Two-Tier Stochastic Frontier Analysis for the Social Sciences

Alecos Papadopoulos • Christopher F. Parmeter

Two-Tier Stochastic Frontier Analysis for the Social Sciences

Springer

Alecos Papadopoulos
Athens University of Economics and Business
Athens, Greece

Christopher F. Parmeter
Department of Economics
University of Miami
Miami, FL, USA

ISBN 978-3-031-81512-6 ISBN 978-3-031-81513-3 (eBook)
https://doi.org/10.1007/978-3-031-81513-3

This Springer imprint is published by the registered company Springer Nature Switzerland AG
The registered company address is: Gewerbestrasse 11, 6330 Cham, Switzerland

Alecos Papadopoulos dedicates this book to the Data Privacy and Protection Laws worldwide.

Christopher Parmeter dedicates this book to the memory of Virginia Grace Berch.

Preface

This book is concerned with measuring the unobservable in human activities. We may live in the "Age of Information," but actual *data*, of the kind, scope, and detail needed for serious scientific work, is another matter. A *lot* of data is there, but never complete. There are always aspects of the situation that are beyond reach, creating unobservables and latent forces—sometimes the very things we want to explore and assess in a quantitative manner.

The challenge is not new, and various statistical methods have been developed to deal with it. What differentiates this book is its singular focus on one such framework of analysis, measurement, and estimation, the "two-tier stochastic frontier model" (2TSF). The model is designed to handle a specific structural situation: latent interactions which produce competing influences on a measurable outcome. But this "specific situation" is a universal and frequent phenomenon characterizing human affairs, not a fringe occurrence. Incomplete information in labor and housing markets, bargaining over prices, international negotiations, conflicting psychological tendencies, healthcare vs. bureaucracy, social pressure vs. individual goals, are all examples where the observed outcome contains more nuance than what the available data can directly explain. The 2TSF model aims to and aids in extracting from the available data at hand the signal of the unobservable influences harboring underneath.

The 2TSF model has a 40-year history, having been applied to many social science milieus, with varying notation and insight. While this demonstrates the depth and breadth of the model, it also discourages researchers in need for consistent and structured guidance transferable across scientific fields, so that they can apply the model to their own research topics. The goal of this book is to operate as a self-contained guide for practitioners and applied research in general with the use of the 2TSF model, its variants, its estimation methods and inference tools.

While both authors are economists with a heavy preference for econometrics, we firmly believe in the applicability of the 2TSF model across social sciences, based on past evidence from the literature but also on our own research endeavors. This belief has guided the structuring of the book: we start in Chap. 1 with a discussion of the fundamental concept of a "stochastic frontier" and the nature of stochastic frontier analysis, contrasted conceptually with other quantitative methods that deal with unobservables (which we present first, for fair-play). We then provide a quick taste of actual situations where the *two-tier* stochastic frontier is the appropriate modeling framework, and we close the chapter with some history and a selective review of past 2TSF applications that can be found in the literature. Chapter 2 introduces a range of 2TSF *metrics*, specific measures that provide tangible results on the unobservables *after* statistical estimation. These metrics are defined mathematically but we also discuss extensively their rationale and conceptual scope and flexibility, since their interpretation may change with the topic and the goals of research. The chapter also contains the first two empirical applications of the book (out of 16 in total), where we showcase what the value-added of 2TSF analysis can be. Rather unsurprisingly, these two chapters form Part I of the book under the title "Concepts and Metrics."

Then comes Part II, "Models and Applications": four Chaps. 3 to 6, nine models and empirical applications to explore matters as diverse as incomplete information in transactions but also management performance, bargaining over wages or military aid, home health care but also extramarital affairs, environmental policy but also primary public education. The focus here is on the real-world situations, the statistical *results* and their interpretation, as well as on introducing many of the different 2TSF models that will be presented later in the book in their technical detail. These chapters stay clear of theoretical elaborations on the tools used. This, as mentioned earlier, is in order to first argue by example our case that the 2TSF framework is as universal in its applicability as it is the structural situation it was designed to address.

Part III includes Chaps. 7 to 11 and is the opposite of Part II: these chapters are full of mathematical formulas. *All* mathematical formulas needed in order to implement 2TSF analysis in its various incarnations: estimation methods like maximum likelihood and generalized method of moments (but also others), tailored for 2TSF models (formulas included). Statistical specifications that are the bridges to the land of unobservables (formulas included). Specification tests and model selection methods (formulas included). The handling of heterogeneous latent forces (Chap. 8). 2TSF panel data models (Chap. 9). Statistical dependence between the antagonistic latent forces, including simulation methods and Copulas (Chap. 10). And yes, all formulas included, together with extensive discussion for particulars and details that often frustrate newcomers and seasoned professional alike. Chapter 11 is a special one, providing specific 2TSF software code in the `R` and `gretl` open-source environments (together with a presentation of an available command in the commercial Stata program that exists for 2TSF analysis). And this is only a taste of the library of scripts and data sets that will be hosted in the dedicated website of this book.[1]

Part IV includes the last three chapters of this work. It is called "Extensions" because the material here is either presented for the first time, or it has been treated in the scientific literature only briefly and in an introductory manner. Chapter 12

[1] https://github.com/Parms23/2TSF

presents the 2TSF "Tug-of-War" model which, on the surface, appears as just a minor modification of the benchmark model, but experience has shown that it may prove handy in many empirical applications. Chapter 13 develops in detail two different 2TSF Binary Response models (whose applications are the subject of Chap. 6). And Chap. 14... well, Chap. 14 includes short presentations (sometimes barely more than an executive summary) of all those additional 2TSF things that two econometricians would love to write hundreds and hundreds of pages about—but then this book would not stay true to its stated purpose: a stand-alone guide for implementing two-tier stochastic frontier empirical analysis across the social sciences.

Acknowledgments

It is simply impossible to thank all of the people who influenced this text. That being said, we would like to acknowledge by name and express our thanks to those who had an overall and profound impact, our mentors and senior colleagues: Subal Kumbhakar, Solomon Polachek, and Mike Tsionas who sadly passed away very prematurely. We also acknowledge the support of our publishing company Springer and especially our managing editor, Lorraine Klimowich. It was Klimowich who gave us complete control of our text. It is therefore the case that if there are any quarrels about the form of the book, these lie solely on us and not our publisher.

But the most important people involved in any text are the immediate family members. They are the ones who have to deal with the induced stress on a daily basis and it is them who bring authors back to reality. Chris Parmeter thanks his loving wife Sandra, and his two children, Elizabeth and Harrison, without whose support the completion of this book would not have been possible. Alecos Papadopoulos thanks them too, since without Chris this book could not and would not have been written.

Athens, Greece — Alecos Papadopoulos
Miami, FL, USA — Christopher F. Parmeter

Contents

Part I Concepts and Metrics

1 Introduction 3
1.1 Accounting for Unobservable Factors 4
1.1.1 Structural Latent-Variable Models 4
1.1.2 Panel Data Individual Effects Models 6
1.1.3 Binary Choice/Response Models 7
1.2 What Is a Frontier? 9
1.2.1 Deciding Which Variables Belong to the Frontier 12
1.3 Why Would There Be Two Frontiers? 13
1.3.1 Incomplete Information 14
1.3.2 Bilateral Bargaining 15
1.3.3 Omitted Variables vs. Latent Forces 16
1.3.4 Rudrian Treatments 19
1.4 Historical Overview of the 2TSF Model 20
1.4.1 Origins 20
1.4.2 Two-Tier Stochastic Frontier Analysis in Action 22

2 Metrics for the Latent Effects 25
2.1 Dependent Variable in Original Measurement Units 26
2.1.1 Sample-Level Metrics 28
2.1.2 Individual Metrics 33
2.2 Application: Personal Computer Pricing, Back in the Day 37
2.3 Dependent Variable Is Transformed 41
2.3.1 Sample-Level Metrics 43
2.3.2 Individual Metrics 47
2.4 Application: The Original 2TSF Earnings Data Set 49

Part II Models and Applications

3 Incomplete Information 57
3.1 A Hedonic Pricing Model with Incomplete Information 58

3.2 Formulating the Empirical Model 63
3.3 Application: A Canadian Housing Market 65

4 Bargaining 73
4.1 The 2TSF Nash Bargaining Model 74
4.1.1 Logarithmic Representation and Estimation 78
4.1.2 Surplus Definition and Measurement in the Nash Bargaining Model 79
4.2 Application: Zambian Wages in the Early 1990s 80
4.2.1 Results on Endogeneity and Intra-error Dependence 84
4.2.2 Bargaining Space and Bargaining Performance 85
4.2.3 Stratification: Bargaining Matters yet Again 87
4.2.4 Bargaining Power 88
4.3 No-Nash Bargaining 89
4.4 Application: No-Nash Bargaining on US Military Aid, 1992–2011 90
4.4.1 Determinants of US Military Aid 91
4.4.2 Data, Specification, and Estimation 92

5 A Jamboree of Latent Forces 97
5.1 The 2TSF Management Model of Production 97
5.1.1 A Production Function with Management................. 98
5.1.2 The 2TSF Stochastic Specification 99
5.1.3 Metrics of Interest 100
5.2 Application: Measuring Management Across the Globe 100
5.3 Idiosyncratic Forces in Secondary Education...................... 104
5.4 Application: Race and Income in Boston's Public Schools 105
5.5 Environmental Policy and Social Pressure: Do Businesses Become Greener? 109
5.6 Application: Going Green in China 2010–2020 110
5.7 2TSF Analysis of Subjective Welfare 115
5.8 Application: World Happiness Report 2020 117

6 Binary Responses and Treatment Effects 125
6.1 Latent Forces in a Latent Regression.............................. 125
6.2 Application: Probability of Extramarital Affairs.................... 126
6.3 The 2TSF Treatment of Treatment Effects 132
6.4 Application: Home Healthcare and US Hospital Readmissions 133

Part III Tools for Estimation and Inference

7 Estimation Methods and Statistical Specifications..................... 145
7.1 Estimation Methods 146
7.1.1 Maximum Likelihood 148
7.1.2 Corrected OLS 150

7.1.3 Corrected Nonlinear Least Squares ... 152
7.1.4 Generalized Method of Moments ... 153
7.2 Statistical Specifications ... 156
7.2.1 Normal-Exponential (NE) ... 157
7.2.2 Normal-Half Normal (NHN) ... 161
7.3 Model Selection ... 165
7.3.1 Forensic Diagnostics ... 166
7.3.2 Formal Statistical Tests ... 167
7.4 Application: Detecting and Testing 2TSF Models ... 169

8 Incorporating Heterogeneity in the 2TSF Model ... 173
8.1 The Importance of Allowing Heterogeneity ... 173
8.2 Moment Dependence and Maximum Likelihood ... 175
8.2.1 Moments and Metrics ... 177
8.3 The Scaling Property and NLS ... 179
8.3.1 The Scaling Property ... 179
8.3.2 NLS Estimation ... 180
8.3.3 Metrics ... 182
8.3.4 Identification Perils and Traps ... 184
8.4 Application: The Housing Market in the USA ... 185

9 2TSF Panel Data Models ... 195
9.1 Why Is Access to a Panel Important? ... 196
9.2 The Generalized 2TSF Panel Data Model ... 196
9.2.1 Fixed-Effects Two-Stage Iterated Estimation ... 199
9.3 Random-Effects Two-Stage Estimation ... 204
9.3.1 Incidental Parameters and Sample Size ... 208
9.3.2 Method-of-Moments Estimation ... 208
9.3.3 Modeling Correlation Between Regressors and Heterogeneity ... 210
9.4 Computing and Interpreting Metrics in the 2TSF Panel Data Model ... 211
9.5 Heterogeneity in the 2TSF Panel Data Model ... 212
9.5.1 Heterogeneity in the Time-Varying Latent Forces ... 212
9.5.2 Nonlinear Heterogeneity ... 213
9.6 Application: The Original 2TSF Panel Data Study Extended ... 215

10 Allowing for Intra-error Dependence ... 223
10.1 The Correlated Exponential 2TSF Model ... 224
10.1.1 The Distribution of the Composite Error Term ... 225
10.1.2 Identification ... 226
10.1.3 Prediction of w and u ... 227
10.1.4 The Distribution of the Difference $z = w - u$ ... 228
10.1.5 Model Diagnostics and a Specification Test ... 229
10.2 The Correlated Truncated Normal 2TSF Model ... 231

10.2.1 The Marginal Distributions of w and u 232
10.2.2 Conditional Expectations for w and u 233
10.2.3 The Distribution of the Difference $z = w - u$ 235
10.3 Using Copulas to Model Intra-error Dependence.................. 236
10.3.1 Simulated Maximum Likelihood Estimation............. 237
10.3.2 Choosing the Copula 239
10.3.3 Conditional Expectations for w and u 241
10.3.4 The Distribution of the Difference $z = w - u$ 242
10.4 Conditioning Out the Source of Dependence 245
10.4.1 Prediction and Conditional Expectations of Latent Forces... 248
10.4.2 Application: Bilateral Wage Bargaining in Ghana 249

11 Software for 2TSF Analysis 253
11.1 2TSF Estimation in `R`... 253
11.1.1 Basic Implementation 253
11.1.2 Calculating Individual Measures.......................... 260
11.1.3 Heterogeneity 262
11.1.4 Allowing Dependence: Closed-Form Specification 267
11.1.5 Allowing Dependence: Copulas and Simulated Methods 270
11.1.6 Calculation of Individual Metrics via Simulation......... 274
11.2 2TSF Estimation in `gretl`... 277
11.2.1 Basic Maximum Likelihood Estimation.................. 278
11.2.2 Individual Metrics 280
11.2.3 Heterogeneity 282
11.2.4 COLS and GMM Estimation.............................. 283
11.2.5 The Tug-of-War Model (with Copula).................... 288
11.2.6 Using the Conditional Mode as Predictor of Latent Forces... 291
11.3 The `sftt` Command in Stata for 2TSF Estimation 294
11.3.1 Basic Maximum Likelihood Estimation.................. 294
11.3.2 NLS Estimation... 294
11.3.3 Presenting Estimation Results 295
11.3.4 Individual Metrics 296

Part IV Extensions

12 Extensions I: 2TSF Tug-of-War Models 299
12.1 Tug-of-War Distributions of Independent Latent Forces 300
12.1.1 The Asymmetric Double Exponential Specification 300
12.1.2 The Asymmetric Double Half-Normal Specification 303

12.2 A Copula Tug-of-War Model for Dependent Latent Forces........ 305
12.2.1 The Likelihood Function 306
12.2.2 Individual Metrics 308
12.2.3 Choosing the Marginal Distributions and the Copula.. 308
12.3 Application: The Greek Depression 2009–2018.................... 309
12.3.1 Results Related to Management and Efficiency........... 310

13 Extensions II: 2TSF Binary Response Models 315
13.1 Unobservable Effects on Binary Outcomes 315
13.1.1 Derivation of the Binary Response Model 316
13.1.2 The 2TSF Probit Exponential Model 319
13.1.3 Estimation... 321
13.1.4 Inference 321
13.1.5 Heterogeneity in the 2TSF Probit Exponential Model 325
13.2 Decomposing Conflicting Treatment Effects 326
13.2.1 The 2TSF Random Treatment Effects Model 326
13.2.2 Estimation and Inference.................................. 330

14 Extensions III: Honey Pots and Epilogue.............................. 337
14.1 More Estimation Methods 337
14.1.1 Nonparametric Estimation 338
14.1.2 Fast Fourier Transform.................................... 342
14.1.3 Bayesian Estimation of the 2TSF Model................. 345
14.1.4 Regressor Endogeneity.................................... 345
14.2 More Distributional Specifications................................ 347
14.2.1 The Generalized Exponential Specification 347
14.2.2 Heterogeneous Distributions for the Latent Forces 348
14.2.3 Latent Class Modeling 349
14.2.4 Sample Selection Bias in the 2TSF Model 352
14.3 Terra Incognita 353
14.3.1 Non-Normal and Asymmetric Noise Component......... 353
14.3.2 Quantile Methods for 2TSF Models 354
14.3.3 Dynamic Models... 354
14.4 Afterword.. 354

References.. 357

12.2 A Copula Tug-of-War Model for Dependent Latent Forces ... 305
12.2.1 The Likelihood Function ... 306
12.2.2 Individual Metrics ... 306
12.2.3 Choosing the Marginal Distributions and the Copula ... 308
12.3 Application: The Greek Depression 2009–2013 ... 309
12.3.1 Results Related to Management and Recovery ... 310

13 Extensions III: 2TSF Binary Response Models ... 315
13.1 Unobservable Effects in Binary Outcomes ... 315
13.1.1 Derivation of the Binary Response Model ... 316
13.1.2 The 2TSF Probit Exponential Model ... 320
13.1.3 Estimation ... 321
13.1.4 Inference ... 322
13.1.5 Heterogeneity in the 2TSF Probit Exponential Model ... 325
13.2 Decomposing Conflicting Treatment Effects ... 326
13.2.1 The 2TSF Random Treatment Effects Model ... 326
13.2.2 [illegible] ... 327

[illegible]

14.1.4 [illegible]
14.2 [illegible]
14.2.1 [illegible]

14.3 [illegible]
14.4 [illegible]

References [illegible]

List of Applications

1. Chapter 2: Personal computer market, USA, 1995. **2TSF focus**: Exploiting the hype while fearing the competition.
2. Chapter 2: Labor market, USA, 1980. **2TSF focus**: Wages and informational deficiencies of employers and workers.
3. Chapter 3: Housing market, Ontario Canada, 1987. **2TSF focus**: Hedonic analysis and informational deficiencies.
4. Chapter 4: Labor market, Zambia, 1993–1995. **2TSF focus**: Nash equilibrium and bargaining performance.
5. Chapter 4: US military aid globally, 1992–2011. **2TSF focus**: No-Nash bargaining performance per geopolitical region.
6. Chapter 5: Management in production globally, 2006. **2TSF focus**: Improving operations while battling external inefficiency.
7. Chapter 5: Public education, Boston USA, 2001. **2TSF focus**: Idiosyncratic forces in primary education.
8. Chapter 5: Business sustainability, China, 2010–2020. **2TSF focus**: Social pressure and business evasion.
9. Chapter 5: World happiness globally, 2006–2018. **2TSF focus**: Optimism and pessimism at the national level.
10. Chapter 6: Extramarital affairs, USA, 1969. **2TSF focus**: Desire for variety.
11. Chapter 6: Home healthcare and hospital readmissions, California USA, 2015–2017. **2TSF focus**: Antagonistic treatment effects.
12. Chapter 7: Wage earnings, USA, 1982. **2TSF focus**: Choosing the right model specification.
13. Chapter 8: Housing market, USA, 1986–1993. **2TSF focus**: Determinants of information deficiencies.
14. Chapter 9: Labor market, USA, 1968–1984. **2TSF focus**: Panel data model to analyze persistent and time-varying latent forces.
15. Chapter 10: Wage bargaining, Ghana, 1992–1994. **2TSF focus**: Dependent latent forces and disentanglement of bargaining power from private information effects.
16. Chapter 12: Sectoral production functions, Greece, 2008/2019. **2TSF focus**: Management performance before and after the Greek depression.

Part I
Concepts and Metrics

Chapter 1
Introduction

With the onset of the twenty-first century, we have entered an era where data are generated with dizzying speeds and information becomes available to an unprecedented degree. This, to scholars of the social sciences that want to engage in quantitative analysis, is what water is to the thirsty. And yet, every time we focus on a specific phenomenon of the real world, on an outcome and the forces that influence it, and we analyze it causally and structurally in order to come up with a theoretical model that we could then use for quantitative analysis, we invariably discover that some of the causal forces, the factors that are expected to influence the outcome, are not available as data.

There are many reasons why this may happen even though we are currently swimming in an ocean of data. To name the most common among them: inability (natural, technical, or social) to measure some of the conceived factors at play, mundane collection-cost reasons, organizational absent-mindedness that keeps missing the proper design for data collection projects, cultural constraints, privacy concerns, ethical red lines, and even "national" barriers to entry (microlevel datasets of statistical agencies and other data-collecting institutions often are available only to localized researchers). All contribute to the existence of "unobservables."[1]

Whatever the reasons, applied researchers constantly need to deal with unobservable variables in their empirical studies. Sometimes these can be "assumed out of the picture," ignored, with the hope (or argumented belief) that this omission does not destroy the validity of the results and the reliability of the conclusions. But there are many cases where the unobservables are clearly important for the determination of the outcome under study and of great interest on their own.

[1] Strictly speaking, the most general description would be "unavailable for the study at hand." This would cover cases where we could, given the constraints, obtain measurements but did not or even cases where measurements exist but they are unavailable to us. So the proper shorthand label would be "unavailables," but we will stick with established terminology.

A. Papadopoulos, C. F. Parmeter, *Two-Tier Stochastic Frontier Analysis for the Social Sciences*, https://doi.org/10.1007/978-3-031-81513-3_1

Statistics is uniquely positioned to deal with unobservables, because its existence starts with an admission of ignorance, of incomplete information. The fundamental concept of probability and the fundamental tool of a random variable were formulated and are useful because of our ignorance of the actual realization of a real-world phenomenon. While this is intuitively understood as relating to a realization in the future, it makes no difference to consider probabilities of events passed or concurrent, and we can model these events as random variables, as long as their actual realization remains unknown. By applying such statistical concepts on the real world, we aim to envelop such phenomena of unknown realization into a structure and hence contain our ignorance and obtain useful results for our purposes, such as bounds, anticipated *likely* values, indicators of volatility, as well as reliable estimates of the strength of covariance and the direction of causal effects.

This book is a detailed exposition and a guide for implementation of a specific statistical model that has been developed to measure the effects that unobservable factors have on an outcome. The model initially appeared in the field of econometrics, but it will become obvious that it is universally applicable across social sciences. It is called the two-tier stochastic frontier model (2TSF), but it is more of a framework for statistical modeling, estimation, and inference.[2] The *defining property* of the model is that it can handle *two opposing* latent forces on the outcome and estimate their effects separately. But before presenting its fundamental premises in some detail, we discuss briefly some other statistical approaches that account for the existence of unobservable factors or latent forces without using a frontier methodology, so that the reader has some tangible background for contrast.

1.1 Accounting for Unobservable Factors

1.1.1 Structural Latent-Variable Models

Structural latent-variable models arise in the context of factor analysis (see Bollen, 1989; Kim & Mueller, 1978), in particular confirmatory factor analysis (as contrasted to exploratory), where a particular structure of relationships is assumed and imposed on the data, instead of letting the data indicate the existing relationships. Let y be the outcome of interest, assumed to be influenced by a vector of unobservable variables $\boldsymbol{z}$. We have data on variables $\boldsymbol{x}$ that are treated as depending, in a noisy way, on the unobservables. Then a system of equations can be formed

$$
\begin{aligned}
y &= g(\boldsymbol{z}; \boldsymbol{\gamma}) + v \\
\boldsymbol{x} &= h(\boldsymbol{z}; \boldsymbol{\lambda}) + \boldsymbol{u}.
\end{aligned}
$$

[2] The creators of this model had initially used "two-tiered" rather than "two-tier." That would imply that we have a *single* frontier with "two tiers." But our view is that the model represents situations where in reality there are *two* frontiers, and so "two-tier" becomes a more accurate label.

The unknown coefficient vectors $\boldsymbol{\lambda}$ and $\boldsymbol{\gamma}$ are the main targets of estimation. To make this implementable, in practice the functions g and h are specified as linear.[3] Suppose also for illustrative purposes that $\boldsymbol{z} = (z_1, z_2)$ and $\boldsymbol{x} = (x_1, x_2)$. Then for every observation i, we have the explicit system of equations

$$\begin{aligned} y_i &= \gamma_1 z_{1i} + \gamma_2 z_{2i} + v_i &&= \boldsymbol{z}_i' \boldsymbol{\gamma} + v_i, \\ x_{1i} &= \lambda_{11} z_{1i} + \lambda_{12} z_{2i} + u_{1i} &&= \boldsymbol{z}_i' \boldsymbol{\lambda}_1 + u_{1i}, \\ x_{2i} &= \lambda_{21} z_{1i} + \lambda_{22} z_{2i} + u_{2i} &&= \boldsymbol{z}_i' \boldsymbol{\lambda}_2 + u_{2i}. \end{aligned}$$

From this formulation, one can show that the variance-covariance matrix of the observables $(y, \boldsymbol{x})$, that are usually centered to have zero mean, can be written in terms of the unknowns of the model, which, apart from the coefficient vectors $\boldsymbol{\gamma}$, $\boldsymbol{\lambda}_1$, and $\boldsymbol{\lambda}_2$, include also the variances of the error terms and the variance-covariance matrix of the unobservables. This leads to a method-of-moments estimation strategy, where sample moments are equated to their population counterparts, forming a system of equations to be solved for its unknowns (in a statistical context, so as to obtain also standard errors for the estimates). Estimation can be done by Gaussian quasi-maximum likelihood (i.e., assuming that the distribution that governs the model is multivariate Normal), but also using other methods like generalized least squares. In Chap. 7 of Bollen (1989), the model is presented in every detail, including its standard statistical assumptions. In particular, the author presents the restrictions that must be imposed in order to achieve identification (e.g. a relative scale assumption, where one of the unknown coefficients becomes essentially the *numéraire*), and describes several methods to examine whether the constructed model allows for identification of all the remaining unknowns. A formal theoretical treatment can be found in Anderson (1989).

Structural latent-variable modeling is a more elaborate version of "modeling with proxy variables." In proxy-variable models, we use the observable $\boldsymbol{x}$ variables as proxies for $\boldsymbol{z}$, and we model y as a function of $\boldsymbol{x}$, directly. Here, the system-of-equations modeling separates the relation between the observables $\boldsymbol{x}$ and the unobservables $\boldsymbol{z}$, from the effects that the latter have on the outcome y. This comes at a price: the nexus of causal or covariation links among $(y, \boldsymbol{x}, \boldsymbol{z})$ is more complicated and hence more challenging to establish through structural arguments.

[3] Throughout this book, we will use the symbols g, G to represent generic functions or other entities, the symbol f to represent a probability density function, or just density, and the symbol F to indicate a cumulative distribution function (CDF), or just distribution function. The symbols h, H, will be jokers, fulfilling notational needs per case. In the same vein, y and Y will denote the dependent variable in a regression equation, x and X the core explanatory variables, while z and Z will take on various roles.

In stochastic frontier models, we also encounter situations where we have available data that relate to the unobservables: the so-called determinants of inefficiency (namely, determinants of the unobservable factors, since in the benchmark stochastic frontier model, we examine production and cost matters). But there is a critical conceptual difference with the structural latent-variable models: there, the unobservable z's are seen to be confounders, affecting all observables $(y, \boldsymbol{x})$. In frontier analysis, these observable determinants of latent forces carry causal effects toward the unobservable factors. Hence, they are used to individualize the effects of the latent forces per observation, and so frontier models with such covariates are categorized as models that address the issue of heterogeneity. In Chap. 8, we present in detail different 2TSF models that incorporate heterogeneity together with an empirical application.

1.1.2 Panel Data Individual Effects Models

A latent factor, an unobservable, is what apparently inspired Yair Mundlak to pioneer the individual effects of panel data model: management in the production process.[4] Mundlak (1961) showed how with a panel dataset of just two observations along the time dimension, and a constant-returns-to-scale assumption on the production function, one could obtain a measure of management and of its output elasticity. He modeled the unobservable management variable as another input in a Cobb-Douglas function with its own elasticity, $Q = B_0 X^{b_1} M^c$ (we use a single additional input for simplicity of exposition). Using lowercase letters for variables in logarithms and adding a random disturbance, the panel data specification becomes

$$q_{it} = b_0 + b_1 x_{it} + c m_i + v_{it}, \qquad i = 1, \dots, n, \quad t = 1, \dots T .$$

So the management variable m_i was mapped to the time-invariant individual effect for each firm, conceptually multiplied by the management output elasticity c. Mundlak re-specified the equation to express the management variable as a deviation from its sample mean $\bar{m}$:

$$q_{it} = (b_0 + c\bar{m}) + b_1 x_{it} + c\,(m_i - \bar{m}) + v_{it} .$$

This allowed him to apply what is now called the least-squares dummy-variables (LSDV) estimator, and he obtained estimates for the slope coefficients (here, just the b_1 coefficient) using the within estimator. Then, pooling the data, he obtained an estimate for $(b_0 + c\bar{m})$. This in turn leads to an estimated series for $c\,(m_i - \bar{m})$. The assumption of constant returns to scale in inputs *including* management, $b_1 + c = 1$,

[4] The measurement of management and its effects on output has produced a large empirical literature, a representative portion of which has been surveyed in Papadopoulos (2021a).

permitted him to estimate the management output elasticity $\hat{c} = 1 - \hat{b}_1$ and then divide $c\,(m_i - \bar{m})$ by $\hat{c}$ to obtain a data series on management in mean-deviation form. The unobservable had been revealed and its effects estimated.

We presented the model in terms of a specific area of application. In general terms this model assumes that, for every observation in a data sample (which represents a person, a group, a business, a country), the observable outcome is partly determined by unobservables that are grouped in the individual effect. A critical assumption that allows this methodology to measure the effects of these unobservables is that this effect remains constant throughout the time length spanned by the sample. This may not be realistic if the time horizon is large, say, more than 2 or 3 years or even more than a few quarters in other cases. Moreover, the model cannot distinguish between positive and negative latent forces, since the effects of all are netted in a single time-invariant variable. Both limitations can be overcome by the 2TSF model for panel data. This is developed in Chap. 9.

1.1.3 Binary Choice/Response Models

It is often the case that observed actions, or discontinuous changes in status, depend on reaching a critical threshold: it is a discontinuous change of status, when two countries formerly at peace are now at war. But in order for this to happen, tensions were built gradually—until a threshold of animosity has been reached and war ensues. Being admitted to a hospital (which turns a person into a patient) is an action that depends on pain or other markers of health status reaching a critical threshold. Buying a second car, which is a discrete action in the sense of buying an additional whole unit of an indivisible product, depends on family income, size, and maturity (kids grow) reaching a certain combined threshold. In practice researchers observe such discrete choices or more generally discrete responses but they do not know how various characteristics of the agent that took the action have contributed to it.

These kinds of situations can be handled by postulating the existence of an unobservable variable y^{ℓ} that is a function of these characteristics, say

$$y^{\ell} = g(\boldsymbol{x}) + v,$$

where v is a random variable having distribution function conditional on the regressors, $F_{v|\boldsymbol{x}}$.[5] Since the value of y^{ℓ} is not known, this relation cannot be estimated, even if we have data on $\boldsymbol{x}$. What we do have data on is whether a discrete action was taken. In the simplest case, the action (or change of status) can be represented as a binary $\{0, 1\}$ variable, an indicator function, say y, where the value

[5] Because the 2TSF model operates in the context of (generalized) regression analysis, we will use the term "regressors" throughout this book to refer to what are traditionally known as "explanatory variables." It also economizes on words.

1 indicates that the action was taken (the change of status has happened), while the value 0 indicates that it was not. And because we assume that in order for the action to be taken, the unobservable y^ℓ must have reached or exceeded some (unknown) threshold value, say c, we have the equivalence

$$y = 1 \Leftrightarrow y^\ell \geq c, \qquad y = 0 \Leftrightarrow y^\ell < c.$$

Due to this equivalence relation, we can form a probabilistic equality statement, conditional on the regressors $\boldsymbol{x}$ whose effects we are interested in estimating while using the relation that involves the latent variable. We have

$$\begin{aligned}
\Pr(y = 1 \mid \boldsymbol{x}) &= \Pr(y^\ell \geq c \mid \boldsymbol{x}) \\
&= \Pr(g(\boldsymbol{x}) + v \geq c \mid \boldsymbol{x}) \\
&= 1 - \Pr(v \leq c - g(\boldsymbol{x}) \mid \boldsymbol{x}) \\
&= 1 - F_{v|\boldsymbol{x}}(c - g(\boldsymbol{x}) \mid \boldsymbol{x}).
\end{aligned}$$

We see that the conditional probability of the binary response given the regressors can be expressed through the latent relation as a probability that is a function of these regressors, embedded in the distribution function of the random disturbance v. This is the conceptual foundation of the celebrated and widely used logistic regression, where the name logistic comes from assuming that $F_{v|\boldsymbol{x}}$ is the distribution function of the logistic distribution.[6] Evidently, by making a different distributional assumption, we arrive at a model with a different name (assuming the Normal distribution gives us, e.g., the probit model), but they all fall into the category of binary response models.

Here, the latent force behind the observed response is the unobservable y^ℓ, but our main concern is not so much to estimate it but rather to estimate how the regressors affect the observed outcome through it. The threshold value itself is a convention, and in models where $g(\boldsymbol{x}) = a + h(\boldsymbol{x})$ the postulated threshold c is indistinguishable from the constant term a. These models provide estimates of how the regressors affect the probability of observing the discrete response. Essentially, the latent variable is an intermediary that summarizes the effects of the regressors.

There are two ways that the binary choice/response model can be extended to become a 2TSF one. First, when used to model a binary *choice* (as in the decision to buy a car or not), the regressors in the latent structural relation are explanatory variables in their own right. Then we may have opposing unobservables that affect the latent variable y^ℓ. In such a case, the only thing that changes in the standard derivation of a binary choice model is the distribution function $F_{v|\boldsymbol{x}}$. We present a related 2TSF application in Chap. 6, where we measure the strength of latent forces that tend to affect the occurrence of extramarital affairs.

[6] This underlying causal and probabilistic structure for logistic regression is usually mentioned in econometric circles, but not so much in other disciplines.

A second, more subtle application of the 2TSF logic arises if we use such a model to describe a binary *response*, as in a treatment effect model. Here, the 2TSF approach will allow us to decompose *dualistic* treatments, where a treatment variable in reality represents the application of *two opposing/antagonistic treatments* on the outcome. A standard treatment effects model would estimate the net effect of them. A 2TSF model is able to estimate them separately. We present one such application of the binary response 2TSF model in Chap. 6, relating to healthcare. The full theory for both models is developed in Chap. 13.

1.2 What Is a Frontier?

The word "frontier" belongs to the standard vocabulary of the English language, and its meanings all relate, in one way or another, to a "boundary," a "limit," and an "extreme." But in order to correctly understand the concept of a frontier as used in this book, it is important to distinguish it and then properly relate it to the mathematical concept of an extreme value, a minimum or a maximum.

To begin with, the mathematical concepts of minimum and maximum are defined as the absolute smallest and largest value that an outcome *can* take, without connecting these extreme values with any one force that influences it. Essentially they are the minimum and maximum values in the range of the outcome, viewing the latter as a numerical variable. Then, once we postulate a functional relation between an outcome y and a collection of variables $\boldsymbol{x}$, we encounter the minimum and the maximum again, now as the minimized and the maximized value of the outcome after engaging in the mathematical procedure of function minimization (or maximization; compactly, optimization). Associated with this optimum is the corresponding optimizer vector: the values that the $\boldsymbol{x}$-variables *must* take jointly, in order for the outcome to reach its optimum value. This are all abstract mathematical operations. If y and $\boldsymbol{x}$ are observed variables, there is no guarantee that the observed values of $\boldsymbol{x}$ constitute the optimizer vector (under the assumption that the assumed functional relation reflects indeed the real world situation). The realized values that the regressors may have actually taken in a real-world study do not enter the picture here. Moreover, rarely if ever do we have a priori knowledge of the exact form that this functional relation takes. In fact, statistical estimation is exactly about quantifying the relation between y and $\boldsymbol{x}$.

The frontier value (in our case, the upper or the lower frontier) does not coincide with these mathematical concepts. First, the frontier is defined *conditional on the regressors* and is estimated conditional on the realized values of the regressors. But, one could ask, if the realizations of the forces that determine the outcome are observational givens, in all likelihood they will not coincide with the optimizer vector, so in what sense does the frontier relate to some boundary or extreme value?

Let y denote the outcome, $\boldsymbol{x}$ the influencing factors/regressors, and G the relation that describes how the regressors jointly determine the outcome, $y = G(\boldsymbol{x})$. Then we have the contrast

$$\text{maximized } y = \max_{x} G(x) \text{ given } G, \quad \text{minimized } y = \min_{x} G(x) \text{ given } G$$

$$\text{upper frontier of } y = \max_{G} G(x) \text{ given } x, \text{ lower frontier of } y = \min_{G} G(x) \text{ given } x.$$

We see that the frontier is a minimum or a maximum, but not along the dimension that function optimization would take us: for the usual minimum and maximum, we take as given the functional relationship G that represents how the regressors jointly determine the outcome, and we optimize with respect to the values of these regressors, the arguments of the function G. In contrast, to obtain the frontier values, we optimize with respect to the functional relationship G given its arguments. Let G^U denote the function for the upper frontier and G^L the function for the lower frontier. Suppose that the realized regressor vector is x_1. Then,

$$\begin{aligned} \text{upper frontier of } y \equiv G^U(x_1) &= \text{maximum } y \text{ given } x_1, \\ \text{lower frontier of } y \equiv G^L(x_1) &= \text{minimum } y \text{ given } x_1. \end{aligned}$$

If we change x_1 to some other vector x_2, the frontier value of the outcome will likely change, but always through the frontier functions G^U or G^L. When we have available a whole series of vectors $\{y_i, x_i\}$, we can attempt to determine the functions G^U and G^L. This kind of optimization of a functional form would be a rather daunting undertaking. But in a statistical parametric approach, as the one we will be concerned with in this book, the actual mathematical form of G is one of convenience and of some very basic logical properties that it must possess in order to reflect properties of the real-world phenomenon under study. It is the *parameters* that actually *characterize* G that are the targets of statistical estimation. By estimating these, we estimate how the regressors affect the outcome in the context of the frontier relation (that the frontier relation may not be what holds in reality is what assigns to stochastic frontier analysis a double duty: to estimate the frontier but also to estimate the distance from it. More on that in a while).

Why are we interested in such conditional bounds like frontiers? Because in social sciences, we mostly observe *ex post* and attempt to reveal the structure after the fact; we do not design a priori what will happen in economic and other social interactions. Certainly, through such investigation of events that have occurred, we obtain specific quantitative relationships, which in turn may be taken into account in designing policies. But no entity controls all the forces that determine an outcome, so the frontier values become the important bounds into which an outcome will be expected to range. Also, it is the frontier values that are the realistic targets to achieve by setting the subset of policy variables (if this is entailed by the goals of analysis), and not some ideal fully planned social engineering scheme that is never going to be implemented as designed on paper, if history is any indication on the matter.

There are many cases where only one of the two frontiers appears meaningful and important to study: for example, in production analysis, we are interested in the upper frontier, in the maximum we can produce given the available resources (production inputs like labor hours, amount of materials, equipment, and their flow of services). The lower frontier is inherently zero: we can engage the inputs and then let them sit idle. Technically, the unconditional lower bound of the outcome coincides with the bound conditional on the regressors, so the latter stops being interesting to study (but see later what happens when an important production input is unobservable itself). Consider another example, from healthcare this time: suppose we have constructed an index that compacts and quantifies the side effects of an administered drug, increasing as the side effects worsen. The regressors for this index-outcome are the chemical compounds of the drug, the pre-medicated health markers of the recipient of the drug, and perhaps other lifestyle characteristics. The upper frontier here is equivalent to biological death. What we want to study is the lower frontier, the minimum degree of side effects given the regressors, and how each one of them affects it.

Such situations can be modeled and studied by a "single-tier" stochastic frontier model.[7] But we have not yet explained the "stochastic" adjective.

Continuing with the production setting, where the upper frontier is of interest, we are certainly interested in estimating the parameters that characterize the upper frontier, but we are also aware, from historical experience and observation, that real-world businesses do not operate on their frontier, given the production inputs that they employ. There are learning ladders to climb and internal vested interests to accommodate, and such forces lead to a not-fully *efficient* use of the available resources (where full efficiency is the equivalent of operating on the frontier). So apart from estimating the upper frontier output function, we want also to estimate the distance of realized output from it. Measuring this distance quantifies how far we are from the ideal and provides tangible motive to reorganize a production process so as to become more efficient. Moreover, there may be forces that co-determine the output beyond the production inputs. These forces may be beneficial (output-increasing) or detrimental (output-decreasing). So they enter the relation with a "plus" sign, with the understanding that their effects may take positive or negative values (through, say, the sign of the coefficient that accompanies them). These are what are usually called control (or, elsewhere, "environmental") variables. But as regards the distance from the frontier, we adopt from mathematics the convention of measuring it in absolute terms (always as a positive magnitude), and here, where we are examining the upper frontier, the variable that represents this distance enters the relationship with a minus sign (it would enter with a plus sign if we were examining the lower frontier). So the actual production output of a business can be modeled as the upper production frontier, plus some other, "external," output-affecting variables minus a term that measures the distance from it:

$$\text{actual } y \equiv \text{upper frontier of } y \ (+) \text{ other forces } (-) \text{ distance from upper frontier},$$

$$y \equiv G(\boldsymbol{x}) \ + \ [H(z) \ + \ v] \ - \ u.$$

[7] In the literature, the "single-tier" qualifier is not used, but here, we need to make a clear distinction with the two-tier stochastic frontier.

In the second line, $G(\boldsymbol{x})$ is the frontier function (here the upper frontier). Note that those "other forces" are represented by two entities, $H(z)$ and v. $H(z)$ is a function of observable control variables, while v represents unobservable forces. An econometrician would simply call v the "regression error term," but in honesty, it stands for those forces that we deem *not* systematically (or if you wish, *systemically*) important for the determination of outcome. The symbol u represents the distance from the frontier due to failures in organization and other tendencies toward inefficiency. It is also unobservable. Just to be clear, we measure u in the units of measurement of the dependent variable, not in some own measurement scale. Hence we measure the effects of u on y, not the "size" of u.

The frontier that we have been talking about, $G(\boldsymbol{x})$, is what we call the "deterministic" frontier, even though the regressors $\boldsymbol{x}$ are treated as random variables. The word "deterministic" is used either to convey the idea that the regressors are decision variables specifically set so as to determine the outcome or to reflect that our frontier analysis on any given sample is done *conditional* on these variables, and the conditioning variables are in practice treated as fixed/deterministic (since they constitute the sample itself, together with the dependent variable). The *stochastic* frontier is defined as $G(\boldsymbol{x}) + v$. This has a historical justification: initially frontier analysis was done without a v component included. This did not allow those early models to invoke probability and statistical theory proper, and so their results could not be validly generalized outside the sample.[8]

1.2.1 Deciding Which Variables Belong to the Frontier

It should be clear by now that, since the frontier is a concept conditional on certain variables and not some absolute boundary, one could conceive of a different frontier for each set of different covariates. Why not include some of the controls z together with the $\boldsymbol{x}$ in the frontier function? In other words, an important issue arises: how do we select the variables conditional on which we will define and subsequently estimate the frontier?

There is no single answer to this question, and in each discipline, and even in each subfield or even in each empirical study, different considerations may give rise to conceptually different sets of frontier-defining regressors. But we argue that there are two clear conceptual criteria that should be used in order to select the variables that belong to the frontier function. Keeping with our example from production theory, the frontier has come to be defined as the boundary with respect to production inputs in the narrow sense (labor, capital, materials), while any other force that may determine the output (from constraints on supply due to regulation, to consumer preferences and the demand curve, to macroeconomic shocks) is treated

[8] From a philosophical point of view, adding an unobservable variable in the relation is a tactful way to admit that we do not know everything.

as control/environmental variables. This consensus can be traced to an apparently influential suggestion by Samuelson (1947, p. 84), but it can also be argued that production inputs are direct *decision variables* of the firm that produces the output, while the other factors are partially or totally out of its control. Also, in many (but not all) cases, *the purpose or the direct effect of these other variables is not to affect output*: they do affect it, but as a side effect. So it makes sense to separate the two sets of forces that determine the firm's output and examine the frontier with respect to the main decision variables that a firm has in its disposition.

These two general criteria can be used to select the frontier-defining variables: they should be under the control of the decision-making unit that attempts to determine the outcome, *and* they should have a direct output-affecting purpose.

So, for example, adversarial marketing and advertisement spending from competitors does have the purpose of influencing (negatively) the output of the firm under consideration, but it is not a variable under the firm's control. A relaxation of the regulatory regime, that can be numerically represented by an index, may have the purpose of (beneficially) influencing the output, but again, it is not under the firm's control. To consider a controlled-experiment example, the pharmaceutical company that examines the side effects of a drug has certainly decided which employees to hire and has certainly selected those that will monitor the clinical trial. These are decisions under the company's control. But they are decisions whose purpose is not to influence the outcome "side effects of a drug," so the professional and/or personal attributes of these employees would not belong to the frontier function here (but maybe they would belong to the environmental variables).

1.3 Why Would There Be Two Frontiers?

We can describe the structural setup that gives rise to the 2TSF model in very generic and abstract terms. Suppose that we have an outcome/dependent variable y for which we postulate that it is a function $g(\cdot)$ of a vector of regressors $\boldsymbol{x}$ on which we have data. The dependent variable is further affected by a stochastic noise/disturbance/shock/error denoted v. Suppose now that we know (or that we can adequately argue) that apart from $\boldsymbol{x}$ and v, there exist two forces that affect y each in the opposite direction, but for which we possess no data. Denote the positive influence w and the negative influence u. Since we have no data on them, it is natural to treat them both as one-sided positive random variables attaching a negative sign to the one representing the negative influence. Then the regression function for the dependent variable becomes

$$y = g(\boldsymbol{x}) + \varepsilon, \quad \varepsilon = v + w - u\,.$$

This is the 2TSF model, characterized by its three-component error term $\varepsilon = v + w - u,\ w, u \geq 0$.[9] To provide stronger intuition, in this section we present different situations where the two-tier stochastic frontier arises as the appropriate modeling choice. These situations will be developed further in Chaps. 3 through 6, where we will present empirical applications of the 2TSF model, occasionally pivoting off published studies which we perturb and also enhance with the metrics that we present in Chap. 2. In the literature review of the next section that closes this chapter, many more structural examples can be found that lead to a 2TSF formulation. This speaks to the versatility and general applicability of the two-tier stochastic frontier model.

1.3.1 Incomplete Information

Consider a buy-sell transaction, for a product, an asset, or the provision of a service. We observe a consummated transaction, one that has actually happened. But why did this particular buyer purchase from this particular seller? Do we expect that the buyer was *fully informed* as regards all the sellers offering this product, so that they went to the seller with the lowest selling price? This is unlikely, because information is costly, in time and maybe in monetary terms. Suppose that we examine the sale of a house. Was this particular seller *fully informed* as regards all the buyers wanting to buy a house like their house, and so they reached out to the buyer willing to pay the highest price? Again, it seems unlikely. In fact we can say with confidence that the transactions that actually happen are characterized by *the ubiquitous reality of incomplete information*: if sellers and buyers operated under full information, we would have observed very different matches and market outcomes.

If the observed (actual) price in the transaction is P, we can say that it would be higher by $u \geq 0$ if the seller had full information. So $P_s \equiv P + u$ can be seen as the *price frontier* for the seller, which the seller does not reach due to their informational deficiency. And we can also say that if the buyer had full information the price could be lower by $w \geq 0$. So $P_b \equiv P - w$ is the price frontier for the buyer, and the distance from it is w. We can combine both distances to express the *full information* price as

$$P^{FI} = P + u - w \quad \Longrightarrow \quad P = P^{FI} + w - u.$$

Note that while the full-information price is unobservable, it can be reasonably expressed as a function $h(\boldsymbol{x})$ of objective, commonly accepted characteristics of what is sold: a function of a vector of regressors. Adding also a random disturbance, we arrive at

[9] From a mathematical point of view, we could equivalently specify $u \leq 0$ and enter it with a plus sign. But it would impoverish visual information and intuition as well as create inconvenience later on.

$$P = h(\boldsymbol{x}) + v + w - u.$$

This is an estimable regression specification that has the three-component regression error of the two-tier stochastic frontier model, $\varepsilon \equiv v + w - u$, where $w, u \geq 0$ are two opposing latent forces. This is one example of how the two-tier stochastic frontier model arises structurally. Why two-tier?

Inserting this expression for P into the seller frontier P_s, we have

$$P_s = h(\boldsymbol{x}) + v + w - u + u = h(\boldsymbol{x}) + v + w.$$

This is the seller's *stochastic frontier*.

Analogously for the buyer's frontier P_b, we have

$$P_b = h(\boldsymbol{x}) + v + w - u - w = h(\boldsymbol{x}) + v - u.$$

This is the buyer's stochastic frontier. So embedded in the expression for the actual price, there are *two* frontiers, one for each side of the transaction.

Several things should be noted here: the variable w expresses the distance of the buyers from their frontier, and at the same time it co-determines the frontier of the *seller*. In the same manner, the variable u determines the distance of the sellers from their frontier, and at the same time it co-determines the frontier of the buyer. Moreover, here $h(\boldsymbol{x})$ cannot be interpreted as a deterministic frontier, in the way that it is in the single-tier stochastic frontier model. If it does represent a frontier, it is with respect to *common information*, since it is the price that would prevail if both parties to the transaction had complete information (and so be at the *informational* frontier).

Also note that, causally, here it is the informational deficiency of the seller that creates the frontier of the buyer, and it is the informational deficiency of the buyer that creates the frontier for the seller. This differentiates the above conceptual construction from the *exchange core* of an Edgeworth box, where the boundaries of the core are frontiers, but they emerge independently from each other. We further develop this framework in Chap. 3.

1.3.2 Bilateral Bargaining

Many buy-sell transactions are not intermediated by bargaining: the seller has set a price, and a buyer appears that is willing to pay this price. But in many cases bargaining enters the picture, as an ancient tradition that survives today in flea markets, or when two companies (or countries) negotiate over a prospective collaboration of financial substance or when workers and employers negotiate over the wage.

If we look at a realized buy-sell transaction that involves bilateral bargaining, the terms w and u, but also the observable part, which here we denote by $h(\boldsymbol{x})$, change conceptual content and mapping: here, $w = \eta \cdot g$, where $\eta \in [0, 1]$ is the *relative bargaining power* of the seller, while g represents a premium that the seller imposes on what is to be sold due to their own *private information* and maybe also for strategic reasons. It is an "own-evaluation premium" (or "self-evaluation," if the sellers sell their labor). For the side of the buyer, we have $u = (1 - \eta) \cdot d$, where $(1 - \eta)$ is the relative bargaining power of the buyer and d represents a "prudential discount," based on the private and incomplete information of the buyer, or their bargaining strategy. But "premium" and "discount" on what exactly? The observable part $h(\boldsymbol{x})$ represents the value of what is under exchange, but based on the *actual* common information (not its full information frontier), the set containing all the information that is known to both the seller and the buyer. To a large extent, this information is usually available also to the researcher. But η, g, and d remain unobservables.

It is instructive to contrast how the w and u latent factors relate to the two sides of the transaction here and in the information-deficiency setup discussed in the previous section. There, w represented a weakness of the buyer and u a weakness of the seller (their information-deficiencies). But in the bargaining situation, w represents a strength of the seller, while u a strength of the buyer. This affects what interpretation(s) we can give to the various *metrics* describing the latent forces that we will present in Chap. 2.

On the statistical front, the presence of η creates "intra-error" statistical dependence, and one needs to use estimation tools that are presented in Chap. 10, which is devoted to the handling of such dependence. There, we also present a conditioning technique that allows us to identify and estimate separately the bargaining power η and the private/incomplete information effects g and d.

The above presupposes a "Nash-bargaining" situation, where the bargaining parties follow maximization rules consistent with expected utility theory and economic-theoretic notions like "Pareto optimality." But the real world does not always conform with them. In such a case, we can still use the 2TSF bargaining framework, but now, the opposing latent forces are not connected through the "relative bargaining power" random variable η that makes them statistically dependent, and one can use the benchmark 2TSF models of Chap. 7 for estimation. We develop more deeply the two behavioral models and present applications for both cases in Chap. 4.

1.3.3 Omitted Variables vs. Latent Forces

Doing quantitative research with observational data related to human activity always bears the danger of leaving some important influencing factors out of the theoretical and empirical models. This is due either to imperfect knowledge of the causal mechanisms that characterize the phenomenon under study or because of lack of data. This is often called the "omitted variables issue." It may destroy desirable properties of estimators, and, regardless, it challenges the interpretational validity of the obtained results: if, say, an outcome y is influenced by factors z_1 and z_2, but we do not include z_2 in our model, this may contaminate the estimate for the effect that z_1 has on y, because it may "take on" part of the effect of the omitted variable z_2. We note that when our goal is forecasting, the model that includes only z_1 is still valid as a forecasting tool based on the information on the specific variables that we have included as regressors. Forecasting is always conditional on the information used, so for the same target-outcome, we can form many valid forecast functions that will give different forecasts, based each time on different information sets.[10] But if our goal is to reveal and measure the structural relations between the outcome and the purported influencing factors, then the absence of any variables that should have been present represent a bigger obstacle, not only because we will ignore the effects of them but because we may obtain distorted estimates for the effects of all others.

But omitted variables have come to signify mostly cases where these variables were available as data, and the researcher mistakenly omitted them from their list of explanatory variables. This is a case of wrong design of our empirical study. In this book, we focus on cases where there are latent variables, factors inherently unobservable, but which should not necessarily be part of the regressors forming the frontier.

The examples we have discussed and presented so far belonged to this category. But in some cases, the two aspects coincide, and we have a latent force that *does* belong to the frontier function given the criteria discussed earlier, but it is itself unobservable.

The related example that we will develop in Chap. 5 is that of management in production: Alfred Marshall considered organization to be an "agent of production," alongside land, capital, and labor. But historically management was perceived as a qualitative aspect of production, and no measure of it has been habitually recorded in financial statements or other sources containing information on production matters.[11] Here, the frontier function is $G(\boldsymbol{x}, m)$, where m represents some measure of management. Actual output should be modeled as (omitting for clarity the presence of control variables, but including inefficiency)

$$y = G(\boldsymbol{x}, m) + v - u.$$

But m is unobservable so a way out of this is to envision the approximation $G(\boldsymbol{x}, m) \approx H(\boldsymbol{x}) + h(m)$ and specify

$$y = H(\boldsymbol{x}) + v + h(m) - u,$$

10 We make a distinction here between *validity* and *quality* of a forecast or a series of forecasts. Certainly, forecasting based on a richer information set is likely to give more accurate results.

11 If one wonders what could be such a measure of management, the economist's answer is simple: for starters, report separately from all other labor the number of personnel with supervising duties and/or the cost of their payroll.

that gives rise to a 2TSF model with $h(m)$ taking the role of w. The function H approximates that part of G that does not include m. Unobservable frontier variables can be encountered also in other settings of course, the phenomenon is not confined to production analysis.

Another field where we tend to suspect that some influencing factors are beyond the reach of measurement is education and educational outcomes. Surely, the educational system, the socioeconomic factors, and the family situation are all important in co-determining performance in secondary education—but is that all? Are there not unmeasurable idiosyncratic factors, the pupil's own personality, for example, that play a *quantitatively significant* role? We explore this in the first-ever application of the two-tier stochastic frontier model in the field of education (as far as we were able to determine), again in Chap. 5.

But latent forces keep coming up. Economics may have adopted the rationality/utility maximization paradigm to model how agents arrive at their economic decisions, but no one seriously argues that this is more than an approximation to reality. On top of that, there are many situations where decisions are not of economic nature in the first place. Then, other disciplines take the lead and they tell us that human beings are characterized by internal conflicting tendencies, rotating between opposing views on the same situation, and arriving at assessments and decisions that reflect this (r)evolving subjective reality. Perhaps the most well-known imagery on this matter is the glass being "half-empty/half-full." But this proverbial device is used in order to convey that different people may experience and view very differently the exact same situation; it is also a fact that *the same person* can swing between such opposing views, for reasons that give rise to whole scientific fields.

Life contentment (happiness) and job satisfaction are two examples of situations where people may go back-and-forth in assessing them for their own selves. An index of happiness or of job satisfaction can be constructed, and we can argue that its rational value can be expressed as a function of certain objective and observable characteristics of a person's health, occupation, income, quality of environment, and related aspects. But then the subjective view kicks in, influenced by the psychological makeup and the aspirations of the subject, and we can study its workings by using a 2TSF model, where the two opposing latent forces represent an optimistic and a pessimistic predisposition, respectively. With the 2TSF model, we can estimate their relative strength and understand in more detail how they affect the observed assessment and reaction of the subject that usually deviates from the fully rational one. These matters have indeed been explored by the use of the 2TSF methodology. Emboldened, after a brief mention of them in Chap. 5, we go global for the last application of that chapter, and we apply 2TSF analysis on data associated with the World Happiness Report.

Another situation where a 2TSF situation arises naturally is the *actual compliance* with laws and regulations that obviously touches upon all aspects of social organization and life, since there does not exist a human activity that is not governed by *some* laws and regulations. In abstract terms, it is the difference between "design" and "implementation of design." All agree that no plan is executed exactly as it was conceived and designed, but the reason for this is not just unforeseen situations, i.e., lack of knowledge or informational deficiency. It is also a matter of diligence and perseverance from the part of those entrusted with implementing it and also of informal "pushback" and risk-taking attitudes from the part of those that do not want to see the plan executed as designed. These are important opposing forces for the determination of the actual outcome, the extent of compliance, and they are bound to be unobservable.

To consider a more specific example from criminology, the number of, say, "property crimes" (like theft, burglary, robbery) committed in a municipality, a state, a country, is a complex result of structural socioeconomic factors that create an environment more (or less) conducive to committing such crimes, of the legal framework, of the measurable resources devoted to law enforcement (police and/or the justice system). These are observable aspects of the situation on which we will have data to one or the other degree. But there is also the "intensity" by which law enforcement agencies actually enforce the law, how diligent they are (or not), and how discretionary they are (or not), and there is also the degree of moral hazard characterizing the candidate offenders: how much, at a personal level, acceptance and respect for property rights has deteriorated. Again, these are important opposing factors that determine the outcome, and they are not observable.

While there exist studies that apply the 2TSF model to matters of compliance with environmental regulation, we do not know of any studies that use it in the field of criminology proper, and it certainly appears to be a very fruitful prospect.

1.3.4 Rudrian Treatments

Almost all drugs that are successful in battling a disease have some detrimental side effects on the human body. So if what we want to study is the overall health status of a patient, then the action of "drug administration" is a dualistic treatment that in reality compacts *two* treatments with opposing effects: the treatment of the disease itself and the treatment of the human body, a Rudrian treatment indeed.[12] And while the treatment of the disease is expected to have a beneficial effect on the health status of the patient, the treatment of the patient's body is expected to have a negative effect.

A standard generic formulation of a binary treatment effect model is

$$h_i = a + z_i'\gamma + bT_i + v_i,$$

where here h is an index measuring the health risk of patients (a higher value indicating higher risk/worse health status), T is the binary variable indicating whether patient i received the treatment or not, and z is a vector of control variables.

[12] Rudra is a Vedic deity who "shoots arrows of disease and death" while also being a healer with a thousand medicines.

If the treatment under examination is dualistic, the above model will estimate the average net effect, b, of the two conflicting treatments. In order to decompose them, we consider the treatment effect to be a random variable and not a constant, consisting of two opposing effects, one positive and one negative, possibly different per individual

$$b_i \equiv b_{p,i} - b_{n,i},$$

which leads to the equation

$$h_i = a + z_i'\gamma + b_i T_i + v_i = a + z_i'\gamma + [v_i + T_i(b_{p,i} - b_{n,i})].$$

With some peculiarities, this can be handled as a 2TSF conditional probability model, of the Probit variety. We present in detail such an application in Chap. 6, and we provide the theoretical treatment of the model in Chap. 13.

The "treatment effects" approach is general enough to cover any *purposeful intervention*. And Rudrian treatments certainly exist in the fields of psychoanalysis and education. Treating a patient with mental distress or illness through psychoanalysis can have antagonistic effects: "know thyself" may be accompanied with not being very happy with what you just realized about yourself, and learning to accept certain situations that are out of one's control may offer a measure of calm but also tendencies for depression. Disentangling and estimating separately these effects through a 2TSF model seems natural.

Apart from the operation of latent forces in education, which we previously discussed and will explore empirically in Chap. 5, the subjection of public education systems to "treatments" is a constant phenomenon, as societies grapple with how to prepare their young for the adult life. The subjects being persons at their early stage of intellectual and psychological development, their "response" to these treatments is bound to be very complex, and we expect to observe dualistic treatments here also.

As far as we know (and we looked rather meticulously), applications of the two-tier stochastic frontier model in these fields of study have not yet appeared, and we consider it a very promising avenue for research.

1.4 Historical Overview of the 2TSF Model

1.4.1 Origins

The 2TSF model first appeared in print in volume 69(2) of *The Review of Economics and Statistics*, May 1987, with co-authors Solomon J. Polachek and Bong Joon Yoon, under the title "A two-tiered earnings frontier estimation of employer and employee information in the labor market." Polachek had been thinking about informational effects on economic behavior and market outcomes already from his graduate and PhD years in the late 1960s. He was working on the idea that biases and preconceptions of employers, as regards women eventually dropping out of the labor force for maternal duties, would lead them to favor men over women in the workplace. But the realities of a PhD being what they are (it is not enough to like the topic, your supervisor—none other than Jacob Mincer—has to like it too) led him to abandon this approach as a way to interpret gender discrimination.[13] But the informational angle remained in the back of his mind, and, to use his own words,[14]

> In 1977 when I was at Chapel Hill, Peter Schmidt and Knox Lovell developed the stochastic frontier model which was a statistical version of data envelopment. Knox and Peter applied the technique to production models, namely how to identify situations in which output was less than possible, i.e., inefficiency. Given Knox and Peter's work, I began thinking about the technique and soon realized that it could apply to any situation in which observations are less than what they could be (or the analogue where they are higher than what they could be, as in cost functions). This led me to think about information acquisition going back to the Stigler model in which one pays more for a product than necessary (e.g., when one doesn't find the dealer charging the lowest price for a car) or when one takes a wage less than the maximum he or she could get because one doesn't know which firm is paying the most. This led to Rick Hofler's dissertation which we jointly first published in the Proceedings of the American Statistical Association in 1982.

The "Stigler model" refers to the seminal papers by Stigler (1961, 1962) that kicked off information economics, while the last sentence refers to the first use of a single-tier stochastic frontier model to capture information effects, Hofler and Polachek (1982, 1985), where the authors modeled the deviation of equilibrium wage below its full-information value due to workers having incomplete information as regards labor demand and failing to find the employer that would pay the maximum wage. From there, the clock was ticking for the birth of the two-tier model:

> I remained intrigued with the idea of measuring incomplete information (which we at the time called ignorance) and presented seminars at Binghamton and Cornell. Structural theorists didn't particularly like the approach, I suspect because it did not arise from a rigorous search model. At any rate, while at Binghamton, I thought the original model deficient because it neglects potential incomplete information employers also had about worker reservation wages. In short, the original one-sided frontier could not be used. At first, I thought one could simply run the one-sided frontier model twice, once with a negative one-sided term and then with a positive one-sided term to get at both employee and employer incomplete information, but then I thought that this approach simply looked at both sides independently in an ad hoc way. I discussed this with Bong, and after a bit of back and forth, we were able to work out the statistical formulation that we ended up using in the published article.

Such were the arcs of thought and chance that led to the 2TSF model. The work was presented at an Econometric Society session in the Meetings of the American Economic Association in New York, December 1985. Yoon remembers Polachek saying that not everyone appeared happy with the two-tier approach: "Why two-

[13] The approach would be eventually generalized and formalized in Phelps (1972).

[14] Personal communication with the authors, March 2021.

tier? Why not 10-tier then?" was a criticism heard from stochastic frontier scholars at the time, perhaps because they still had a vast *terra mysteriosa* to explore in production and cost analysis, where single-tier frontiers were the natural model. Polachek is not sure whether this objection was meant seriously or whether it was a slightly humorous invocation of Occam's Razor. Nevertheless overall the reception of the inaugural 2TSF paper must have been positive, since the authors submitted it for publication shortly after its presentation, and then it only took 5 months to pass the review process and gain acceptance (on September 15, 1986). Yoon remembers from the referee reports that the reviewers indeed celebrated "a new idea." He also remembers writing code in "Ancient language. I recall I wrote a Fortran program with the ZXMIN subroutine from the IMSL library." This is how the journey of the 2TSF model begun, originally as a tool to estimate informational deficiencies and to explain wage dispersion in the labor market.

1.4.2 Two-Tier Stochastic Frontier Analysis in Action

Scholarly preoccupation with the effects of incomplete information in the labor market goes back even before Stigler's papers and information economics and was linked to the existence of monopsonistic power in the hands of employers as buyers of labor services. As quoted by Mortensen (2003, p. 4), Bronfenbrenner (1956, p. 577–578) writes "The typical employer in an unorganized labor market is by no means a pure competitor facing market wage rates which he cannot alter. The mobility of the labor force, even between firms located close together, is low by reason of the inability of workers to wait for employment or risk unemployment, plus the inadequacy of the information usually available to them regarding alternative employment opportunities. This low mobility permits each employer to set his own rates and form his own labor market within limits which at some times may be quite wide."

But Polachek and Yoon (1987) (hereafter PY1987) viewed a two-way information effect and were concerned with quantifying the resulting dispersion of observed wages.

The authors pointed to the fact that even in relatively homogeneous competitive labor markets, we observe wage variation and not a single equilibrium wage as standard theory would predict. Based on search theory premises, they attributed this phenomenon to the (optimal) existence of incomplete information: employees searching for work do not know of all the opportunities and work offers (because it would be costly to obtain such information). On the other hand, employers do not know all workers that search for work and what they would be willing to supply at any given wage. And because incomplete information is heterogeneous and varies from employer to employer and from employee to employee, we also observe wage dispersion.

In order to estimate both effects, the authors extended the previous work by Hofler and Polachek (1982, 1985) that used a single-tier SF model with a negative one-sided term, by adding an additional one-sided unobservable term representing the employer "ignorance" and having a positive effect on the realized wage. This is how they arrived at the 2TSF composite error term $\varepsilon = v + w - u$. The systematic part of the regression equation represented the wage under full information but taking into account individual heterogeneity and variation around the observable characteristics of the average firm-worker pair. And thus, the 2TSF model was born, technically as a combination of an upper and a lower frontier in the same equation.

The authors assumed that the two one-sided error terms each followed an Exponential distribution, derived the density of the three-component error term (the third one being the random disturbance, assumed to follow a zero-mean Normal distribution), and applied maximum likelihood estimation. They also stratified their sample and calculated measures of employee and employer "ignorance" for various subsamples partitioned according to characteristics like gender, race, education, and tenure. In all cases, the realized wage was on average below the full-information estimated level.

Many 2TSF studies of the labor market followed the original PY1987 paper.[15] Among them, we point to Groot and Oosterbeek (1994) that extended the model to account for heterogeneity through individual attributes, Polachek and Yoon (1996) that applied their model using panel data in order to disentangle *unobserved* individual heterogeneity from the informational imperfections, Blanco (2017) that focused on the market for job placement services, and Das and Polachek (2017) who developed a new panel data 2TSF model in order to estimate gross flows in and out of the labor market, employment, and unemployment (joiners and leavers, respectively), embedding heterogeneity directly into the composite error term of the 2TSF specification.

Kumbhakar and Parmeter (2010) developed a 2TSF hedonic price framework for the housing market, which was picked up by Rajapaksa (2015) for the housing market in Brisbane, Australia. She found that the information deficiency of the buyer was higher than that of the seller's, leading to a price above the full-information level, in accordance with the argument in Pope (2008). But in the empirical study in Kumbhakar and Parmeter (2010), the selling price was found to be somewhat *below* the full-information one, showing that buyers are in a better position than sellers (in both the standard model as well as in its heteroskedastic variant). We can rationalize this result by observing that sometimes houses are sold under financial distress for the sellers, creating an unfavorable *bargaining* situation for their side that may mitigate and even reverse their informational advantage. We encountered an analogous situation earlier, where information and bargaining effects may compete with each other in the *same* side of the transaction. This creates an interesting identification and measurement challenge for which we will have more to say in Chaps. 4 and 10. Other 2TSF studies for the housing and land market are Xiaojun et al. (2018), Curtiss et al. (2021), and Seifert et al. (2021).

[15] A detailed literature review of the applications of the 2TSF model in a wide variety of situations can be found in Papadopoulos (2021d).

The 2TSF model is particularly suitable to analyze situations where bilateral bargaining takes place. The original 2TSF Nash-bargaining approach is due to Kumbhakar and Parmeter (2009), while Papadopoulos (2018, ch. 5) and Papadopoulos (2021d, 2024a) provided stronger structural foundations. Apart from the labor market, the model has seen applications in highly diverse situations, like the biotechnology market (Kinukawa & Motohashi, 2010, 2016) and international relations (see Wang, 2016, 2018, 2021) but also tourist shopping (Zhang et al., 2018) or the water-rights market in China (Xu et al., 2021).

The 2TSF model has also seen many applications in the health services market, where informational asymmetries and inefficiencies are believed to abound, favoring the supply side of the market. Gaynor and Polachek (1994) applied it to the market for physician services in the USA, while Tomini et al. (2012) estimated the effects of incomplete information on the informal (under-the-table) payments made to physicians, using a sample from Albania. Lu et al. (2011) applied the 2TSF model to data from China for the period 1989–2006. Other similar studies are Xu et al. (2013) and Yang (2018), while Wang and Lei (2017) examine how the informational deficiency and bargaining effects change with the age bracket of the patient.

Already the applications presented exemplify the wide reach of the 2TSF framework, outside the traditional topics of research in stochastic frontier analysis. But there is more: auctions with systematic underbidding and overbidding behavior (Ferona & Tsionas, 2012), over-pricing and under-pricing in the wine market (Fried & Tauer, 2019), the semen market for artificial insemination of bulls (Tauer, 2021), psychometrics, as well as many studies from Chinese scholars that examine financing constraints and agency costs on investment behavior, especially for research and development, the premium and under-pricing effect in initial public offerings (IPOs), the exchange rate of China's currency (the renminbi), the fruit export market, energy demand, and environmental regulations. On the other hand, European researchers appear reluctant to abandon their love for meat consumption and seem to channel their 2TSF explorations to the sausage market (Bonanno et al., 2019) and to the meat market (Petersen et al., 2023), in a "credence/hedonic" (hedonistic?) context.

While the diversity of applications rests on the abstract setting with which we started Sect. 1.3, when the time comes to interpret the estimation results, it is useful to have a defined set of metrics that each captures an interesting and important aspect of the effects of the unobservable factors on the outcome. This is the purpose of the next chapter. It is worth a careful reading in order to have a clear picture of what the 2TSF model actually measures and how. Also, it will help clarify certain misunderstandings that have appeared in the literature on the use and interpretation of the various measures. In order to make the chapter more interesting, we have also included two empirical applications that illustrate the use of these metrics.

Chapter 2
Metrics for the Latent Effects

In this chapter, we present a set of *metrics* which we can obtain once estimation is done, metrics that summarize from various angles the strength of the unobservables but also their interrelations and their effects on the outcome, as well as on the *potential* outcome. Although in actual work, these metrics follow estimation, it is they that carry the value and interpretational power of the 2TSF model in them, and we believe it is important to present them first, long before we deal with the more technical matters of statistical specification and estimation. For each metric, we discuss in some detail it's conceptual meaning, showing how it connects with and leads to its computable mathematical expression. These metrics spring from intuition and logic and are then formalized as a quantitative tool.

Our metrics live in two levels: the sample-level and the individual/observation level. At the sample level, we use the ("unconditional") expected value to compute the metrics, which is universally intuitively linked to the essentially non-statistical concept of "arithmetic average," a metric that people computed long before statistics, in order to get a sense of what happens "on average," and what should be "usually" anticipated.

For the individual metrics, we use the *conditional* expected value (or conditional expectation), which are well-established metrics of random quantities, being optimal in minimizing mean squared error (MSE). Although technically a conditional expectation is constructed in the same way as the unconditional one (it is an integral where we compute a weighted average of a variable over its domain), it operates as a *predictor*; not necessarily of something that has not yet happened, but of something that is unknown to us. It is an estimate of what actually is, and at the individual level, at each observation. So, say, the conditional expectation $E(y_i|\boldsymbol{x}_i)$ can be very different from the unconditional expected value $E(y)$. But by the law of iterated expectations, we know that $E\left[E(y_i|\boldsymbol{x}_i)\right] = E(y)$. So if $E(y)$ provides a point-estimate of the situation "on average," we could think of the collection of the conditional expectations as providing a *predicted distribution* of outcomes around this average situation.

A. Papadopoulos, C. F. Parmeter, *Two-Tier Stochastic Frontier Analysis for the Social Sciences*, https://doi.org/10.1007/978-3-031-81513-3_2

An argument could be made that expected values are more pertinent to characterize repetitive, mid-term phenomena rather than one-off decision-making situations, for which other estimators/predictors like the mode of the density or the median of a distribution may be more appropriate. While that may be true, the main purpose of applying statistics to social sciences is to analyze and uncover what has happened and why, not to form a forecast or to make a one-time decision. So in this book we will stick to the use of the expected value, conditional and unconditional.[1]

Nevertheless, for the individual metrics, where we will obtain a whole series of estimates, it is prudent to explore some quantile properties of these series, like their empirical median (namely, the empirical median of the series of conditional expectations). This guards against estimation inaccuracy that may lead the expected value of some metrics being unreasonably high due to the presence of outliers.

There is another dimension along which our metrics differentiate: the units of measurement at which we carry out the estimation, contrasted to the original measurement units of the dependent variable.

If the equation that we will estimate includes the dependent variable in its original measurement units, the estimated latent effects are also measured in the measurement units of the outcome, and so they can be used directly in the computation of the metrics. But if in the estimating equation the dependent variable has been transformed, the situation changes: the metrics are conceptually similar but they have to be computed using different formulas, since we want them to express their effects on the original outcome, not on the transformed one, while the estimates will be expressed in the transformed measurement units. The transformation we consider is the (natural) logarithmic one, ubiquitous across disciplines.

Some conceptual interpretations of the metrics are common throughout, and so they are discussed the first time the metric is presented, and not again after that. On the other hand, when we examine models with logarithmic dependent variables, some intricacies arise. Therefore, readers are strongly encouraged to read through the whole chapter, without self-selecting some parts of it.

2.1 Dependent Variable in Original Measurement Units

There are *four* values that have special importance in the analysis of an outcome influenced by opposing latent forces, one actual and three counterfactuals. For the case where the estimating equation coincides with the original one, these four values are defined as follows:

The *actual* outcome

$$y = g(\boldsymbol{x}) + \varepsilon = g(\boldsymbol{x}) + v + w - u,$$

[1] Nevertheless, we offer a pathway to mode-based prediction in Sects. 11.2.6 and 14.2.1.

which is also the regression equation itself.

The *balanced* outcome

$$y^B = g(\boldsymbol{x}) + v = y - w + u.$$

The balanced outcome represents a counterfactual where the latent factors have zero effect, or perhaps more interestingly, when their strength is equal and so their effects cancel out.

The *upper frontier*

$$y^{UF} = g(\boldsymbol{x}) + v + w = y + u.$$

The upper frontier captures the stochastic maximum value of the outcome, if it were the case that the negative unobservable was zero. Connecting with the discussion in Chap. 1, note that the *upper* frontier is defined as the value of the outcome in the absence of the force u that drags the outcome *down*, and not as the value of the outcome when the positive force w obtains its maximum value.

Finally, the *lower frontier*

$$y^{LF} = g(\boldsymbol{x}) + v - u = y - w.$$

The lower frontier captures the stochastic minimum value of the outcome, if it were the case that the positive unobservable was zero. Analogously with the upper frontier, the lower frontier is defined as the value of the outcome in the absence of the positive force w, and not as the value of the outcome when the negative force takes its maximum value (in absolute terms).

It is useful to show the interrelations between these concepts. We have

$$y - y^{LF} = y^{UF} - y^B = w, \qquad y^{UF} - y = y^B - y^{LF} = u,$$

$$y - y^B = w - u, \qquad y^{UF} - y^{LF} = w + u.$$

So w can be viewed as the distance by which the actual outcome is pushed up from the lower frontier, but also as the distance between the upper frontier and the balanced outcome. In an analogous manner, u is the (absolute) distance by which the actual outcome is pushed down from the upper frontier, but also the absolute distance between the balanced outcome and the lower frontier. The difference $w - u$ is the (signed) distance of the actual from the balanced outcome. Their sum $w + u$ provides the range of the outcome due to these latent forces, which can be used to assess volatility caused by the unobservable factors.

We warn the reader that when we will look at the case where estimation is carried out using this additive representation of the error term but the original equation is multiplicative, the definition and the resulting relations between these four fundamental outcome values will be different. In short, while the additive structure leads to equivalences that are expressed in differences, in the multiplicative case, the equivalences will be expressed in ratios.

Combinations of these values together with measures of unobservables produce meaningful results, since each of these outcome values can be used as a base of measurement to produce a *relative* metric. We present them next, starting with sample-level metrics where each provides an average characterization for the whole sample, and then continuing with their individual, observation-level counterparts. We note that all these metrics incorporate an assumption that will be used throughout this book, namely, that the noise error component v has expected value zero, $E(v) = 0$.

2.1.1 Sample-Level Metrics

To avoid clutter we do not show in our notation that the expected values will be estimated with the use of the estimated parameters and the assumed distribution and also with the use of sample means when we need the expected value of the dependent variable or of a function of the regressors. In other words, we avoid here the "hat" symbol.

I. Relative Weight in, and Relative Loss of, Outcome

A first meaningful relative metric is what proportion of the actual outcome is attributed to the positive unobservable. Also, what is the loss we experience due to the presence of the negative unobservable. These speak for their individual importance relative to what actually happens in the real world. As an average over the sample, we have for the first one

$$M_1 \equiv \frac{E\left(y - y^{LF}\right)}{E(y)} = \frac{E(w)}{E(y)}.$$

The logic of the M_1 metric (in expected value terms) is the following: we ask "what is the relative weight of w in the actual outcome?"—and we note that it is equivalent to ask "what is the difference/distance between the actual outcome and the counterfactual outcome where $w = 0$?" This counterfactual is y^{LF}. So the contribution of w in the outcome is the difference $y - y^{LF}$. We then divide it by y to obtain a relative metric, which verbally is expressed as "on average the relative contribution of w in the outcome is XX%."

In an analogous manner, we ask "what is the difference/distance between the actual outcome and the counterfactual outcome where $u = 0$?" This counterfactual is y^{UF}. The distance $y^{UF} - y$ is what we *could* have had in addition, if it were not for the negative unobservable u. Divided by y, it is expressed in terms relative to it, giving the M_2 metric:

$$M_2 \equiv \frac{E\left(y^{UF} - y\right)}{E(y)} = \frac{E(u)}{E(y)}.$$

Verbally, this metric says "the loss of y due to u is XX% of the actual value." Note that we express it in absolute terms (i.e., as a positive number), for convenience, but it represents a reduction in the value of the outcome. Certainly, one can have a model where the outcome y represents something "undesirable" (say, the amount of pollution), in which case the w-unobservable would represent the undesirable latent force that tends to increase pollution, and the u-unobservable would represent a desirable force that tends to reduce pollution. To focus ideas we will keep thinking that y is such that "the more the better," but it is easy to make the interpretational switch when need arises.

2.1.1.1 Alternative View of M_1 and M_2 as Producer and Consumer Surplus

Suppose that we are studying a sample of buy/sell transactions. Then we can view the M_1 and M_2 metrics in a different way. Here y is the actual price, and one of the ways that the latent forces can be interpreted is the following: w is a force that tends to increase the price because the buyers/consumers are not fully informed about the opportunities in the market, while u tends to decrease the price because the sellers/producers are not fully informed about the same.

M_1 as Producer Surplus

The quantity $y - y^{LF}$ that is used for the M_1 metric measures that part of the actual price that the seller was able to enjoy, because buyers/consumers were *not* fully informed, while the seller was in both cases information deficient. The counterfactual y^{LF} has the w-factor equal to zero (a case of fully informed buyers), while both y and y^{LF} have a non-zero u (which represents sellers with incomplete information). Now, in standard microeconomic market theory, "producer surplus" is that part of actual revenues that suppliers were able to enjoy, because the other side of the market was *not* a monopsonist imposing full "price-discrimination" and forcing every supplier to sell at its own minimum acceptable price. The correspondence of the two situations is clear: in our example sellers enjoy something because buyers do not have full information. In the standard microeconomic theory setting, sellers enjoy something because buyers do not have full market power. This is why the metric M_1 is analogous to the standard concept of producer surplus (in relative units, since we divide it by actual y).

Carefully note that the formula for the computation of the metric involves w which is associated with the *buyer/consumer*—but the metric gives us in this interpretation the surplus of the *seller/producer*. This is another incarnation of the interdependence of the two sides.

M_2 as Consumer Surplus

In an analogous manner, the quantity $y^{UF} - y$ that is used for the M_2 metric measures *what the buyers avoid paying* because the sellers were *not* fully informed, while buyers in both cases were information deficient. The counterfactual y^{UF} has the u-factor equal to zero (a case of fully informed sellers), while both y and y^{UF} have a non-zero w (which represents buyers with incomplete information). And in standard microeconomic market theory, "consumer surplus" is the amount that consumers avoided to pay because the other side of the market was not a monopolist applying full price discrimination and having every consumer buy at their maximum willingness to pay. Again the correspondence of the two situations is clear: in our example buyers avoid an extra payout because sellers do not have full information. In the standard microeconomic theory setting, buyers avoid higher expense because sellers do not have full market power. So the metric M_2 is analogous to consumer surplus (again in relative units, since we divide it by actual y).

Here too, note that the computation of the metric involves the latent factor u that is associated with the seller/producer, but it measures the surplus of the buyer/consumer.

2.1.1.2 Yet Another Alternative View of M_1 and M_2 as quantifying the Needed Action to Reach the Frontiers

Suppose that the lower frontier is a desirable one, say it is the minimum level of pollution. Then the metric M_1 measures how much we have to reduce current pollution y in percentage terms, to reach this lower frontier Y^{LF}; it is the *mark-down* needed given our current position. Correspondingly, suppose that the upper frontier is a desirable one, say, maximum sales. Then the M_2 metric quantifies how much we should increase our current sales y in order to reach the maximum y^{UF}, it is the *mark-up* needed given our current situation. So these metrics may operate also as quantifiers of *goals*, in a language (percentages) that both businesses and policy makers are comfortable with.

II. Relative Strength

Next, comparing the two unobservables using the ratio of their expected values

$$M_3 \equiv \frac{E(w)}{E(u)},$$

gives us their relative strength. M_3 is a useful measure to gauge which side of the relationship has more influence over variation in y. Note that M_3 does not require that the two latent forces are "similar in nature" in order to be meaningful, because we measure not the latent forces themselves but their effects on the outcome, in the units of the outcome.

III. Volatility

Another useful metric is akin to a coefficient of variation (which is defined as the standard deviation over the mean):

$$M_4 \equiv \frac{E\left(y^{UF} - y^{LF}\right)}{E(y^B)} = \frac{E(w) + E(u)}{E(g(\boldsymbol{x}))}.$$

This metric answers (in relative terms) the question "what is the (average) range that the unobservables create around the *balanced* outcome, the outcome that would have occurred if the unobservables had zero effect?" In other words, what is the volatility (on average) of the outcome due to the latent forces?

IV. Markup/Markdown Effects on the Balanced Outcome Equally useful is to see the effect of the unobservables as a "markup/markdown" but here on the *balanced* outcome, how far we tend to move from it in relative terms due to the latent forces.

$$M_5 \equiv \frac{E\left(y^{UF} - y^{B}\right)}{E(y^B)} = \frac{E(w)}{E(g(\boldsymbol{x}))},$$

$$M_6 \equiv \frac{E\left(y^{B} - y^{LF}\right)}{E(y^B)} = \frac{E(u)}{E(g(\boldsymbol{x}))},$$

$$M_7 \equiv \frac{E\left(y - y^{B}\right)}{E(y^B)} = \frac{E(w) - E(u)}{E(g(\boldsymbol{x}))}.$$

The metrics M_5 and M_6 give us the individual effect of the positive and negative unobservables on the balanced outcome, respectively. Here too, M_6 is expressed as a positive number, but it express a "markdown," a downward force that tends to reduce y^B. It is to be noted that we have the relation $M_4 = M_5 + M_6$. Adding the two markup/markdown effects, gives us the volatility metric.

The metric M_7 gives the net effect on the balanced outcome, the net markup/markdown effect. M_7 is an "after all is said and done" index of how far from the balanced outcome (and in which direction) we have moved because of the workings of the latent forces. M_7 is a signed metric, and it may be positive or negative.

V. Distance from the Frontiers

Distance of the actual outcome from the two frontiers in relative terms should be measured with respect to the frontier value. So,

$$M_8 \equiv \frac{E\left(y^{UF} - y\right)}{E(y^{UF})} = \frac{E(u)}{E(y) + E(u)},$$

$$M_9 \equiv \frac{E\left(y - y^{LF}\right)}{E(y^{LF})} = \frac{E(w)}{E(y) - E(w)}.$$

M_8 measures the distance of the actual outcome from the upper frontier as a percentage of the latter. M_9 measures the distance of the actual outcome from the lower frontier as a percentage of the latter. Computationally, one could use for the denominators the expressions $E(g(\boldsymbol{x})) + E(w) = E(y) + E(u)$ and $E(g(\boldsymbol{x})) - E(u) = E(y) - E(w)$, but those that we use contain fewer elements affected by inaccuracy induced by the estimation process, since $E(y)$ will be computed as the sample mean of y.

The metric M_8 is useful if, for example, we aspire to reach the upper frontier. Compared to metric M_2, the latter as already mentioned provides the markup figure based on our current situation in order to reach the UF, while the M_8 metric gives the proportion of UF that we are still missing out on. Or, if the upper frontier is *not* desirable (say maximum contamination), it gives the proportion of it that we have managed to avoid.

Analogously, the metric M_9 tells us how much of an undesirable lower frontier in proportional terms we still bear. If we think again in terms of pollution, $M_9 = 0.25$ conveys the information "our current level of pollution is 25% higher than minimum pollution attainable." The related metric M_1 (which, in deterministic terms would be equal to 0.2 when $M_9 = 0.25$), tells us how much we have to reduce current pollution in order to reach LF, as a percentage. As another example, suppose now that LF is the break-even (zero profit) point of sales for a firm. Then $M_9 = 0.25$ would send the message "we are still 25% above our break-even point," while the related metric $M_1 = 0.2$ would say "we can lose up to 20% of current sales and remain above our break-even point."

VI. Net Gain

We started with metrics expressed in terms of the actual outcome y, M_1 and M_2. Metric M_3 is concerned with a comparison of the two latent forces. Metrics M_4, M_5, M_6, and M_7 provide information of the effects of the latent forces on the counterfactual balanced outcome y^B, and the metrics M_8 and M_9 offer results in relation to the upper and the lower frontier, respectively. We close our set of metrics by coming full circle and returning to a measure grounded in reality, a final measure relative to the actual outcome y.

This is a metric that measures the difference of the actual outcome from the balanced outcome, but *as a percentage of actual outcome*:

$$M_{10} \equiv \frac{E\left(y - y^B\right)}{E(y)} = \frac{E(w) - E(u)}{E(y)}.$$

The metric M_{10} has the same numerator as M_7, but crucially, it is expressed in terms of the actual outcome y. Why does this matter? Metrics relative to the balanced outcome are in principle of more interest to researchers and policy makers;

on the other hand, metrics relative to the actual outcome are metrics that are more meaningful to the actual agents/decision-makers involved in the real-world phenomenon under study. To give a tangible example, suppose that we examine observed wages and how they are affected by incomplete information. We conduct our research, and we tell a worker "your wage is 10% below the balanced/full-information wage" (that would be the M_7 metric). The reaction of the worker would likely be "who cares about this non-existent 'balanced wage'? This is not a base of measurement I care about. I care about what affects my wage, so I do care about information deficiencies, but what I want to know is what I lost due to these deficiencies as a percentage of what I got. This I understand and can relate to." And that would be the M_{10} metric. The difference is that policy makers have in principle a mindset, or at least a duty, to remain neutral and promote balance and an "equal footing" of all involved in a situation, hence their interest for metrics relative to y^B. But self-interested participants do not want an "equal footing" and do not seek balance in economic or other interactions; they look to promote their own interests.

We call this metric a measure of "net gain," and it must be clearly distinguished from metrics M_1 and M_2 which as we saw can be interpreted as "surpluses" in the specific sense that the word has in microeconomic theory. Apart from their conceptual differences, we note that M_{10} is signed (hence "net"); it can be positive or negative (while M_1 and M_2 are always positive). M_{10} can be usefully decomposed into its negative and positive subsets, when we apply the metrics at the level of each individual observation.

We have completed the set of core metrics related to the two-tier stochastic frontier model. Before presenting the analogous measures at the observation-level, we summarize in Table 2.1 the computing formula for each metric and its interpretation.

2.1.2 Individual Metrics

After estimation, we will obtain a series of residuals $\{\widehat{\varepsilon}_i\}$ that are consistent predictors of the true error term series $\{\varepsilon_i\}$. This series includes (imperfect) information on the unobservable factors w and u for each observation. This information can be summarized by the relevant conditional expectations, $E(w_i|\varepsilon_i)$, $E(u_i|\varepsilon_i)$. With these we can compute the metrics per observation in the sample.

There are three main ways to utilize the individual metrics. At the observation level we will obtain a series for each metric. The first application of these series is to use them in order to *rank* the observations. Which of the metrics provide meaningful ranking criteria depends on the application.

The second is to explore the empirical distribution of each series. To begin with, one should obtain the *median* of the series, as an alternative measure of centrality that is robust to outliers, compared to the mean. But we can proceed further and graph the empirical relative frequencies, examine other empirical quantiles apart from the median, and reflect on skewness and kurtosis properties of the empirical

Table 2.1 Metrics of latent forces. Formulas and interpretation when the dependent variable is in its original measurement units. Metrics M_2 and M_6 represent absolute distances

Metric	Formula	Description
M_1	$\frac{E(w)}{E(y)}$	Proportion of actual outcome y attributed to positive force w / producer surplus / markdown to reach the lower frontier
M_2	$\frac{E(u)}{E(y)}$	Relative reduction in actual outcome y due to negative force u / consumer surplus / markup to reach the upper frontier
M_3	$\frac{E(w)}{E(u)}$	Relative strength of w compared to u
M_4	$\frac{E(w)+E(u)}{E(g(\boldsymbol{x}))}$	Volatility of balanced outcome y^B due to w, u
M_5	$\frac{E(w)}{E(g(\boldsymbol{x}))}$	Markup effect of w on balanced outcome y^B
M_6	$\frac{E(u)}{E(g(\boldsymbol{x}))}$	Mark-down effect of u on balanced outcome y^B
M_7	$\frac{E(w)-E(u)}{E(g(\boldsymbol{x}))}$	Net effect on balanced outcome y^B ($\pm$)
M_8	$\frac{E(u)}{E(y)+E(u)}$	Distance of actual outcome y from upper frontier y^{UF}
M_9	$\frac{E(w)}{E(y)-E(w)}$	Distance of actual outcome y from lower frontier y^{LF}
M_{10}	$\frac{E(w)-E(u)}{E(y)}$	Net gain ($\pm$)

distribution, on dispersion measures like the interquantile range, etc. This goes well beyond the computation of a sample-central measure, and it presents a much more refined picture about what happens with the unobservables in the sample.

The third use is as an informal diagnostic for the existence of missed heterogeneity in the w and u forces or even endogeneity of regressors. Specifically, after an estimation stage of a model where regressors are assumed exogenous and w and u are assumed identically distributed over the sample (see Chap. 7), we first obtain the series of individual metrics based on these estimates, and then we stratify the sample based on a regressor (usually a binary or a categorical variable) and compute subsample means or medians of the metrics, per stratum. If they differ visibly, it is an indication that there exists heterogeneity in w and u (along the characteristic represented by the regressor used for the stratification), and one should consider the possibility of estimating a model that allows for heterogeneity (see Chap. 8). Also, it could be an indication that this regressor is correlated with the composite error term, creating endogeneity that undermines the reliability of the estimates (see Sect. 14.1.4).

The individual-level measures are defined and computed as their sample-level counterparts with one exception: when the dependent variable itself appears, what we use is its actual realization, not a conditional expectation. This is justified because all our estimation results are essentially conditioned on the sample at hand. So, formally we look at $E(y|\boldsymbol{x}, \varepsilon) = E(y|y) = y$.

I. Relative Weight in/Loss of Outcome The contribution of w in the outcome and what we lost due to u as a proportion are defined as

$$\{M_{1i}\} \equiv \frac{E(w_i|\varepsilon_i)}{y_i},$$

$$\{M_{2i}\} \equiv \frac{E(u_i|\varepsilon_i)}{y_i}.$$

We have used curly brackets in the left hand side to stress that these refer to a series with length equal to the sample size.

The interpretations of these metrics as producer/consumer surpluses or as markdown/markup factors to reach the frontiers are valid here too.

II. Relative Strength The relative strength of the latent forces is defined as

$$\{M_{3i}\} \equiv \frac{E(w_i|\varepsilon_i)}{E(u_i|\varepsilon_i)}.$$

Our best prediction of w (in mean squared error sense) is $E(w_i|\varepsilon_i)$ and correspondingly for u. It is natural to consider the ratio of these best predictions to gauge their relative size and strength per observation.

III. Volatility The volatility series at the individual level is defined by

$$\{M_{4i}\} \equiv \frac{E(w_i|\varepsilon_i) + E(u_i|\varepsilon_i)}{y_i - E(w_i|\varepsilon_i) + E(u_i|\varepsilon_i)}.$$

The denominator is an estimation of the balanced outcome per observation that uses the conditional expectation as predictor for the unobservables.

IV. Markup/Markdown Effects Analogously, the markup/markdown effects on the balanced outcome, unilateral and net, use this estimate for the balanced outcome:

$$\{M_{5i}\} \equiv \frac{E(w_i|\varepsilon_i)}{y_i - E(w_i|\varepsilon_i) + E(u_i|\varepsilon_i)},$$

$$\{M_{6i}\} \equiv \frac{E(u_i|\varepsilon_i)}{y_i - E(w_i|\varepsilon_i) + E(u_i|\varepsilon_i)},$$

$$\{M_{7i}\} \equiv \frac{E(w_i|\varepsilon_i) - E(u_i|\varepsilon_i)}{y_i - E(w_i|\varepsilon_i) + E(u_i|\varepsilon_i)}.$$

Of interest is to consider two variants, the positive and negative values of $\{M_{7i}\}$ separately:

$$\{M_{7i}^{(+)}\} \equiv \frac{E(w_i|\varepsilon_i) - E(u_i|\varepsilon_i)}{y_i - E(w_i|\varepsilon_i) + E(u_i|\varepsilon_i)} \Bigg| \forall i : E(w_i|\varepsilon_i) > E(u_i|\varepsilon_i),$$

$$\{M_{7i}^{(-)}\} \equiv \frac{E(w_i|\varepsilon_i) - E(u_i|\varepsilon_i)}{y_i - E(w_i|\varepsilon_i) + E(u_i|\varepsilon_i)} \Bigg| \forall i : E(w_i|\varepsilon_i) < E(u_i|\varepsilon_i).$$

These will answer the questions "if positive, how positive" and "if negative, how negative" and is an example of what kind of sample stratification is proper and can provide additional insights.

V. Distance from the Frontiers The formulas to compute the distance from the frontiers at the individual level incorporate less estimation uncertainty compared to the markup/markdown formulas, because in each case we have to remove, through the use of a conditional expectation, only one of the latent forces from the denominator:

$$\{M_{8i}\} \equiv \frac{E(u_i|\varepsilon_i)}{y_i + E(u_i|\varepsilon_i)},$$

$$\{M_{9i}\} \equiv \frac{E(w_i|\varepsilon_i)}{y_i - E(w_i|\varepsilon_i)}.$$

VI. Net Gain For the final net gain metric, we have

$$\{M_{10i}\} \equiv \frac{E(w_i|\varepsilon_i) - E(u_i|\varepsilon_i)}{y_i}.$$

As is the case with $\{M_{7i}\}$, metric $\{M_{10i}\}$ is also signed so we can usefully partition it in two, but with a twist: because this metric measures actual gains, for its subseries with the negative values, we reverse the sign, so that we see again a positive number, since it is the gain "of the other side":

$$\{M_{10i}^{(+)}\} \equiv \frac{E(w_i|\varepsilon_i) - E(u_i|\varepsilon_i)}{y_i} \Bigg| \forall i : E(w_i|\varepsilon_i) > E(u_i|\varepsilon_i),$$

$$\{M_{10i}^{(-)}\} \equiv \frac{E(u_i|\varepsilon_i) - E(w_i|\varepsilon_i)}{y_i} \Bigg| \forall i : E(w_i|\varepsilon_i) < E(u_i|\varepsilon_i).$$

This will separate the observations according to which side of the transaction enjoyed a gain, and it will answer the respective conditional questions.

2.2 Application: Personal Computer Pricing, Back in the Day

To illustrate the use of the metrics when the dependent variable is in its original form, we employ a subsample of a rich data set that contains the retail price and product characteristics on personal computers from the early days of the personal computer market. The broader data set has been used in Deltas and Zacharias (2004), Stengos and Zacharias (2006), and Deltas et al. (2011).[2] The data set was constructed by scavenging corporate advertisements selling personal computers by mail in a popular monthly magazine of the era ("PC Magazine"—early 1990s). Even though these prices were not negotiable, being advertised prices they first and foremost reflect the price-setting behavior of the firm. So, with such a data set, a hedonic regression would reflect the *beliefs of the firm* about the value that consumers assign to each product characteristic.

We can then wonder what other considerations played a part on the price decided upon by the firm. We construct a 2TSF model by arguing as follows: at the time certainly, and even today, personal computers were never only about value-in-use but also a new device reflecting a modern approach to life. So there was room for a firm to exploit the hype surrounding each new computer offering, by charging a bit more for the privilege. This can be represented by the positive force w. On the other hand, competition was always fierce, and so the component u can represent fear of competition that may make a firm reduce the price of a product. This is also a good example as to how flexible the 2TSF framework is, since the tendencies just described are very intuitive, but this is the first time they are examined with a 2TSF model. We will use data from the first half of 1995 and only related to advertisements by the firm "ZEOS" which for a period was the strongest no-name competitor of the computer brands of the time and the firm that introduced many commercial innovations in the market (like bundling software with the computer hardware). The size of the data set is 1,054 observations (spread over six monthly editions of the magazine), which averages to 175 selling ads per issue.

To understand the historical context, Pentium processors were introduced 1 year before, while 17-inch screens in the second half of 1994. The dependent variable, price, is in nominal US dollars (USD). We use three integer variables, size of hard drive higher than 85 (that would be MB, not GB), size of RAM higher than 2 MB, and clock speed higher than 25 MHz. In this way the constant term represents a basic configuration of 2 MB of RAM, a clock speed of 25MHz, and 85 MB of storage. Also, this basic configuration has a 14" screen. We then include five dummy variables: whether the processor is Pentium, whether the screen is 15" or whether it is 17", whether the product has a CD-drive, and whether it has a multimedia kit (speakers and a sound card). In Table 2.2 we present summary statistics of the variables used.

[2] We thank professor Eleftherios Zacharias for providing access to this data set.

Table 2.2 Summary statistics of the variables used, $n = 1{,}054$

Variable	Unit	Mean	Median	S.D.	Min	Max
Price (dep. variable)	USD	2485.0	2494.0	622.8	1195.0	4190.0
RAM > 2	MB	11.2	14.0	7.4	2.0	22.0
Clock speed > 25	MHz	51.6	50.0	21.4	8.0	95.0
$D_{pentium}$	0,1	0.61		0.49	0.0	1.0
D_{scrn15}	0,1	0.51		0.50	0.0	1.0
D_{scrn17}	0,1	0.10		0.30	0.0	1.0
D_{CD}	0,1	0.78		0.41	0.0	1.0
$D_{multimedia}$	0,1	0.15		0.36	0.0	1.0

Table 2.3 Maximum likelihood estimation of the 2TSF Normal-Exponential model. Period : 1995-S1. $n = 1{,}054$. Dependent variable is advertised price in nominal 1991 USD. Standard errors based on outer products matrix

Variable	Coefficient	Std. Error	p-value
Constant	1070.50	22.43	0.00
Hard drive size > 85 MB	− 0.33	0.04	0.00
RAM > 2 MB	60.33	1.73	0.00
Clock speed > 25 MHz	7.74	0.30	0.00
$D_{pentium}$	394.00	12.83	0.00
D_{scrn15}	59.17	18.06	0.00
D_{scrn17}	310.88	26.90	0.00
D_{CD}	187.56	19.28	0.00
$D_{multimedia}$	11.40	16.85	0.50
σ_v	120.63	11.63	0.00
σ_w	164.03	12.27	0.00
σ_u	55.16	9.17	0.00

For estimation purposes, we will use a linear regression model and the Normal-Exponential 2TSF statistical specification for the error term, which is presented in detail as regards its statistical and technical aspects in Chap. 7. Our purpose here is to show how the various metrics work and can be interpreted, irrespective of how the estimates on which the metrics are based have been obtained. So as regards the model used, it suffices to say that the components of the composed error term $\varepsilon = v + w - u$ are specified as $v \sim \mathrm{N}(0, \sigma_v)$, $w \sim \mathrm{Exp}(\sigma_w)$, $u \sim \mathrm{Exp}(\sigma_u)$. Namely, the noise error component v is assumed to be a zero-mean Normal random variable with standard deviation σ_v, and the two latent forces are both assumed to follow the Exponential distribution with scale parameters σ_w and σ_u, respectively, which also represent their mean and standard deviation. The three components are assumed jointly independent. The sample is assumed to be comprised of identically and independently distributed observations. We apply maximum likelihood estimation and present the results in Table 2.3.

But since this is the first table with estimation results in the book, before looking at the results, a comment is in order as to its informational content: the specific table contains, apart from the estimates themselves, both standard errors *and* p-values.[3] Usually, published tables with estimates include either one or the other, but not both. We believe that each addresses a different situation: standard errors are the core metric of the uncertainty surrounding the estimates, and it is the more suitable when the focus is on scientific explanation and possible further exploration. But p-values are those that clearly reflect the risk that we have to accept in order to treat the associated estimates as usable for conclusions, decisions, and policy. "Usable" here means that we can statistically ascertain the *sign* of the coefficient, and hence the direction of influence that a regressor has on the the dependent variable (which is what the "statistical significance" comes down to in practice). So in subsequent tables, sometimes we will include one of the two, sometimes both.

Turning to the estimation results, interesting conclusions arise: the basic configuration (which includes also every other component and associated cost that may not have direct hedonic value for the consumers) represents on average 43% of the price. What appears counterintuitive is the *negative* coefficient on the size of the hard drive. It can be interpreted as follows: the firm believes that consumers place negative value on large hard disks, because at the time, they did not really have much (perceived) use for them, compared to having a Pentium processor, a larger screen, or a CD drive (which are the three product characteristics that drive the price set).

Turning to the latent forces, they appear to have relatively smaller effects on the price, with the average positive force (σ_w), the hype and novelty aspect, weighing much more prominently in the mind of the firm compared to its fear for the competition (σ_u).

We move now to the metrics defined previously. We note that for the metrics M_1 and M_2, the most appropriate interpretation is the more abstract one, as "relative contribution" and "relative loss," respectively, while the last "net gain" metric M_{10} does not appear meaningful here since what we are examining is internal conflicting forces in the decision-makers of the firm. For the other nine metrics, we first present side-by-side the sample-level mean values (which are computed using sample means and the estimated σ_w and σ_u) and the sample averages of the series with the individual metrics (Table 2.4).

In theory, by the law of iterated expectations, we have $E[E(x|z)] = E(x)$, so at the limit these two magnitudes should be equal for all metrics. But of course, finite sample size and variability as well as estimation inaccuracy get in the way. Still, the estimates are pretty close and essentially convey the same information for all practical purposes, except in one case, M_3, the relative strength metric. The first number is computed as $\hat{E}(w)/\hat{E}(u) = \hat{\sigma}_w/\hat{\sigma}_u$, while the second number is computed as $n^{-1}\sum_{i=1}^{n}[\hat{E}(w_i|\varepsilon_i)/\hat{E}(u_i|\varepsilon_i)]$. It is no surprise that a ratio metric like this could be furthest from the limiting result.

Interpreting the metrics, we see that their percentage relation to the actual price (metrics M_1 and M_2) is small. The average volatility of the hedonic price (the "balanced outcome" in this case) is no more than 10% (M_4 metric), while the

[3] It does not contain "significance asterisks," and no table in this book will include such.

Table 2.4 Metrics of latent forces—mean values

Metric	Description	Sample-level mean	Series mean
M_1	Relative contribution to outcome of positive force w	0.066	0.066
M_2	Relative loss in outcome due to negative force u	0.022	0.024
M_3	Relative strength w/u	2.974	3.590
M_4	Volatility of balanced outcome due to w, u	0.092	0.097
M_5	Markup effect of w on balanced outcome	0.069	0.072
M_6	Mark-down effect of u on balanced outcome	0.023	0.025
M_7	Net effect on balanced outcome	0.046	0.047
M_8	Distance from upper frontier	0.022	0.024
M_9	Distance from lower frontier	0.071	0.073

Table 2.5 Deciles of individual metrics

	Cumulative Probability								
Metric	0.1	0.2	0.3	0.4	0.5	0.6	0.7	0.8	0.9
M_1	0.026	0.031	0.038	0.044	0.051	0.059	0.071	0.094	0.128
M_2	0.013	0.015	0.017	0.019	0.021	0.024	0.026	0.031	0.039
M_3	0.759	1.073	1.506	2.035	2.467	3.033	3.789	5.584	8.36
M_4	0.051	0.057	0.064	0.072	0.079	0.087	0.102	0.122	0.165
M_5	0.026	0.031	0.038	0.044	0.052	0.062	0.075	0.102	0.144
M_6	0.014	0.016	0.018	0.020	0.022	0.024	0.027	0.032	0.04
M_7	−0.009	0.002	0.012	0.023	0.030	0.040	0.054	0.082	0.127
M_8	0.013	0.015	0.017	0.019	0.021	0.023	0.025	0.030	0.038
M_9	0.026	0.032	0.039	0.046	0.053	0.063	0.077	0.104	0.147

markup and markdown effects are $\sim$7% and $\sim$2.5%, respectively (metrics M_5 and M_6), leading to an overall push-up effect of the hedonic price by just 4.6% (M_7). But also the distance from the frontiers of the set price is small (metrics M_8 and M_9).

Some intuition as to why these forces appear to affect little the price set by the firm can be gained by going back to the summary statistics of the regressors and noting that there is not a strong novelty feature included in this period: the Pentium processor is already present in 61% of the products and the CD-drive in 78% of them. The 17" screen is present only in 10%, but even today a large screen remains one of the most expensive components of a desktop computer, so already this would have pushed the total price high enough, to be exploited for its novelty value. As for the price-depressing competition effect, ZEOS relied on affordability to counter the brand power of its competitors, so its prices were generally already lower, and therefore only for relatively few of its products it could feel threatened by pure price competition.

More nuance can be obtained from the quantiles of the series with the individual metrics. We provide the deciles in Table 2.5.

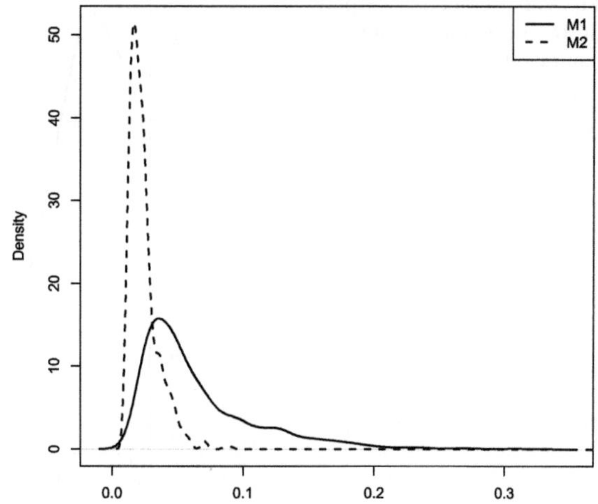

Fig. 2.1 Kernel density plot of M_1 and M_2 using rule of thumb bandwidth

Inspecting the deciles, we see that there is a small proportion of offerings where the fear of competition overtakes marginally the tendency of the firm to exploit the hype (the negative quantile value in the 0.1-decile of M_7). But we also see that in 90% of the products, the pricing above the hedonic price does not exceed 13% (the 0.9-decile of M_7). Other usual measures like the standard deviation, skewness, and excess kurtosis or the interquantile range (IQR) can also be useful in gleaning information and conclusions from the individual metrics.

Lastly, we can visually inspect our metrics. Figure 2.1 plots the kernel density estimates using a rule-of-thumb bandwidth for both M_1 and M_2. What the density plot clearly reveals is the much longer right tail of M_1 and the asymmetry in both estimated densities. Consistent with the average values, the density plots reveal that M_1 and M_2 are typically small, though, M_1 can on occasion be larger than M_2. This is consistent with the idea that the relative strength of the positive force outweighs the relative strength of the negative force for these new personal computers. It is likely that a lot of price uncertainty pertaining to these new devices was present in the mid-1990s.

We also plot the kernel density for M_7 in Fig. 2.2. This score is kept separate since it has a much smaller scale than either M_1 or M_2. Remember that M_7 is the net effect on a balanced outcome. Here we see that the density is quite compact and skewed with a long right tail. The mode arises near 0.03, so the most likely outcome is a net balance of about 3% above the hedonic price.

2.3 Dependent Variable Is Transformed

We now consider the case where the dependent variable of the model is in natural logarithms, a widely used practice across social sciences. This transformation has

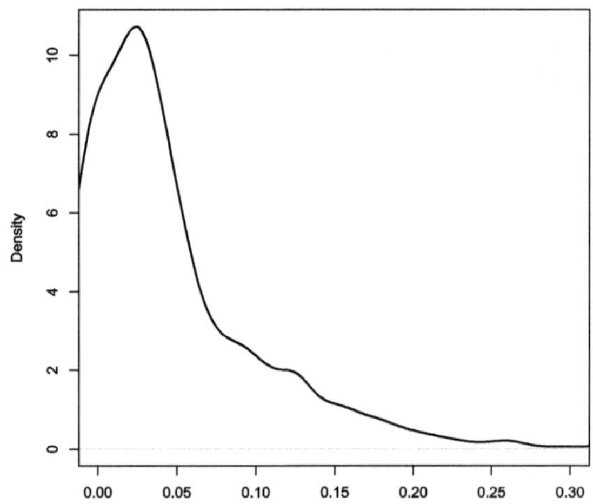

Fig. 2.2 Kernel density plot of M_7 using rule of thumb bandwidth

the limitation that it can be applied only on dependent variables that take strictly positive values, but most outcomes are measured using strictly positive scales (price, e.g., as in our previous example). Here, the four concepts of the outcome are defined as follows. The *actual* outcome is

$$Y = G(\boldsymbol{x}) \cdot \exp\{\varepsilon\} = G(\boldsymbol{x}) \cdot \exp\{v + w - u\}.$$

Uppercase Y denotes the outcome in its original measurement scale. The error term enters multiplicatively and exponentiated, so that it becomes additive when we take logarithms. The estimating equation here would be $\ln Y = \ln G(\boldsymbol{x}) + \varepsilon$, and $G(\boldsymbol{x})$ can take any form that guarantees strictly positive values. It is *not* the case that this formulation requires necessarily that $\ln G(\boldsymbol{x})$ contains the logarithms of the regressors. For example we may have, say for a single regressor, $\ln G(\boldsymbol{x}) = \beta x$, which would imply that in the original units, $Y = \exp\{\beta x\} \exp\{\varepsilon\}$. This could be the case for binary (zero-one) variables.

For the outcome counterfactuals, we have:

The *balanced* outcome

$$Y^B = G(\boldsymbol{x}) \cdot \exp\{v\},$$

the *upper frontier*

$$Y^{UF} = G(\boldsymbol{x}) \cdot \exp\{v + w\},$$

and the *lower frontier*

$$Y^{LF} = G(\boldsymbol{x}) \cdot \exp\{v - u\}.$$

The meaning of these is as in the previous section. But their interrelations are now revealed in terms of ratios, rather than differences, although with exact correspondence. We have

$$\frac{Y}{Y^{LF}} = \frac{Y^{UF}}{Y^{B}} = \exp\{w\}, \qquad \frac{Y^{LF}}{Y} = \exp\{-w\},$$

$$\frac{Y^{UF}}{Y} = \frac{Y^{B}}{Y^{LF}} = \exp\{u\}, \qquad \frac{Y}{Y^{UF}} = \exp\{-u\},$$

$$\frac{Y}{Y^{B}} = \exp\{w - u\}, \qquad \frac{Y^{UF}}{Y^{LF}} = \exp\{w + u\}.$$

The reader should compare these relations with those that hold when the outcome is measured in its original measurement units.

Here $\exp\{w\}$ can be viewed as the markup factor to go from the lower frontier to the actual outcome, which is equivalent to go from the *balanced* outcome to the upper frontier. In an analogous manner, we can describe $\exp\{u\}$. Inverse relations also become meaningful in the multiplicative setting: $Y^{LF}/Y = \exp\{-w\}$ is the lower frontier as a percentage of the actual outcome (and it will appear in the metrics below), while $Y/Y^{UF} = \exp\{-u\}$ is the actual outcome as a percentage of the upper frontier, so it stands for the actual achievement toward the maximum. The factor difference, $\exp\{w - u\}$, is the net effect of the unobservables on the actual outcome, as a gross markup/markdown on the balanced outcome. The factor $\exp\{w + u\}$ expresses the distance between the upper and the lower frontier in relative terms of the latter (how many times is the upper frontier larger than the lower). For the net gain metric M_{10}, we will also encounter the term $\exp\{u - w\}$.

2.3.1 Sample-Level Metrics

The abstractness of the metrics and the question that each one answers is basically the same as in the previous section, but due to the multiplicative nature of the relation, some things change. One of them is that here we define the metrics (all of which are ratios) as the expected value of the ratio rather than as a ratio of expected values. To actually compute the metrics, we will need the expected values of $\exp\{\pm w\}$, $\exp\{\pm u\}$ and $\exp\{\pm(w - u)\}$, conditional and unconditional. The expressions for these will be provided in those chapters where statistical specifications are presented (the benchmark ones in Chap. 7 and others more specialized in subsequent chapters).

I. Relative Weight in/loss of Outcome As regards what proportion of the actual outcome is attributed to w and how much we lost due to u, we have

$$M_1 \equiv E\left(\frac{Y - Y^{LF}}{Y}\right) = E\left(\frac{G(\boldsymbol{x}) \cdot \exp\{v + w - u\} - G(\boldsymbol{x}) \cdot \exp\{v - u\}}{G(\boldsymbol{x}) \cdot \exp\{v + w - u\}}\right)$$
$$= 1 - E\left(\exp\{-w\}\right),$$

$$M_2 \equiv E\left(\frac{Y^{UF} - Y}{Y}\right) = E\left(\frac{G(\boldsymbol{x}) \cdot \exp\{v + w\} - G(\boldsymbol{x}) \cdot \exp\{v + w - u\}}{G(\boldsymbol{x}) \cdot \exp\{v + w - u\}}\right)$$
$$= E\left(\exp\{u\}\right) - 1.$$

Note that the simplifications do not depend on any assumption about the statistical dependence or independence between regressors and the error term or the components of the error term in between them. The same holds for all subsequent metrics.

We note that the possible interpretation of these metrics as producer/consumer surplus or as markdown/markup factors to achieve the corresponding frontier (see the previous part of the chapter) are also valid here.

II. Relative Strength Here we have

$$M_3 \equiv E\left(\frac{\exp\{w\}}{\exp\{u\}}\right) = E\left(\exp\{w - u\}\right).$$

If w and u are assumed statistically independent, then this expected value can be decomposed:

$$w \perp\!\!\!\perp u \implies M_3 = E\left(\exp\{w\}\right) \cdot E\left(\exp\{-u\}\right).$$

This will not be the case for the individual metric.

III. Volatility

$$M_4 \equiv E\left(\frac{Y^{UF} - Y^{LF}}{Y^B}\right) = E\left(\frac{G(\boldsymbol{x}) \cdot \exp\{v + w\} - G(\boldsymbol{x}) \cdot \exp\{v - u\}}{G(\boldsymbol{x}) \cdot \exp\{v\}}\right)$$
$$= E\left(\exp\{w\}\right) - E\left(\exp\{-u\}\right).$$

Note the asymmetry here: $\exp\{w\}$ ranges in $(1, \infty)$, while $\exp\{-u\}$ ranges in $(0, 1)$. In the case where the distribution of w is slowly decaying, its effect on volatility will not be moderated by $\exp\{-u\}$ even if the latter decays fast. So the use of this metric requires some care, depending on the estimated parameters related to w and u.

IV. Markup/Markdown Gross and Net Effects The effects of unobservables as a "markup/markdown," always on the *balanced* outcome, are here:

$$
\begin{aligned}
M_5 \equiv E\left(\frac{Y^{UF} - Y^B}{Y^B}\right) &= E\left(\frac{G(\boldsymbol{x}) \cdot \exp\{v + w\} - G(\boldsymbol{x}) \cdot \exp\{v\}}{G(\boldsymbol{x}) \cdot \exp\{v\}}\right) \\
&= E\left(\exp\{w\}\right) - 1,
\end{aligned}
$$

$$
\begin{aligned}
M_6 \equiv E\left(\frac{Y^B - Y^{LF}}{Y^B}\right) &= E\left(\frac{G(\boldsymbol{x}) \cdot \exp\{v\} - G(\boldsymbol{x}) \cdot \exp\{v - u\}}{G(\boldsymbol{x}) \cdot \exp\{v\}}\right) \\
&= 1 - E\left(\exp\{-u\}\right),
\end{aligned}
$$

$$
\begin{aligned}
M_7 \equiv E\left(\frac{Y - Y^B}{Y^B}\right) &= E\left(\frac{G(\boldsymbol{x}) \cdot \exp\{v + w - u\} - G(\boldsymbol{x}) \cdot \exp\{v\}}{G(\boldsymbol{x}) \cdot \exp\{v\}}\right) \\
&= E\left(\exp\{w - u\}\right) - 1, \\
\text{and if } w \perp\!\!\!\perp u \implies M_7 &= E\left(\exp\{w\}\right) \cdot E\left(\exp\{-u\}\right) - 1.
\end{aligned}
$$

It is important to note that the metrics M_5 and M_6 have subtle but crucially different interpretations than before. When the dependent variable is in its original measurement units, we have the intuitive relation that M_7 is just the difference between M_5 and M_6. This *does not hold here*. In fact, the net effect metric M_7 does not have an obvious relation with the M_5 and M_6 metrics. This happens because M_5 and M_6 ignore the interactive effect that one force has on the other (even if they are statistically independent). Such interaction is not present when the dependent variable is in its original units and the relation between the two is additive. But when we use the logarithmically transformed dependent variable, it is the multiplicative relation that reflects the real-world situation, and it is captured by the M_7 metric (which as with M_3, can be decomposed into the product of two expected values if (w, u) are independent). Therefore the metrics M_5 and M_6 should be interpreted as *counterfactual* tendencies that would occur had the other force not been present, rather than actual effects that co-determine as-is the net effect. This interpretation is also consistent with the fact that these two continue to satisfy the relationship $M_5 + M_6 = M_4$ (volatility).

Also, note that $M_7 = M_3 - 1$. The net markup/markdown effect in this multiplicative setting is equal to the relative strength metric M_3 minus one. This relationship does not hold when the estimating equation has the dependent variable in its original measurement units.

2.3.1.1 Interpretation Alert: Beware of Misleading Pairing of Metrics

Reconsider now the pairs of metrics (M_1, M_2) and (M_5, M_6):

$$M_1 = E\left(\frac{Y - Y^{LF}}{Y}\right) = 1 - E\left(\exp\{-w\}\right),$$

$$M_2 = E\left(\frac{Y^{UF} - Y}{Y}\right) = E\left(\exp\{u\}\right) - 1$$

$$M_5 = E\left(\frac{Y^{UF} - Y^B}{Y^B}\right) = E\left(\exp\{w\}\right) - 1,$$

$$M_6 = E\left(\frac{Y^B - Y^{LF}}{Y^B}\right) = 1 - E\left(\exp\{-u\}\right)$$

A mistake that has appeared often in the literature is to pair and compare M_1 and M_6, perhaps due to the symmetry in their mathematical expressions. For example, in studies of wage equations, they are compared as being the "surpluses" extracted by the two sides of the market. But while M_1 can indeed be interpreted as the "worker (producer/supplier)" surplus, M_6 is *not* the buyer/firm surplus. Already from a mathematical point of view, it should be clear why it is wrong to compare/contrast M_1 and M_6: M_1 provides a measure relative to the actual outcome Y, while M_6 provides a measure relative to the *balanced* outcome Y^B. But this makes all the difference, since (M_1, M_2) give us information about the effects of the unobservables relative to a magnitude that has already been affected by them (the actual outcome Y), while (M_5, M_6) measure the effects of the unobservables relative to a magnitude that presumes that these effects are absent (the balanced outcome Y^B). So the meaningful thing to do is to compare M_1 to M_2 (which can be interpreted as surpluses) and M_5 to M_6 (which can be interpreted as markup/downs).

V. Distance from the Frontiers We show now that, with a logarithmically transformed outcome, the distance metrics M_8 and M_9 are equivalent to the metrics M_6 and M_5, respectively.

$$M_8 \equiv E\left(\frac{Y^{UF} - Y}{Y^{UF}}\right) = E\left(\frac{G(\boldsymbol{x}) \cdot \exp\{v + w\} - G(\boldsymbol{x}) \cdot \exp\{v + w - u\}}{G(\boldsymbol{x}) \cdot \exp\{v + w\}}\right)$$
$$= 1 - E\left(\exp\{-u\}\right) \implies M_8 = M_6.$$

Also,

$$M_9 \equiv E\left(\frac{Y - Y^{LF}}{Y^{LF}}\right) = E\left(\frac{G(\boldsymbol{x}) \cdot \exp\{v + w - u\} - G(\boldsymbol{x}) \cdot \exp\{v - u\}}{G(\boldsymbol{x}) \cdot \exp\{v - u\}}\right)$$
$$= E\left(\exp\{w\}\right) - 1 \implies M_9 = M_5.$$

These equivalences spring from the multiplicative nature of the relations. They also strengthen the understanding as to why M_5 and M_6 should be considered as counterfactual measures for the case the other factor is absent.

Table 2.6 Metrics of latent forces. Formulas and interpretation when the dependent variable is in natural logarithms. These formulas measure the effects of w, u in the original measurement units of the outcome Y. Metrics M_2 and M_6 represent absolute distances

Metric	Formula	Description
M_1	$1 - E(\exp\{-w\})$	Relative contribution in Y of positive force w / producer surplus / markdown factor to reach Y^{LF}
M_2	$E(\exp\{u\}) - 1$	Relative loss in Y due to negative force u / consumer surplus / markup factor to reach Y^{UF}
M_3	$E(\exp\{w-u\})$	Relative strength of w compared to u
M_4	$E(\exp\{w\}) - E(\exp\{-u\})$	Volatility of Y^B due to w, u
$M_5 = M_9$	$E(\exp\{w\}) - 1$	Counterfactual markup effect of w on Y^B
$M_6 = M_8$	$1 - E(\exp\{-u\})$	Counterfactual markdown effect of u on Y^B
$M_7 = M_3 - 1$	$E(\exp\{w-u\}) - 1$	Net effect on balanced outcome Y^B ($\pm$)
M_{10}	$1 - E(\exp\{u-w\})$	Net gain relative to Y ($\pm$)

VI. Net Gain For the net gain metric, we have

$$M_{10} \equiv E\left(\frac{Y - Y^B}{Y}\right) = E\left(\frac{G(\boldsymbol{x}) \cdot \exp\{v + w - u\} - G(\boldsymbol{x}) \cdot \exp\{v\}}{G(\boldsymbol{x}) \cdot \exp\{v + w - u\}}\right)$$
$$= 1 - E(\exp\{u - w\}).$$

Before presenting the analogous measures at the observation-level, we summarize in Table 2.6 the computing formulas for each metric and its interpretation.

2.3.2 Individual Metrics

The individual metrics are the conditional versions of their sample-level counterparts. So we obtain their formulas by substituting conditional expectations for unconditional expectations. This leads to only one complication, as we will see.

I. Relative Weight in/loss of Outcome We have

$$\{M_{1i}\} \equiv E\left(\frac{Y - Y^{LF}}{Y} \middle| \varepsilon_i\right) = 1 - E(\exp\{-w_i\}|\varepsilon_i),$$

$$\{M_{2i}\} \equiv E\left(\frac{Y^{UF} - Y}{Y} \middle| \varepsilon_i\right) = E(\exp\{u_i\}|\varepsilon_i) - 1.$$

Here too the simplifications inside the conditional expectation do not depend on any statistical assumptions about the existence or not of any dependence relations regarding w and u.

II. Relative Strength

$$\{M_{3i}\} \equiv E\left(\frac{\exp\{w_i\}}{\exp\{u_i\}}\middle|\varepsilon_i\right) = E\left(\exp\{w_i - u_i\}|\varepsilon_i\right).$$

Unlike the corresponding sample-level metric, here this conditional expected value cannot be separated into the product of the two:

$$E\left(\exp\{w_i - u_i\}|\varepsilon_i\right) \neq E\left(\exp\{w_i\}|\varepsilon_i\right) \cdot E\left(\exp\{-u_i\}|\varepsilon_i\right),$$

This is because, even when w are u are assumed independent with respect to each other, they stop being independent conditional on ε, and due to the multiplicative nature of the relation, conditional dependence cannot be ignored.

III. Volatility The expected range of the balanced outcome due the latent forces at the observation level can be predicted as

$$\{M_{4i}\} \equiv E\left(\frac{Y^{UF} - Y^{LF}}{Y^B}\middle|\varepsilon_i\right) = E\left(\exp\{w_i\}|\varepsilon_i\right) - E\left(\exp\{-u_i\}|\varepsilon_i\right).$$

In practice, we will use the estimation residuals to compute these expressions, and not the true errors. This introduces some additional estimation uncertainty, and when outliers are present, they may produce unreasonably high values. This is a problem that may occur in any metric that includes $\exp\{w\}$ or $\exp\{u\}$.

IV. Markup/Markdown Effects As with their sample-level selves, metrics M_{5i} and M_{6i} are to be interpreted as counterfactuals, and they are computed as

$$\{M_{5i}\} \equiv E\left(\frac{Y^{UF} - Y^B}{Y^B}\middle|\varepsilon_i\right) = E\left(\exp\{w_i\}|\varepsilon_i\right) - 1,$$

$$\{M_{6i}\} \equiv E\left(\frac{Y^B - Y^{LF}}{Y^B}\middle|\varepsilon_i\right) = 1 - E\left(\exp\{-u_i\}|\varepsilon_i\right).$$

The net multiplicative effect on the balanced outcome per observation is

$$\{M_{7i}\} \equiv E\left(\frac{Y - Y^B}{Y^B}\middle|\varepsilon_i\right) = E\left(\exp\{w_i - u_i\}|\varepsilon_i\right) - 1.$$

V. Distance from the Frontiers We skip over $\{M_{8i}\}$ and $\{M_{9i}\}$ since they are the same as $\{M_{6i}\}$ and $\{M_{5i}\}$, respectively. Despite this equivalence, bear in mind that "distance from the frontiers" may be the most appropriate interpretation when we operate in the multiplicative setting.

VI. Net Gain Relative to Y Here we have

$$\{M_{10i}\} \equiv E\left(\frac{Y - Y^B}{Y}\Big|\varepsilon_i\right) = 1 - E\left(\exp\{u_i - w_i\}|\varepsilon_i\right).$$

The two signed metrics, $\{M_{7i}\}$ and $\{M_{10i}\}$, can be separated into their positive and negative subsets, as before, with the $\{M_{10i}^{(-)}\}$ metric presented as a positive number, to align with the conceptual intuition ("gain of the other side").

2.4 Application: The Original 2TSF Earnings Data Set

We close this chapter by going back to the origin: the Polachek and Yoon (1987) article that introduced the two-tier stochastic frontier model. The authors wanted to measure how incomplete information of workers and employers affects the observed wage. They used data from the USA Panel Study of Income Dynamics (PSID) of 1981 that includes information on the year 1980. As is almost universal in earnings studies, they used a semi-log regression equation, where the dependent variable (yearly earnings) enters in logarithms while the regressors are in levels (education, experience, tenure, and a constant), plus experience squared and tenure squared to capture nonlinear effects of human capital on earnings. We contacted the authors but they were not able to provide us with the data files they used back then. So we manually extracted data from the PSID for the same period. We were not able to compile an equivalent data set, probably because adjustments and corrections were made over the years. But our purpose here is not to replicate the study, but to showcase the application of the 2TSF model and its metrics when the dependent variable is transformed. Summary statistics of the variables used are in Table 2.7.

We note that the earnings series has an upper truncation point (earnings of six figures are recorded as 99,999), but this affects only ten observations. Here too we will use the Normal-Exponential 2TSF specification for the error term (which was also the one used in the original study).

Table 2.7 Summary statistics of the PSID data set, $n = 3{,}856$

Variable	Unit	Mean	Median	S.D.	Min	Max
Earnings 1980	USD	17,433.0	15,486.0	12,545.0	4.00	99,999.0
Log-earnings 1980 (dep. var.)		9.47	9.65	0.932	1.39	11.5
Tenure	Months	75.6	36.0	96.9	0.0	540.0
Experience	Years	14.6	10.0	12.1	0.0	64.0
Education	Years	12.2	12.0	2.90	0.0	17.0

Table 2.8 Maximum likelihood estimation, 2TSF Normal-Exponential model. The dependent variable is the logarithm of yearly earnings in 1981. Standard errors are based on the outer products matrix. The sample size is $n = 3{,}856$

Variable	Coefficient	Std. Error
Constant	8.263	0.045
Education	0.091	0.003
Experience	0.040	0.002
Experience^2	−0.001	0.000
Tenure	0.003	0.000
Tenure^2	−0.000	0.000
σ_v	0.184	0.022
σ_w	0.295	0.014
σ_u	0.622	0.013

The positive latent force w represents the employer's "ignorance" (incomplete information or information deficiency) that pushes the wage up from its full-information level, while the negative latent force u represents the worker's incomplete information of the market that pushes the wage down from the full-information level. But unlike the empirical illustration previously, here the dependent variable used is in logarithms. The estimation results are presented in Table 2.8.

Adopting the usual interpretation of a semi-log regression, the coefficients of the regressors approximate percentage effects on the dependent variable in levels. So we see that a 1-year increase in education tends to increase earnings by 9.1% while a 1-year increase in professional experience by 4%. On the other hand, tenure (which we remind is measured in months) tends to increase the wage by 3.6% per year. The very small but negative coefficients on the square variables indicate the existence of a weak nonlinear relationship between earnings and experience or tenure, where there is a maximum point reached after which more experience or tenure would tend to slightly decrease earnings.

The magnitudes of the sigma parameters, the scale parameters of the three components of the error term, are not directly interpretable. We should revert back to the original units. We do that in Table 2.9 where we present the sample-level means of the various metrics (computed using the estimated sigma parameters), but also the average and the median of the series of the individual metrics. We remind the reader that when the dependent variable is transformed, metrics M_8 and M_9 (distance from the frontier) are identical to M_6 and M_5, respectively, which in turn here measure counterfactual effects in the absence of the antagonist.

Prior to discussing the substantive results, some important remarks of a methodological nature are in order. First, note that for six out of the eight metrics the series mean is fairly close to the sample-level mean. In theory, they should be identical; in practice, we anticipate that they should be close, especially with a sample size that is not small. But for M_2 and M_{10}, they differ markedly. This is clear indication that there are extreme outliers in these two series. In fact just one outlier would suffice to severely distort the series mean. These are outliers that depend on the

Table 2.9 Metrics of latent forces—mean and median values

Metric	Description	Sample-level mean	Series mean	Series median
M_1	Worker surplus	0.228	0.228	0.174
M_2	Employer surplus	1.642	2.705	0.533
M_3	Relative strength w/u	0.874	0.881	0.845
M_4	Volatility of Y^B due to w, u	0.801	0.792	0.679
$M_5 = M_9$	Markup effect of w on Y^B	0.418	0.416	0.263
$M_6 = M_8$	Markdown effect of u on Y^B	0.383	0.376	0.304
M_7	Net effect on balanced outcome	−0.126	−0.119	−0.155
M_{10}	Net gain	−1.048	−1.891	−0.217
$M_7^{(+)}$	Positive effects on Y^B (32%)	–	0.374	0.201
$M_{10}^{(+)}$	Worker gain (30%)	–	0.208	0.161
$M_7^{(-)}$	Negative effects on Y^B (68%)	–	−0.355	−0.309
$M_{10}^{(-)}$	Employer gain (70%)	–	2.779	0.470

estimation results; they are not outliers from the data directly, but outliers in the series of residuals. And a very large residual could arise from a single mistake in recording the value of a data point (say, of the dependent variable). In such cases there are two remedies often encountered in the literature: the first is to identify the outliers, perhaps review them in an attempt to understand why they are outliers, and then in any case discard them and re-compute the empirical mean without them. The alternative, which we have already adopted here, is to look instead at the series *median*, as a robust measure of centrality that is unaffected by outliers. We see that in both cases, the series median is much smaller in absolute value not only from the series mean but also from the sample-level mean (which uses only the estimated sigma parameters, not the residuals). So the median does not stand-in as a "corrected mean," but as an autonomous measure that identifies the center of probability mass and, in combination with the sample-level mean, indicates that there is a long right tail in the distribution of M_2 and a long left tail in the distribution of M_{10}.

A second methodological note is to observe that the proportion of the sample where $M_7 > 0$ (32%) is not equal to the corresponding proportion where $M_{10} > 0$ (30%). Also it is not necessary that for the same observation they will have the same sign. Identifying the observations where the two have opposite signs and reviewing them may provide interesting insights, but we do not do that here.

A third methodological note is that the differences between the M_7 and the M_{10} metrics justify the usefulness of their separate existence. We remind the reader that in their abstract definition, their numerator is the same, but they differ in the denominator, the base of measurement. Surely, in a roughly symmetric situation as regards the latent forces, where $Y \approx Y^B$, the two are bound to be quite close and convey largely the same information. But if this is not the case (the "case" being to have this symmetry observation-per-observation throughout the sample, not on average over the sample), each has something different and useful to offer.

Table 2.10 Deciles of individual metrics

	Cumulative probability								
Metric	0.1	0.2	0.3	0.4	0.5	0.6	0.7	0.8	0.9
M_1	0.167	0.167	0.167	0.168	0.174	0.188	0.216	0.266	0.375
M_2	0.253	0.277	0.326	0.411	0.533	0.718	0.982	1.436	2.605
M_3	0.359	0.531	0.652	0.752	0.845	0.929	1.022	1.138	1.372
M_4	0.542	0.558	0.586	0.628	0.679	0.742	0.821	0.930	1.120
$M_5 = M_9$	0.250	0.250	0.250	0.253	0.263	0.287	0.340	0.445	0.718
$M_6 = M_8$	0.168	0.182	0.209	0.251	0.304	0.375	0.457	0.558	0.701
M_7	−0.641	−0.469	−0.348	−0.248	−0.155	−0.071	0.022	0.138	0.372
M_{10}	−1.884	−0.948	−0.585	−0.372	−0.217	−0.104	−0.002	0.098	0.248

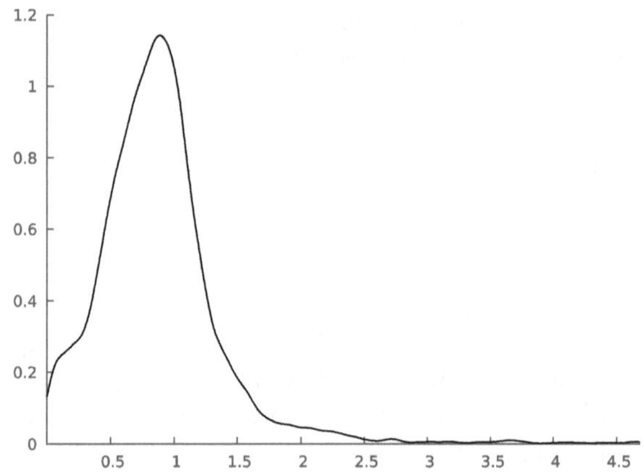

Fig. 2.3 Kernel density plot of M_3 using rule of thumb bandwidth

Before commenting on the results themselves, we also present in Table 2.10 the deciles of the metrics, so that they are available in the discussion that will follow.

Turning to the actual results and working mainly with the medians except otherwise noted, we see that employer surplus (M_2) is much higher than worker surplus (M_1): the wage cost employers avoided paying due to information deficiencies of the workers is much higher than what the workers received in addition, due to employers' information deficiencies (we highlight here that consumer/producer surpluses are not antagonistic zero-sum measures).

The M_3 metric tells us that on average, the informational deficiencies of the employers (the w variable) are only $\approx 85\%$ in strength as those of the workers (the u variable). Looking at the deciles of the series, we see that in almost 70% of cases w is predicted as being smaller than u. Figure 2.3 shows the kernel density of M_3. The mode of this density lies at about the value 0.9, which is the "most likely" to occur.[4]

[4] Readers may notice that Fig. 2.3 has a slightly different visual style than the previous ones. True. We navigate freely between the software platforms R and `gretl` whose numerical and visual outputs differ somewhat in style, and we see no reason for absolute uniformity here.

The M_4 metric reveals that the average volatility is about 34-40% above or below the counterfactual full-information (balanced) outcome, which indicates a visible role on the determination of the wage by the latent information deficiencies.

Turning to the effects on the balanced outcome, we note that the markup effect of w on the balanced outcome (full-information wage), the metric M_5, appears to be stronger than the markdown effect of the negative force u (metric M_6): the first has an average uplifting effect of ≈42%, while the second has a reducing effect of ≈38%. Instinctively, one would proceed to conclude that "so the net effect must be something like 4% *above* the full information wage." But what we observe is that the net effect is ≈−12%, namely 12% *below* the full information wage (metric M_7). This is also corroborated by the metric M_3, which is below unity, indicating that the effect of exponentiated w on the dependent variable is weaker than the effect of exponentiated u. We also note that in terms of *median* values, the M_5 metric is below the M_6 metric, and this also helps to explain why the M_7 metric is negative.

Still, are we perhaps getting conflicting, irreconcilable results here? No, because, as we mentioned earlier, the metrics M_5 and M_6 represent *counterfactuals*, the strength of a latent force if the other one were absent. So they must be interpreted *alone and not in combination*: M_5 tells us that with fully informed workers, employers' information deficiency would tend to push the wage 42% above the *common* full-information wage. And M_6 tells us that with fully informed employers, workers' information deficiency would tend on average to reduce the wage 38% below the common full-information wage. This is why we warned earlier that these formulas are better interpreted as the metrics M_9 and M_8, respectively (distance from the frontiers): 42% is how much the actual wage is above the lower frontier (minimum wage), and 38% is how much the actual wage lies below the upper frontier (maximum wage). How the actual wage is positioned with respect to the common full information wage is estimated directly by the M_7 metric that takes into account the realized *pairs* of employer-employee information deficiency. Figure 2.4 presents the kernel density plot of M_7.

We see both positive and negative values with a long right tail and a mode near −0.1. The long right tail corresponds to cases where the wage was *above* the full information level, but the deciles indicate that 2/3 of probability mass lies in the negative values. So we have a more frequent occurrence of the event $\{Y < Y^B\}$, but when the opposite holds, it may hold strongly. This is intuitive in that firms, being structured organizations, are expected to manage better their information deficiencies than workers, while for workers it is reserved the rare exceptional beneficial case due to diligence and luck. This robust performance by employers is also indicated by the decomposition of M_7 into its negative and positive subseries: the median of the negative subseries $M_7^{(-)}$ (actual wage below the balanced one) is −31%, while the median of the positive subseries $M_7^{(+)}$is only 20%.

In this empirical exercise we also encounter the net gain metric M_{10}, which, seen as a whole, is mostly in the negative, indicating gains enjoyed by the employer's side. Its decomposition into its negative and positive subseries reveals even more:

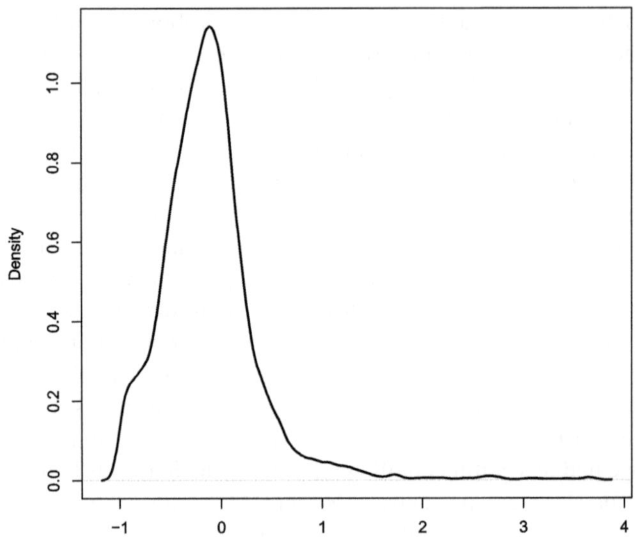

Fig. 2.4 Kernel density plot of M_7 using rule of thumb bandwidth

when workers gain, their gain is equal to 16% of actual wage. When employers gain (cost avoidance), their gain is equal to 47% of actual wage—a powerful insight into the workings of the market.

With this we conclude this empirical illustration, this chapter, and indeed Part I of this book, where we have introduced the two-tier stochastic frontier model, discussed its origins, browsed its applications up to now, and presented the heart of its contribution, the core metrics that we can compute after we estimate it. In the next part of the book, we present applications of the model in various topics and fields, so that the reader is provided with a diverse and detailed exposition of the applied power of the 2TSF approach.

Part II
Models and Applications

Chapter 3
Incomplete Information

In presenting our metrics in Chap. 2, we finished the discussion with the empirical application from the foundational 2TSF paper of Polachek and Yoon (1987). Their concern was with wage variation, which they attributed to the existence of incomplete information on both sides of the labor market. This, as confessed by Polachek himself (see Chap. 1), was inspired by the seminal paper of Stigler (1961) who attributed price variation to two overarching phenomena: heterogeneity and "ignorance" (lack of information).[1]

The wage is a price: the price of labor services. But price dispersion is also observed in the goods and services market. As an example, consider the Amazon marketplace. Different sellers sell the same product on Amazon at various prices. The product being sold is identical and yet different prices are observed. Thus, a buyer pays a higher price (or a seller receives a lower price) for a homogeneous product due to his/her ignorance. Another example is the retail gasoline market. Here there are numerous gas stations along a travel corridor and each charge slightly different prices for an identical good.

One could expect that if buyers knew of the seller(s) with the *highest* "willingness to accept" (WTA), namely, the seller with the *cheapest* price (since the seller has "high willingness" to accept a transaction and so is more prone to facilitate it by accepting a lower price), as would be the case if the market was located at one central location and perfectly competitive, there would be no price variation except for the inherent randomness of the market and seller heterogeneity. But empirical studies consistently dispute that this is what characterizes actual markets, even those that could be considered as being close to the ideal of "perfect competition." For example, Pratt et al. (1979) found significant price dispersion, in terms of the coefficient of variation, for 39 goods in what many would consider competitive markets. Other studies such as Sorensen (2000) and Lach (2002) find

[1] See (Papadopoulos, 2021d, pp. 171–173), for a compact as well as graphical presentation of how the 2TSF model arises from the labor supply and demand curves.

A. Papadopoulos, C. F. Parmeter, *Two-Tier Stochastic Frontier Analysis for the Social Sciences*, https://doi.org/10.1007/978-3-031-81513-3_3

price dispersion in "brick and mortar" retail markets, and Brown and Goolsbee (2002) analyze price dispersion in the market for life insurance, while Häring (2003) and Cooper (2006) find price dispersion in online markets. Whatever its cause, price dispersion can be taken as an empirical regularity. And as such, it demands that we examine its causes.

Note that many of the reasons for price dispersion are wrapped up in information and obtaining information is costly. When purchasing goods, buyers face search costs in obtaining price information from different sellers. These search costs lead the buyer to search optimally in the sense that they will stop searching when the marginal gain from another search is equal to its marginal cost (what we could call the "shadow price of search"). Analogously, inferring the price that each prospective consumer would be willing to pay for a product is costly for the seller. They will attempt such inference, but up to a point. So, given that search costs exist, market participants most likely will not become fully informed, and price variations due to incomplete information will exist even after controlling for product characteristics. So, incomplete information can be seen as the outcome of a rational optimizing behavior. For this reason we will call buyers and sellers who lack full information, "deficient." As opposed to "ignorant," deficient means that buyers and sellers may want to gather more information through search, but further search is costly and time-consuming, and thus they end up entering the market with incomplete information. This approach to information deficiency is intimately connected with the recent literature on rational or selective inattention, see, for example, Maćkowiak et al. (2023).

There are different ways to exploit and model incompleteness of information. Recent work of Pennerstorfer et al. (2020) develops a theoretical model of price dispersion where consumers differ in their degree of informedness. For some consumers, it is costly to obtain additional information, while others are fully informed. Intuitively, their model implies that average prices decline as the share of informed consumers increases. Compare this to the work of Kaplan et al. (2019), who construct a model of relative price dispersion where sellers listing different prices is a consequence of pricing strategies carried out to discriminate between high-valuation buyers who need to make all of their purchases in one location and low-valuation buyers who are able to purchase different items in different locations. For this strategy to be successful, sellers need to have some information on the types of buyers they are selling to and when these strategies will be optimal for buyers. But it is time now to develop a general model that encompasses information deficiency of both buyers and sellers.

3.1 A Hedonic Pricing Model with Incomplete Information

We base our presentation on the work of Parmeter (2006), which Kumbhakar and Parmeter (2010) extend, who generalizes the hedonic price framework of Griliches (1971) and Rosen (1974) while accounting also for information deficiency. In the standard hedonic price model, a good is composed of a set of characteristics. The affine transformation between these characteristics and their implicit values (known as shadow prices) determine the market value of the good for both buyers and sellers. The tangency of the loci (between buyers and sellers) traces out the hedonic price function.

To help formalize the hedonic regression, Rosen (1974) uses the idea of a bid function. Bid functions are constructed to discern the price that a buyer would pay for a good with certain attributes for a fixed level of utility and income, indicating the highest price a buyer is willing to pay for the good (or simply "wilingness-to-pay" – WTP). Utility is maximized when the bid function is equal and tangent to the market price function, the minimum price a buyer must pay in the market. Rosen (1974) also develops the idea of an offer function, which plays a role analogous to that of the bid function, for sellers. Firms have offer functions for the good in question, given a certain level of profit. The offer function (or just willingness to accept, "WTA") indicates the lowest price a seller will accept for the product up for sale. Sales take place where an individual seller's offer function becomes tangent to an individual buyer's bid function. The locus of the points of tangency between the offer functions $\varphi(z)$ (which differ for different levels of profit, π) and bid functions $\psi(z)$ (for different levels of income, I and utility, U) traces the hedonic price function $P(z)$, of the good with scalar characteristic z (holding other characteristics fixed). This is shown in Fig. 3.1.

However, this model implicitly assumes that there is full information in the market, as well as perfect competition (i.e., markets are "thick"). In this setup, there would be no rationale for price variation except for that which is due to inherent

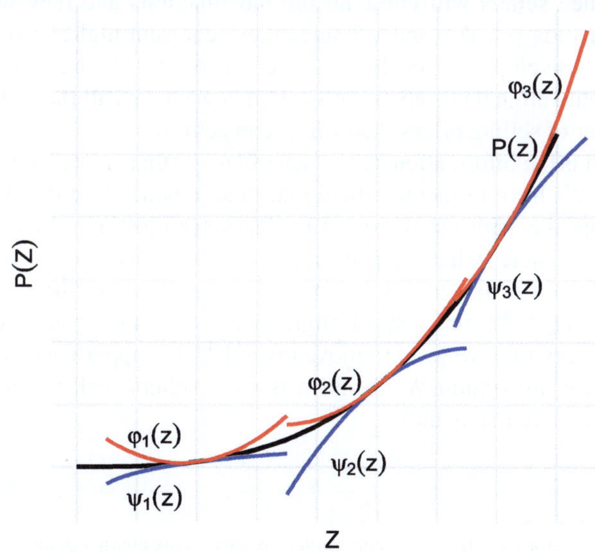

Fig. 3.1 Hedonic price equilibrium with full information

heterogeneity of the good.[2] With full information, every buyer knows of the highest WTA and every seller knows of the highest WTP. This forces the market to generate a unique price for each z in the market, given also that it assumes that all bid (offer) functions at a given point are exactly the same for all buyers (sellers), a homogeneity premise inherent in a perfect competition model.

But in reality, with incomplete information the market price is affected by the levels of information that buyers and sellers possess specifically because there is no reason for all bid and offer functions to be the same at a given point. That is, for a given z there are numerous bid and offer functions due to differing levels of information. The multiplicity of bid and offer functions for a specific characteristic results in a gap between the highest bidding buyer and the lowest offering seller. This gap is the source of price variations that are observed in the market, after controlling for product characteristics. A transaction occurs when a buyer-seller pair align themselves such that their levels of information translate into tangency between their bid and offer functions as in the traditional hedonic setup with full information.[3]

The above account of how information affects price variation is detailed below. For clarity we describe each side of the market separately before continuing on to equilibrium.

A hedonic market with incomplete information on the buyers' side only is shown in Fig. 3.2. Each ψ-curve represents a buyer's bid function, indicating different WTP. We see that there are buyers in the market that are looking to purchase the good at the relatively lower bids ψ^1, ψ^2, and ψ^3 and others that have higher WTP (bid functions ψ^4, ψ^5, and ψ^6), with bid function ψ^7 somewhere in the middle between ψ^4 and ψ^1.

Fully informed sellers will know all the bid functions and they will approach buyers with the bids ψ^4, ψ^5, and ψ^6 since they represent higher profits for them. It is this upper envelope, which here we denote by $P_s(z)$, that sellers with full information would align themselves with, and it is what we call the *sellers'* frontier, the upper cluster of selling prices that sellers can achieve.

Notice that it is the information deficiency of *buyers* that creates the price frontier for the *sellers*. We have to mention here that this nomenclature deviates, in fact it is the exact opposite with the one used by Parmeter (2006) and Kumbhakar and Parmeter (2010). There, what we just discussed was called the "buyer's frontier" (because it is linked to buyers' WTP). But we have come to the conclusion that calling it the sellers' frontier instead aligns better with the abstract concept of a frontier, a boundary that we want to move toward.[4] This upper envelope in Fig. 3.2 represents buyers' maximum WTP, and it is what sellers seek to reach (and will benefit from if it were to expand).

---

[2] This would be the error term from the regression of prices on the characteristics.

[3] It is as though information is being directly incorporated into the utility and profit functions.

[4] This is also fully consistent with what a frontier implicates in the single-tier setting.

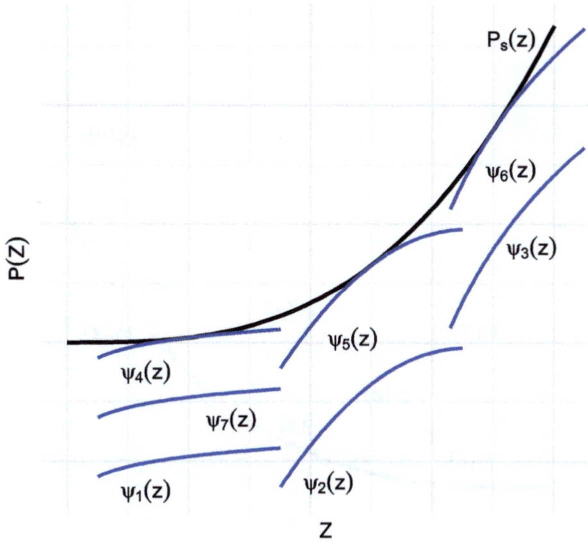

Fig. 3.2 Sellers' viewpoint of the buyers' side of the market with incomplete information

Now consider Fig. 3.3, which depicts the case of incomplete information on the seller side only. Here we have the expensive sellers with offer functions φ^4, φ^5, φ^6 (low WTA), the relatively cheaper ones with offer functions φ^1, φ^2, and φ^3 (high WTA), and φ^7 somewhere in the middle. Fully informed buyers will know of the cheaper offers and will opt for them since it minimizes their expenditure. The envelope of these cheapest prices, $P_b(z)$, is the *buyers'* frontier, and it emerges because sellers are characterized by incomplete information, while buyers are not. Again, this frontier will benefit buyers as they approach it (and it is buyers that will benefit even more if it is pushed "further away," i.e., if it moves downward).

In Rosen's model, we would have $P_b(z) = P_s(z)$, and price fluctuations would only exist due to heterogeneity. However, with incomplete information, $P_b(z) \neq P_s(z)$. Equilibrium in the differentiated goods market with incomplete information is depicted in Fig. 3.4. This figure captures the essence of Figs. 3.2 and 3.3 by placing them on top of one another. Since it is assumed that buyers and sellers search optimally, tangency of bid and offer functions occurs where the marginal cost of searching, along with the marginal bid and offers, are equivalent for both parties. We depict three different equilibrium points (i.e., realized transactions), $P^1(z')$, $P^2(z')$, and $P(z'')$.

It can be seen from this representation that instead of having a unique price for a given level of a characteristic (as in Fig. 3.1), we can have more than one: in our example, $P^1(z') > P^2(z')$ even though they relate to the same value z' of the z-attribute. One of them, $P^2(z')$, lives on the buyers' frontier $P_b(z)$; the other is in between the buyers' and sellers' frontiers. We realize that there does not exist an equilibrium price *function* $P(z)$ anymore, but a price *surface* (the area between

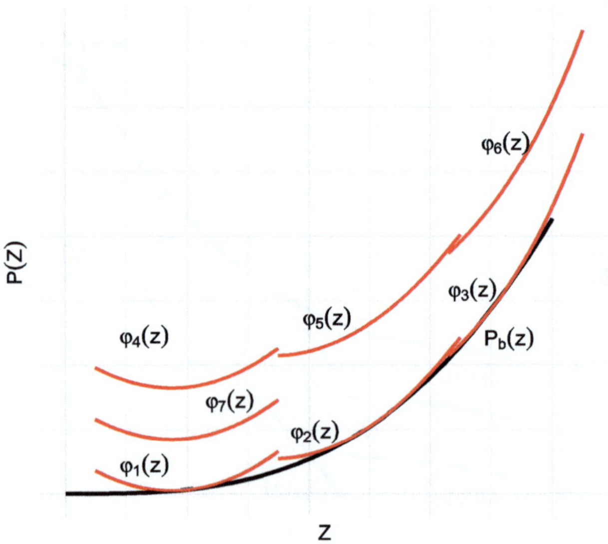

Fig. 3.3 Buyers' viewpoint of the sellers' side of the market with incomplete information

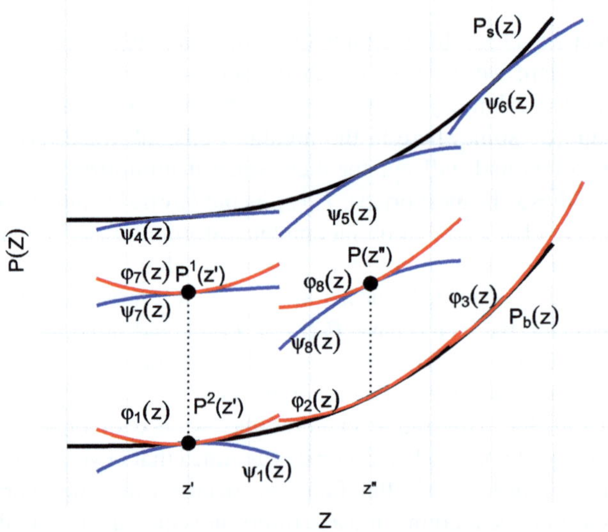

Fig. 3.4 Structure of a differentiated good market when incomplete information exists

the frontiers $P_s(z)$ and $P_b(z)$ including these boundaries), where transactions can happen. The observed prices will depend on the characteristics of the good, the levels of information possessed by both the buyer and seller, as well as any inherent randomness (unexplained product heterogeneity) that may be present.

The sellers frontier $P_s(z)$ indicates that observed prices cannot be higher than it, otherwise no sale will result. Sellers would wish to charge prices corresponding to this upper envelope but incomplete information prevents them from doing so: they do not know its location. The same story holds on the buyers' side. Here buyers attempt to find the "lowest prices" available in the market, i.e., prices corresponding to points along $P_b(z)$. The presence of search costs limits their ability to find the location of $P_b(z)$.[5] Additional information will change the observed transaction prices and the pattern of variation, depending on who is acquiring the information and how pertinent it is. These basic remarks point out that the effects of information deficiency are the distance of a realized price from $P_s(z)$ and from $P_b(z)$ *at the same time*. But these boundaries are unobservable not only to the sellers and buyers but also to researchers. Therefore, we need a method to measure them both individually but also as regards their combined effect while acknowledging that they are latent forces, and this is how the 2TSF approach arises naturally as a modeling strategy.[6]

3.2 Formulating the Empirical Model

Armed with robust theoretical foundations, we turn to the standard hedonic price equation that is typically specified as

$$P^{FI} = \beta_0 + h(z) + v, \tag{3.1}$$

where P^{FI} is the hedonic price (or the logarithm of the hedonic price), z is a vector of product characteristics influencing the overall value of the good in the market, and v represents random noise and measurement error in price. The constant β_0 represents those components of the price that do not relate to the hedonic value but are necessary costs in order for the consumers to be able to actually acquire access to it (costs related to administrative costs, logistics, packaging, retail margins etc., that could all be characterized abstractly as "access costs"). The hedonic price function in (3.1) corresponds to the full-information model developed by Rosen (1974), hence the "FI" superscript. So it maps to the "balanced outcome" presented abstractly in Chap. 2, since nowhere in this setup are the buyers with the highest WTPs or the sellers with the highest WTAs accounted for.

If we recast the hedonic price model to take account of these price bounds, then from a seller's standpoint, the price received (P_r) can be represented as

---

[5] But the occasional individual market agent may be fully informed as point $P^2(z')$ in Fig. 3.4 depicts.

[6] Before moving on, we also mention that the "tunnel" that exists due to information deficiencies in the hedonic model was also studied by Pope (2008), though in his setup he uses various information disclosures (such as a sex offender registry or a noise ordinance from a change in an airport runway) to understand how this impacts bargaining. But bargaining is not our focus here (it will become so in Chap. 4).

$$P_r = P_s - u = P^{UF} - u, \tag{3.2}$$

where P^{UF} is self-evidently identified with the highest WTP in the market (i.e., the upper frontier of the price, the Seller frontier) and $u \geq 0$ can be viewed as the loss to a seller for being information deficient. From a buyer's point of view, the price paid (P_p) can be written as

$$P_p = P^{LF} + w = P_b + w, \tag{3.3}$$

where P^{LF} is identified with the highest WTA (lowest price) in the market (the lower/buyer frontier) and $w \geq 0$ represents the cost the buyer bears for being information-deficient.

For a market transaction to occur, it must be the case that the market price, P, equals both the price paid by a buyer and the price received by a seller, $P_p = P_r$. This yields the following equality:

$$P^{LF} + w = P^{UF} - u = P. \tag{3.4}$$

This equality was obtained by logical description and by the fact that each transaction has a unique price. At the same time, it can be immediately obtained from the definitions of P^{LF}, P^{UF}, and P in Chap. 2, where we had

$$P^{LF} \equiv P - w, \quad P^{UF} \equiv P + u.$$

Then we can apply the whole framework of Chap. 2, in particular, the definition for the full-information outcome P^{FI}, and combine it with the hedonic expression (3.1),

$$P - w + u \equiv P^{FI} = \beta_0 + h(z) + v \tag{3.5}$$

which leads us to the expression of the market price P as

$$P = \beta_0 + h(z) + \varepsilon, \quad \varepsilon = v + w - u. \tag{3.6}$$

Equation (3.6) shows that the market price of a good is composed of (i) the implied value of the characteristics $h(z)$, (ii) inherent heterogeneity and random noise (v), and (iii) the effects of incomplete information to the buyers (w) and sellers (u), measured in terms of higher or lower prices, respectively (plus the constant term β_0 that takes on the part of the price that does not relate to the hedonic characteristics). With this representation, one can estimate not only marginal values of attributes but the impact of incomplete information on prices (perceived as a deviation associated with information deficiency). The market price should be the same as the hedonic price (aside from v) when (i) there is no information deficiency on the part of either buyers or sellers or (ii) the effect of buyers' incomplete information completely offsets the effect of sellers' incomplete information (i.e., $w - u = 0$). The standard hedonic price models take the market price as the true valuation of the good. In doing so the asymmetric informational incompleteness characterizing the buyers and sellers is ignored, and consequently the only explanation of price dispersion is heterogeneity.

Our model (3.6) incorporates information deficiency from both sides of the market, as suggested by Rothschild (1973), which should help characterize the nature of price dispersion better than the standard model that assumes away information deficiency. At the same time, its three-component composite error term ε defines a two-tier stochastic frontier model, readily estimable. In the next section, we do just that.

3.3 Application: A Canadian Housing Market

Pope (2008) remarked that it is reasonable to expect incomplete and asymmetric information in the housing market, since sellers have lower search costs and are more informed than buyers by virtue of knowing the property beforehand and/or actually living in the area. This argument, plausible indeed, anticipates that actual house prices should be above the balanced, full-information price, at least on average. The 2TSF model is exactly the model that can test such hypotheses. To show this, we present an application using data originally curated by Anglin and Gençay (1996). They estimated a hedonic price regression using 546 observations taken from the multiple listing service for houses sold in Windsor, Ontario, Canada, from July through September 1987.[7] The data set contains information on the sale price as well as 11 physical characteristics of the home. For our purposes we estimate a hedonic price regression function in semi-logarithmic form in a two-tier frontier setting to assess information deficiency on both sides of the market.

Our empirical specification is

$$\ln P_i = \beta_0 + \boldsymbol{x}_i'\boldsymbol{\beta} + \varepsilon_i, \qquad \varepsilon_i = v_i + w_i - u_i \qquad i = 1, \dots n, \tag{3.7}$$

where $\boldsymbol{x}$ contains the logarithm of the lot size of the home, as well as dummy variables for air conditioning, gas heat, a finished basement, a driveway, a recreation room, and location in a preferred neighborhood of Windsor as well as count variables for the number of bedrooms, bathrooms, garage places, and stories of the home. All variables except the lot size enter in level form. Here, the constant term β_0 acquires a different interpretation: it is that part of the value of the house not accounted for by the included hedonic characteristics. We implement the Normal-Exponential 2TSF model (see Chap. 7) which is estimated through maximum likelihood. Here the error term is decomposed as $\varepsilon = v + w - u$, where v is treated

[7] The data can be accessed in R through the AER package (Kleiber and Zeileis (2008)).

Table 3.1 Two-tier estimation of the hedonic price function

Regressors	OLS	2TSF
Constant	7.745	7.869
	(0.216)	(0.208)
$D_{Driveway}$	0.110	0.114
	(0.028)	(0.027)
$D_{RecRoom}$	0.058	0.057
	(0.026)	(0.025)
$D_{FinBasement}$	0.104	0.102
	(0.022)	(0.021)
$D_{GasHeat}$	0.179	0.160
	(0.044)	(0.046)
$D_{AirConditioning}$	0.166	0.155
	(0.021)	(0.020)
Garage Places	0.048	0.045
	(0.011)	(0.011)
$D_{PreferredNeighborhood}$	0.132	0.125
	(0.023)	(0.021)
ln Lot Size	0.303	0.293
	(0.027)	(0.026)
# Bedrooms	0.034	0.032
	(0.014)	(0.014)
# Bathrooms	0.166	0.166
	(0.020)	(0.019)
# Stories	0.092	0.095
	(0.013)	(0.011)
Scale parameters		
σ_ε	0.208	0.211
σ_v	–	0.114
σ_w	–	0.110
σ_u	–	0.139

as noise and w represents the effects of information deficiency of buyers (deficiency that increases the price) and u the effects of information deficiency of sellers (that reduces the price). Table 3.1 contains the estimation results, with standard errors in parentheses. As a benchmark, we present also the OLS estimates.

The first thing to note from Table 3.1 is that the OLS and 2TSF estimates for the hedonic price function are quite similar. Moreover, all of the coefficients have positive signs. Given that we have estimated a semi-logarithmic hedonic price function, these estimates are interpreted as "quasi-elasticities," which means that they give directly the percentage change in the dependent variable for a one-unit change in the regressor. For example, having a driveway leads to a roughly 11% increase in the sale price of a home (ceteris paribus), while an additional garage place increases sales price by just under 5%.

But mostly, we are after the decomposition of variance that happens as we move from the standard hedonic price model that ignores the presence of information deficiencies to the 2TSF model. Note that the overall variation in the composite error term is estimated very closely by the two models: In the OLS case, where we do not (and cannot) decompose the error term, we obtain as a direct estimate an overall standard deviation of the error equal to $\sigma_\varepsilon = 0.208$. In the Normal-Exponential 2TSF model, the three sigma coefficients each represent the standard deviation of the respective error component. The standard deviation of the composite error term is computed after estimation as $\sigma_\varepsilon = (\sigma_v^2 + \sigma_w^2 + \sigma_u^2)^{1/2} = 0.211$. This cross-validates the models. Now it is time to compute some of the metrics we presented in Chap. 2.

First, note that under the Exponential distribution specification for w and u, σ_w and σ_u estimate also their mean values, and we see that on average the information deficiency of the sellers, $\sigma_u = E(u)$, has a somewhat stronger effect than the information deficiency of the buyers $\sigma_w = E(w)$. But this is at the logarithmic level, and we will want to know what happens with actual prices.[8]

We first look at how information deficiency has affected the actual, realized price. These are the metrics $M_1 = 1 - E(\exp\{-w\})$ (measuring the effect of the buyer's information deficiency) and $M_2 = E(\exp\{u\}) - 1$ (measuring the effect of the seller's information deficiency). The average actual price information deficiency is 9.8% for buyers and 16% for sellers. Namely, on average 9.8% of the actual price was due to the information deficiency of the buyers (the M_1 measure), while at the same time sellers lost a premium equal to 16% of the actual price, due to their information deficiency (the M_2 measure). So our measures suggest that there is information deficiency on both sides of the market and, on average, this deficiency is economically relevant.

We get more detailed results if we look at the metrics defined at the individual level, and to help visualize the range of effects, we plot the estimated kernel densities for M_1 compared to M_2, in Fig. 3.5.

We see from Fig. 3.5 that the M_1 density (buyers' deficiency—solid line) is much more compact with a thin right tail while the M_2 density (seller's deficiency—dashed line) is more spread out with a fatter right tail. This comports with our averages: buyers and sellers both have information deficiency, but sellers appear to lose more on account of it than buyers, in this sample. This could be an indication of other forces at play, for example, the effect of employing real-estate agents or that "sales under duress" comprise a not negligible part of the transactions in the sample, implying a bargaining undercurrent to the situation that we will explore in detail in Chap. 4.

Next, we turn to view these information deficiencies as forces operating on the full-information, "balanced" price. The relevant *unilateral* measure for buyers is $M_5 = E(\exp\{w\} - 1)$. Recall from Chap. 2 that this measures the effect of w on the

[8] Applied experience has indicated that sometimes, Jensen's inequality for expected values may reverse the inequality when going from the logarithmic level to the original measurement units.

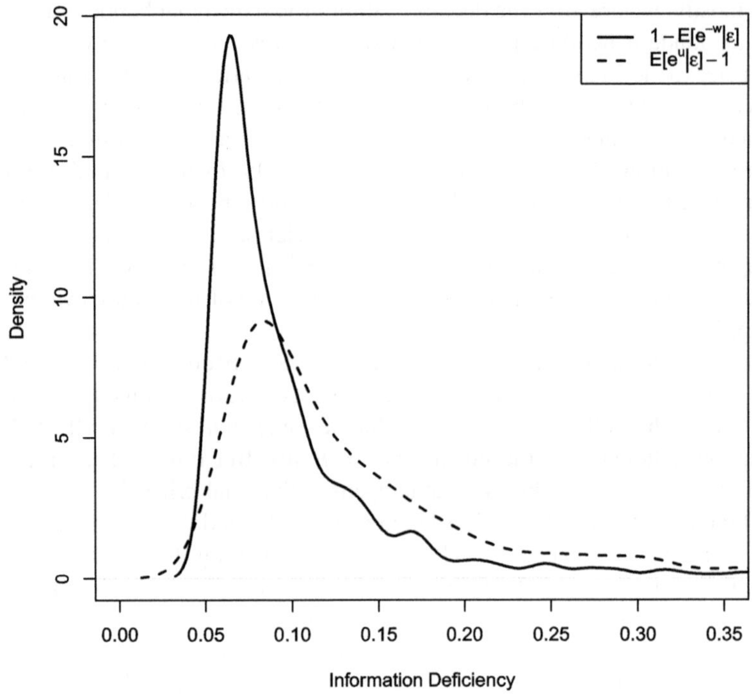

Fig. 3.5 Kernel densities of M_1 and M_2 metrics

balanced price, under the assumption that information deficiency does not exist on the opposite side of the market. The average value of M_5 is 12%. The corresponding counterfactual measure for sellers is $M_6 = 1 - E(\exp\{-u\})$, and it has mean value equal to 9.8%. This implies that the information deficiency of buyers tends to increase *on its own*, the full-information price by 12%, while the information deficiency of sellers tends to decrease *on its own* the full-information, balanced price, by 9.8%. The related kernel density estimates are presented in Fig. 3.6.

Both the average values obtained, but also the eventual first-order stochastic dominance of buyers' M_5 density over the sellers' M_6 density, as we see in Fig. 3.6, indicate that the information deficiency of buyers appears here to have a stronger effect than that of the sellers (but on the balanced price). Does this contradict the previous results that related to the actual price? Even if it should be clear that the M_1/M_2 pair measures different aspects than the pair M_5/M_6, still, at an intuitive level, some dissonance may arise, since one could expect that if the one latent force is "stronger" than the other, its effects should everywhere appear "greater," for logical consistency. But this is false, it is an "intuition trap."

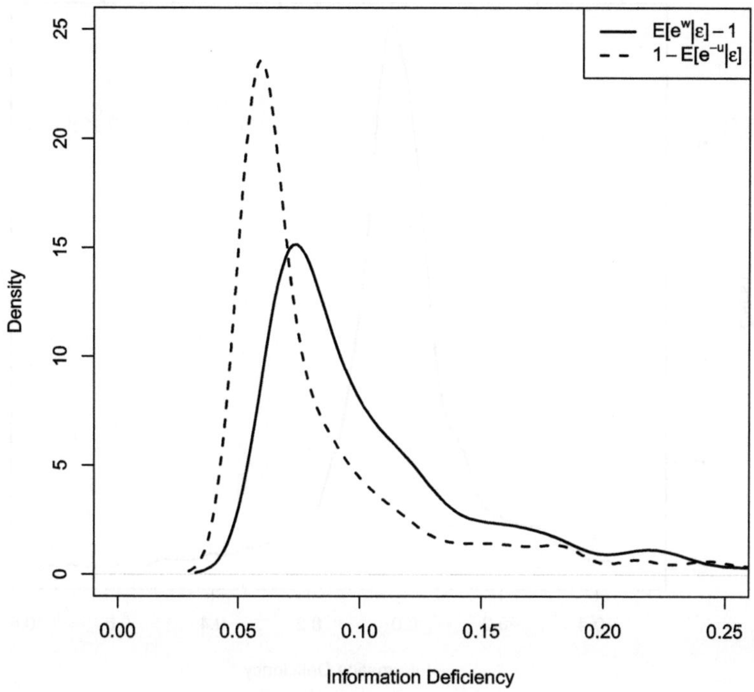

Fig. 3.6 Kernel densities of M_5 and M_6 metrics

It is instructive here to repeat the conceptual definitions of the metrics:

$$M_1 = E\left(\frac{Y - Y^{LF}}{Y}\right), \qquad M_2 = E\left(\frac{Y^{UF} - Y}{Y}\right)$$

$$M_5 = E\left(\frac{Y^{UF} - Y^B}{Y^B}\right), \qquad M_6 = E\left(\frac{Y^B - Y^{LF}}{Y^B}\right)$$

Ignoring for clarity the expected value operator, we obtain empirically

$$M_1 < M_2 \Longrightarrow Y < \frac{Y^{LF} + Y^{UF}}{2}, \qquad M_5 > M_6 \Longrightarrow Y^B < \frac{Y^{LF} + Y^{UF}}{2}.$$

So we found that both the average actual price and the average balanced, full-information price lie below the midpoint of the range determined by the lower and upper price frontier. The result related to actual outcome Y indicates informally that the negative force u "must" be stronger than the positive force w, since it pulls Y toward its own lower frontier more and Y includes the effects of both w and u. Note also that the effects of w are incorporated in Y^{UF} and the effects of u in Y^{LF}.

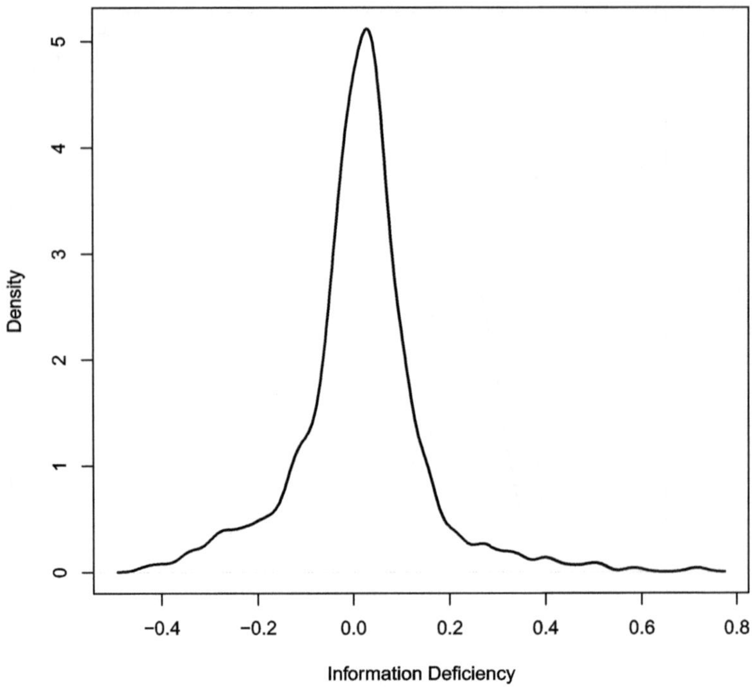

Fig. 3.7 Kernel density of the $M_7 = E(\exp\{w - u\}\text{mid}\varepsilon) - 1$ metric

The result that $M_5 > M_6$ and that the *balanced* outcome Y^B (where the latent forces are absent) is also below the midpoint does not reflect the relative strength of the two forces, exactly because Y^B, as a counterfactual, is not affected by them. The higher unilateral *proportional* pull that w exerts on Y^B compared to u (as reflected by the inequality $M_5 > M_6$) is equivalent to Y^B being below the midpoint, but what this indicates is a low Y^B (which is not determined by the latent forces), not a w stronger than u.

Moreover, the M_5 and M_6 metrics are counterfactuals, each measured under the assumption that the other side has zero information deficiency. To complete the picture and the interpretation, one needs also the metric M_7 which measures the joint effect of w and u on the balanced price. This is a measure that takes into account the individual *pairs* of buyers and sellers that were realized and may give us yet a different picture. We have $M_7 = E(\exp\{w - u\}) - 1$, which has an average value of 0.015. This means that, in the end, on average the housing price is 1.5% higher than the full-information, balanced price, a small informational advantage benefiting the sellers. More illuminating is its kernel density plot, shown in Fig. 3.7.

We note the thicker "shoulder" in the negative values: buyer/seller pairs with informational advantage for the buyers are relatively concentrated to small effects of such advantage. The quartiles of this distribution are

$$Q1 = -0.038, \quad \text{median} = 0.020, \quad Q2 = 0.065,$$

indicating that in more than half of the transactions, the actual price was above the balanced price.

Which set of metrics is more pertinent depends on who is conducting the research and for what purpose. The (M_1, M_2) metrics speak to the two sides of the market individually, as an ex post assessment of the effects of information deficiencies on the sample. On the other hand, the metrics (M_5, M_6, M_7) are closer to a more structural estimation of the consequences of incomplete information and more suitable for discussing policy issues toward improving the degree to which market participants do make informed decisions.

But if policy is in our minds, it must be pointed out that nowhere in this chapter did we name the *sources* of information deficiency. They could be attributed to the actual characteristics of the product, characteristics of the seller, characteristics of the buyer, or a combination of the three. Understandably, this information would be important in discussing policy—and we can gain access to it if we include covariates that purport to explain the information deficiency itself. This setup is developed in Chap. 8, of the book.

Chapter 4
Bargaining

Business folk wisdom has it that "In life you don't get what you deserve; you get what you negotiate." While in the era of mass consumer markets, it is true that many products and services are not up for negotiation, in the sense that their characteristics and their prices are set by the firms and the consumer faces a strict "buy/not buy" choice, there are still many settings where negotiation between parties takes place. The labor market is one of the prominent examples in relation to economic activity, and here we observe both individual bargaining (between employer and worker) and also collective bargaining (between labor unions and trade associations or large firms). Bargaining still happens between suppliers and consumers in flea markets, between buyers and sellers in business-to-business transactions, but also in political settings, between political parties in a government coalition or in international relations between sovereign countries. We could argue that any human interaction that involves some degree of interdependence and common action and fate involves some "bargaining" but, to avoid hubris, we stick to cases where a mostly economic matter is at stake.

The issue with a bargaining situation is that informational asymmetries and private information are crucial to the outcome. What one party knows that the other does not affects critically the positions and the offers and counteroffers that this party may make or accept. These pieces of information are also unobservable to a researcher that uses observational data. Moreover, bargaining power, namely, the extent to which one party can "hold out" or the degree to which it can confidently make high demands, impacts the outcome. A model that does not handle informational asymmetry and bargaining power cannot claim to be a good device to examine and quantitatively assess bargaining situations.

In this chapter, we present two variants of the 2TSF bargaining model that recognize the existence of latent forces on the bargaining outcome from both sides of the negotiation and provide estimated results for the strength and effects of these forces. The two variants share most everything but how one models bargaining power. The common core brings together elements from both axiomatic/normative

A. Papadopoulos, C. F. Parmeter, *Two-Tier Stochastic Frontier Analysis for the Social Sciences*, https://doi.org/10.1007/978-3-031-81513-3_4

and strategic game theory, exploits the dynamic nature of bargaining, and is applicable to any bargaining situation where one party can be designated as a "seller" (even in a metaphorical sense) and the other as a "buyer" and which also has the following characteristics:

1. The two parties share some common knowledge but also possess asymmetric and heterogeneous information.
2. The parties formulate guiding targets, to structure their negotiation actions.
3. There exists uncertainty about the total gain over which bargaining takes place.
4. There exists systematic mark-up from the "seller" and systematic discount from the "buyer," pivoting off a commonly accepted (but not final) base assessment of the value at stake.

The first two characteristics exist in all cases of bargaining, to one or another degree. The third characteristic can simply spring from the inherent uncertainty of the future, but it may also reflect that the two negotiating parties arrive at different conclusions as to the potential gains of collaboration/transaction that still allow for a common ground to be reached. The final characteristic is also very familiar: either "honestly" or just as a bargaining tactic, sellers tend to exaggerate positively, and buyers tend to exaggerate negatively.

The model is developed in detail in Papadopoulos (2024a), which focuses on the Nash bargaining variant. We present it summarily here and offer also an empirical application related to the labor market. But the Nash bargaining model assumes a specific interconnection between the two bargaining parties, an interconnection that may not always be present in a bargaining situation. This is something that a researcher can argue from a priori premises, or the data may indicate so, rendering the Nash bargaining model inappropriate. To cover this case also, in the second part of this chapter, we present the no-Nash 2TSF bargaining model, where we model bargaining power in a different way that in turn leads to different statistical properties and different specifications. We illustrate this variant of the 2TSF bargaining model with another application, this time from the field of international relations.

4.1 The 2TSF Nash Bargaining Model

The core setup of the 2TSF Nash bargaining model is as follows: we have two parties, A and B. If they collaborate, a gain will be realized, and the issue is how this gain will be split. For example, in a wage negotiation, collaboration will result in the production of a good that has revenue-generating value for the firm; the issue here is how much of this value will the worker receive, in the form of a wage. Here, the uncertainty relates to the revenue potential of the output that will be produced, and the wage is the object of negotiation. As another example, in a political setting over negotiations for a tax/subsidies package, party B of the coalition may be interested in raising taxes to fund certain social services to the general population, while party

A of the coalition wants to secure part of the increased tax revenue as targeted subsidies to a professional class that is part of its core voting base. Here, the "seller" is party A, demanding a "wage" (the targeted subsidy) in order to provide the service "support for the tax hike," while party B is the "buyer," looking to "pay" as little as possible in order to secure as much as possible of the additional tax revenue to realize its own political agenda. The actual level of additional tax revenues is the uncertain overall gain, and the parties have to form expectations about it.

Let ω denote the value under negotiation between two parties A and B, and p denote the uncertain overall gain from the collaboration. Metaphorically (or not), A will represent the seller, while B has the role of the buyer. So if collaboration happens, A will receive ω, and B will receive $(p - \omega)$.

The situation is also characterized by the "no-agreement payoffs," say b^A and b^B, namely, what each party will receive if agreement is not reached. These no-agreement payoffs may be positive, zero, or negative. For the wage negotiation example, the no-agreement payoff of the worker (the seller of labor services) could be unemployment benefits or the wage of another job that they have already secured. For the employer (the buyer of labor services), it could be zero, positive if they have other candidates with whom they have reached a tentative agreement, or even negative, if the firm is pressed to hire fast to fulfill contractual obligations which may carry a monetary penalty if they are not completed on the agreed-upon time. For the political negotiation example, the no-agreement payoff for the party interested in the tax revenue for broad social services could be some leeway in re-purposing existing tax revenues, while for the other party, it could be already-marked subsidies from the current government budget. As mentioned, the "seller" here is the party demanding the subsidies, selling its voting support to pass the tax legislation and perhaps also the services of its media-propaganda machine to reduce tendencies for tax evasion.

The *surplus* function of party A is $S_A = \omega - b^A$, while the surplus function of party B is $S_B = (p - \omega) - b^B$.[1] Finally denote by $\eta \in (0, 1)$ the relative bargaining power of party A. This is a critical aspect of the Nash bargaining model: here "bargaining power" is modeled in relative terms from *two* points of view: it is relative because it ranges in the interval (0,1), but also, it is relative because the magnitudes characterizing the two parties *add up* to unity. This creates a strong interconnectedness between them.

Note that for the agreement to be incentive-compatible, it must be the case that $p > b^A + b^B$. The overall gain should be greater than the sum of the no-agreement payoffs. Re-written, this condition becomes more intuitive: we require $p - b^B > b^A$. This says that the maximum surplus for party B should exceed the no-agreement payoff of party A. The maximum surplus for party B is what party B would at most be willing to give to party A. If this was smaller than what party A could secure

[1] In the bargaining context, the word "surplus" has a somewhat different meaning than the one we encountered in previous chapters, as "supplier/consumer surplus" in a market equilibrium context. Readers are invited to ponder their differences and how they could be reconciled ... or wait just a little bit longer.

without agreement, then party A would not be interested in collaborating. So the condition $p > b^A + b^B$ is a basic requirement to make the bargaining situation feasible.

Nash (1950) introduced the *normative* solution to this problem (the "Nash bargaining solution"), while Roth (1979) provided the formal mathematical treatment: we can obtain the solution if we maximize the product (or the weighted product) of the two surplus functions. The solution is normative because it maximizes the overall gain while at the same time determines the split between the parties, in a "Pareto-optimal" way (meaning that any higher overall gain will be accompanied by a split that will make one of the parties worse-off compared to the Nash bargaining solution). In mathematical terms as a solution over ω, we have

$$\omega^* = \text{argmax}_\omega\{S_A^\eta \cdot S_B^{1-\eta}\} = \left(\omega - b^A\right)^\eta \cdot \left(p - \omega - b^B\right)^{1-\eta}.$$

Due to the logarithm being a monotonic transformation, we can equivalently consider the maximizer of the logarithm of the above expression. Computing the first-order condition leads to

$$\omega^* = \eta(p - b^B) + (1 - \eta)b^A. \tag{4.1}$$

The optimal level of the variable whose value must be agreed-upon is a convex combination (weighted by relative bargaining power η) of the maximum possible surplus of party B (buyer), $p - b^B$, and of the no-agreement payoff of party A (seller).

While the fundamental intuition of this model carries through, it does not take into account the dynamics inside the bargaining process. It envisions a one-step process where rational parties move directly to the above solution, given the involved factors. But issues arise: under asymmetric and private information, how could we reasonably argue that the expectation on the uncertain p is the same for both parties? Also, during negotiations, once a party makes a concession, then this becomes a fall-back option for the other party. And when these concessions exceed the initial no-agreement payoffs, the latter become irrelevant, and it is not rational to fall back on them since the parties could be better off if they would agree to what was eventually on offer.

Papadopoulos (2024a) dealt with these two important issues by applying the following transformations: first, instead of formulating the surplus functions as surpluses over the no-agreement payoff, he formulated them as surpluses over the current offer of the other party. Second, he postulated that these offers during a bargaining process are rationally linked to a *target* each party has. This approach led first to the following result as regards the expression for the maximizer, the optimal ω^*,

$$\omega^* = \eta\omega_A^T + (1 - \eta)\omega_B^T, \tag{4.2}$$

where ω_A^T is the equilibrium target that party A had at the time of agreement, while ω_B^T is the equilibrium target that party B had at the time of agreement.

Further, these targets can be conceptually decomposed as follows: denote I_c the common information set of the parties, namely, the information set that includes only the information that is known to both. Per our earlier discussion, the full individual information sets of the two parties are bigger than I_c.

Define the expectation on p conditional on the common information set as $E(p|I_c)$. This does not coincide with the individual expectations of each party on p, because these expectations are based on different information sets. The target of the seller can then be decomposed into

$$\omega_A^T = E(p|I_c) + g + v, \qquad g \geq 0,\ v \in \mathbb{R}, \tag{4.3}$$

while for the buyer, it is

$$\omega_B^T = E(p|I_c) - d + v, \qquad d \geq 0. \tag{4.4}$$

The above formulation incorporates the assumption that sellers always ask for a mark-up over some "common-information" value (the $g \geq 0$ factor) and also the assumption that buyers always demand a mark-down on the same commonly accepted value (the $d \geq 0$ factor entering with a minus sign in front). These tendencies exist either due to normative reasons or as a strategic tool to extract as much surplus as possible. The factor g is called the "own-evaluation premium," while d is called the "prudential discount." The variable v represents changes in the common-information expected value during negotiations.

Inserting these into the expression for ω^*, we obtain

$$\omega^* = E(p|I_c) + (v + \eta g - (1 - \eta)d). \tag{4.5}$$

Keeping with the wage bargaining example, ω^* would be the observed wage as the dependent variable, while the common information set can be reasonably represented by a set of variables $\boldsymbol{x}$ that measure worker, firm, and market characteristics, which all influence and co-determine the expected value of p. So $E(p|I_c)$ can be mapped to the systematic part of a regression equation $h(\boldsymbol{x})$, and the variables in $\boldsymbol{x}$ are often available to the researcher as data.

On the other hand, the composite term $(v+\eta g-(1-\eta)d)$ is full of unobservables and has the 2TSF structure, $\varepsilon = v+w-u$, with v being v, $w \equiv \eta g$ and $u \equiv (1-\eta)d$. It includes those unobservable factors that have an important effect on the outcome (wage): the relative bargaining power η and the effects of private information on both sides (g, d).

A conceptual difference of the bargaining framework with the incomplete information setup of Chap. 3 that is useful to keep in mind, is that there, the w latent force represented "information deficiency *of the buyer*," while here it represents "bargaining performance *of the seller*." Mirror-like, the latent force u

in the incomplete information framework represents informational deficiency of the seller, while in the bargaining framework, it represents bargaining performance of the buyer.

A critical technical issue is that by construction, the two one-sided components (w, u) are dependent (and with *negative* covariance/correlation coefficient), because, looking now at a sample of data over many agreements, it is not reasonable to argue that relative bargaining power η is a constant, the same for every seller. But if it is a random variable, then $w_i = \eta_i g_i$ is dependent with $u_i = (1 - \eta_i)d_i$. This will require the use of statistical specifications discussed in Chap. 10.

4.1.1 Logarithmic Representation and Estimation

It is often the case that in actual estimation, we use the dependent variable in logarithmic terms. For example, if we examine a wage-bargaining situation, the wage enters in logarithms in the left-hand side of the estimating equation.

This can be accommodated by the 2TSF Nash bargaining model. First, we rewrite the optimization problem as

$$\begin{aligned}\ln \omega^* &= \arg\max \left(\ln \omega - \ln \omega_f^T\right)^{\eta} \left(\ln \omega_e^T - \ln \omega\right)^{1-\eta} \\ &= \arg\max \left\{ \left[\ln \left(\frac{\omega}{\omega_f^T} \right) \right]^{\eta} \left[\ln \left(\frac{\omega_e^T}{\omega} \right) \right]^{1-\eta} \right\}.\end{aligned}$$

Here, the gains are measured as mark-ups (or mark-downs), which is a proportional approach often encountered in actual transactions. Solving for the maximum, we obtain

$$\ln \omega^* = \eta \ln \omega_e^T + (1 - \eta) \ln \omega_f^T \implies \omega^* = \left(\omega_e^T\right)^{\eta} \cdot \left(\omega_f^T\right)^{1-\eta}.$$

The optimal wage is a weighted geometric average instead of a weighted arithmetic average. Next, we define the targets in a multiplicative fashion

$$\omega_e^T \equiv E\left(p|I_c\right) e^{v} e^{g} \implies \ln \omega_e^T \equiv \ln E\left(p|I_c\right) + v + g,$$

$$\omega_f^T \equiv E\left(p|I_c\right) e^{v} e^{-d} \implies \ln \omega_f^T \equiv \ln E\left(p|I_c\right) + v - d.$$

Note that we took the logarithm of the conditional expectation and not the conditional expectation of the logarithm because, prior to deciding on the evaluation scale, the players have to face the inherent uncertainty of the situation. So it is first $E\left(p|I_c\right)$ instead of p, and then we consider the logarithms.

From these we see that the components of the error term (v, g, d) now refer to the logarithmic scale. Then, we obtain the expression for the wage

$$\begin{aligned}\omega^* &= \left[E\left(p|I_c\right) e^v e^g\right]^\eta \cdot \left[E\left(p|I_c\right) e^v e^{-d}\right]^{1-\eta} \\ &= E\left(p|I_c\right) \cdot e^v \cdot e^{\eta g} \cdot e^{-(1-\eta)d} \\ \Longrightarrow \ln \omega^* &= \ln E\left(p|I_c\right) + v + \eta g - (1-\eta)d. \end{aligned} \tag{4.6}$$

This means that if $h(\boldsymbol{x}; \boldsymbol{\beta})$ is the observable part of a regression function, it estimates $\ln E\left(p|I_c\right)$, and so if we want to obtain an estimate for the balanced outcome in the original measurement scale, we need

$$\widehat{E}\left(p|I_c\right) = \exp\{h(\boldsymbol{x}; \widehat{\boldsymbol{\beta}})\}. \tag{4.7}$$

4.1.2 Surplus Definition and Measurement in the Nash Bargaining Model

We have already footnoted that the word "surplus" as used in the context of a bargaining framework is not equivalent to how the word is used in the standard microeconomic market theory that we have encountered in Chap. 3 and were able to map to information deficiencies and to the metrics M_1 (for producers/sellers) and M_2 (for consumers/buyers).

Here, the traditional definition of seller surplus is "what did I get above my no-agreement payoff," and for the 2TSF Nash bargaining model, we have adjusted it to mean "what did I get above the target that the other side had at the time of agreement." And for the buyer side, it is "how much did I avoid paying below the target that the other side had at the time of agreement." So it is a more antagonistic concept, judging the outcome against the other side's intentions. As analyzed earlier, this qualified concept of surplus leads necessarily to a quantity that is less than the initial bargaining concept (benefit above the no-agreement payoff). But this latter quantity cannot be estimated inside the model, because we always lack any information on it.[2]

Our qualified concept of surplus, relative to the actual outcome (here, the actual wage) and in a logarithmic formulation, is fundamentally defined as

[2] In territories where unemployment benefits exist, one could think that they could proxy for the no-agreement payoff for the worker. But a worker's no-agreement payoff may be another job offer that they have secured or parallel streams of income without claiming unemployment benefits. And we would still lack information of what the no-agreement payoff for the employer side might be.

$$\text{Seller/worker Surplus (SS)} \equiv \frac{\omega - \omega_f^T}{\omega} = 1 - \exp\{-g\},$$

$$\text{Buyer/employer Surplus (BS)} \equiv \frac{\omega_e^T - \omega}{\omega} = \exp\{d\} - 1.$$

But these expressions are *not* what we can directly estimate with the 2TSF Nash bargaining model, since we do not deal with g and d, the own-evaluation premium, and the prudential discount factors, but with $w = \eta g$ and $u = (1 - \eta)d$, i.e., with g and d adjusted for relative bargaining power η. To wit, for the metrics M_1 and M_2, we have, putting everything in average terms,

$$M_1 = 1 - E\left(e^{-w}\right) = 1 - E\left(e^{-\eta g}\right) < 1 - E\left(e^{-g}\right) = SS,$$
$$M_2 = E\left(e^{u}\right) - 1 = E\left(e^{(1-\eta)d}\right) - 1 < E\left(e^{d}\right) - 1 = BS.$$

So our M_1 and M_2 metrics measure "surplus" qualified for relative bargaining power (which enters autonomously in the model, and this is why it does not appear in the core definition of surplus). These are still useful metrics, so we can talk about "surplus extraction" in the above sense. Moreover, as we will see in Sect. 10.4, if we are prepared to make a (very reasonable) distributional assumption for the relative bargaining power η, we can apply a conditioning technique that will allow us to estimate $E(e^{-g})$ and $E(e^{d})$, at least as averages over the sample.

We turn now to present a specific application of the 2TSF Nash bargaining model.

4.2 Application: Zambian Wages in the Early 1990s

We use part of a matched employer-employee data set that has been originally used by Bigsten et al. (2000).[3] Their purpose was to examine physical and human capital returns in the manufacturing sector of five African countries. Data were collected in three survey-waves during the early 1990s. We will use the data set relating to Zambia for the years 1993–1995, with sample size n = 1,545. Its composition as regards number of employees and firms per survey wave is presented in Table 4.1. The composition exhibits some variability across survey waves.

[3] The data was collected as part of the Regional Program on Enterprise Development (RPED), organized by the World Bank. The project received support from the Swedish, Norwegian, United Kingdom, Canadian, and Dutch governments. The full data set was downloaded from the website of the Centre for the Study of African Economies at the University of Oxford (CSAE): http://www.csae.ox.ac.uk.

Table 4.1 Composition of the Zambia sample

Wave	Number of firms	No of employees (observations)	Distribution of firm-sizes (%)			
			up to 10	11–50	51–250	>250
1 (1993)	144	698	14.2	33.5	40.5	11.7
2 (1994)	106	492	15.7	37.8	36.2	10.4
3 (1995)	85	355	14.1	42.5	23.9	19.4
	Sample size	1,545	14.6	37.0	35.3	13.1

Table 4.2 Descriptive statistics

Label	Description	Mean	S.D.	Min	Max
LMEARNPP	Dep var: log monthly earnings in PPP-USD	4.55	0.97	– 1.88	10.22
Employee variables					
MALE	Binary sex dummy (1 = male)	0.81	0.39		
TENURE	Tenure in the firm in years	6.54	6.60	0.00	36.00
EDUC	Formal education in years	10.02	3.22	0.00	16.00
PEXP	Potential Experience (maximum possible)	17.89	10.48	1.00	58.00
MGMT	Binary hierarchical dummy (1 = management)	0.10	0.30		
SUPER	Binary hierarchical dummy (1 = supervisor)	0.13	0.34		
ADMIN	Binary functional dummy (1 = administration)	0.07	0.26		
SALES	Binary functional dummy (1 = sales)	0.14	0.34		
TECH	Binary functional dummy (1 = technician)	0.18	0.38		
Firm variables					
CAPCITY	Socioeconomic dummy (1 = HQ in the capital)	0.43	0.50		
WOOD1	Industry Sector (1 = wood sector)	0.15	0.36		
TEXTILE1	Industry Sector (1 = textile-clothing sectors)	0.27	0.44		
METAL1	Industry Sector (1 = metal-machinery sectors)	0.28	0.45		
ANYFOR	Ownership dummy (1 = some foreign equity)	0.18	0.38		
ANYSTAT	Ownership dummy (1 = some State equity)	0.12	0.32		
LEMP	log of number of employees	3.86	1.39	0.69	8.01
LVADEMPPPP	log of Value-added/employee in PPP-USD	7.84	1.48	– 0.69	11.34
LCAPEMPPPP	log of Physical capital/employee in PPP-USD	8.84	1.68	0.11	12.64

We pooled the data and used a linear trend $t = 0, 1, 2$, to capture possible time effects. The dependent variable is the logarithm of monthly earnings (basic wages plus allowances) expressed in US-PPP dollars as calculated in the original study. Descriptive statistics of the dependent variable and the explanatory variables are presented in Table 4.2.

Except for the covariates appearing in Table 4.2, the regression specification also included a constant term and the time trend mentioned earlier, as well as the squares of the variables tenure and education, to capture a possibly nonlinear relation. To model the intra-error dependence between w and u, we use the 2TSF-Bivariate Truncated Normal (BTN) specification presented in Sect. 10.2 and implement maximum likelihood estimation. In this specification, v is assumed to follow a Normal distribution, while (w, u) follow jointly a BTN distribution.

We must also consider the possibility of regressor endogeneity. We will treat as potentially endogenous two variables: education, to acknowledge its possible correlation with the unobservable "ability" and also "potential experience" since it is by design an inaccurate, maximal proxy for actual experience and so the related measurement error resides in the error term. To account for endogeneity, we attach a (Gaussian) copula density in the likelihood.[4] With this method we do not need to find instruments for the endogenous variables.

The estimation results are presented in Table 4.3, where, for comparison, we have also included estimates from the simpler 2TSF Normal-Half Normal (NHN) specification (see Sect. 7.2.2) that ignores intra-error dependence and regressor endogeneity. We provide either the direct estimates when they are comparable or compute magnitudes based on them, when they are not. This is the case with the σ_w and σ_u scale parameters that have different roles in the two specifications and their magnitudes differ wildly by design. But the resulting mean values for w and u can be meaningfully compared. In brackets we present p-values rather than standard errors.

The estimation results in Table 4.3 show that the two specifications lead to different conclusions in many aspects. Before discussing several diagnostics that lead us to consider the BTN model as more appropriate, we point out some of these differences.

As regards the regressors, we have certain variables whose coefficient or estimation accuracy changes visibly (most appear in bold letters). With the NHN model we would likely conclude that male workers do not enjoy a premium due to their sex; not so with the BTN model. The education variables have unusual signs in their coefficients in both models, and we will discuss this in a while. Being in the wood industry sector as well as working in administration appear as having stronger effects in the BTN model. But overall, the results validate the anticipation that the wage is affected by the triad "worker-firm-socioeconomic environment."

For the residuals, the differences are perhaps stronger, already as regards their basic statistics: the NHN model does not have room for the noise component v estimating its standard deviation σ_v as zero, while the BTN model estimates it as of mediocre strength but existent. It is not unusual with 2TSF models to have $\sigma_v = 0$ (this is the topic of the "Tug-of-War," Chap. 12), but the point here is that the two models clearly do not agree. The residuals in the BTN model have much larger variance, positive skewness instead of roughly zero, and much lower excess kurtosis (hence fewer outliers). Moreover, the mean value of the self-evaluation premium w and of the prudential discount u is considerably lower in the NHN model (although we note that the *ratio* of these two expected values is essentially the same in the two models).

To assess the acceptability of the BTN model, we implemented the Doornik and Hansen (2008) test (see Sect. 7.3), in order to assess whether the (numerically evaluated) distribution function of the error term, that is based on the assumed

[4] See the brief discussion in Sect. 14.1.4 and for every detail Papadopoulos (2022).

Table 4.3 Zambian worker-firm matched data, n=1,545. Dependent variable is the logarithm of monthly earnings

Regressor	2TSF NHN	2TSF-BTN with copula	Regressor	2TSF NHN	2TSF-BTN with copula
MALE	0.0376 [0.435]	0.0714 [0.086]	CAPCITY	0.0950 [0.005]	0.1353 [0.002]
TENURE	0.0242 [0.001]	0.0275 [0.032]	**WOOD1**	0.1788 [0.001]	0.2949 [0.000]
TENSQ	−0.0008 [0.005]	−0.0009 [0.030]	TEXTILE1	0.0951 [0.033]	0.1283 [0.014]
EDUC	−0.0172 [0.397]	−0.1226 [0.0753]	METAL1	0.2789 [0.000]	0.2741 [0.000]
EDUCSQ	0.0053 [0.000]	−0.0077 [0.040]	ANYFOR	0.1671 [0.000]	0.2192 [0.000]
PEXP	0.0082 [0.000]	0.0059 [0.251]	ANYSTAT	0.2080 [0.000]	0.2458 [0.000]
MGMT	0.9023 [0.000]	0.8984 [0.000]	LEMP	0.1235 [0.000]	0.0969 [0.000]
SUPER	0.2199 [0.000]	0.2630 [0.000]	LVADEMPPPP	0.0851 [0.000]	0.0785 [0.000]
ADMIN	0.3830 [0.000]	0.4891 [0.000]	**LCAPEMPPPP**	−0.0243 [0.011]	−0.0054 [0.771]
SALES	0.1644 [0.005]	0.2088 [0.009]	Trend	0.0834 [0.000]	0.0769 [0.001]
TECH	0.0545 [0.245]	0.1487 [0.420]	constant	2.1933 [0.000]	4.5732 [0.000]
σ_ε	0.6485	1.5274	$E(w)$	0.7433	1.3314
Resid. Skew	−0.0396	0.1541	$E(u)$	0.4292	0.7896
Resid. Excess Kurt	15.373	2.4867			
σ_v	0.000 [1.000]	0.3455 [0.062]	Corr(EDUC, ε)	–	0.8964 [0.000]
ρ	–	−0.9958 [0.000]	Corr(PEXP, ε)	–	−0.2823 [0.000]

Asymptotic p-values are given in brackets, based on the outer-product matrix for 2TSF-NHN model and on the sandwich VCV estimator for 2TSF-BTN

distributional specification, behaves like a Uniform(0,1) when viewed as a random variable or equivalently whether $\Phi^{-1}(F_\varepsilon(\varepsilon))$ is distributed as Standard Normal. The test statistic is asymptotically distributed as $\chi^2(2)$. For rejection of the null, the obtained value of the statistic was 2.807 that gives a p-value of 0.245, which does not reject the chosen distributional specification at the conventional statistical confidence levels. We discuss in detail the implementation of the test together with additional formal model selection and "goodness-of-fit" tests in Sect. 7.3.2. Next, we present direct evidence that the BTN model better represents the data.

4.2.1 Results on Endogeneity and Intra-error Dependence

In the BTN model, the correlation coefficients between the endogenous regressors and the error term (both doubly transformed) are non-zero and high for the education variable. This is a direct formal test for whether the suspected regressors are indeed endogenous, and in this case endogeneity is supported, undermining the reliability of the results from the NHN model.

Further, in the NHN model, the education coefficients both have the opposite signs from those usually obtained, which furthers compromises its reliability. In the BTN model *both* coefficients have negative signs, which appears counterintuitive since it implies that formal education tends to have a negative effect on the wage (and the more the years of education, the stronger the negative marginal effect). To make sense of this result, we must interpret it together with the high positive correlation between education and the error term. The correlation link is assumed to be the unobservable "ability," and ability is expected to have a direct positive effect on the wage, and we do obtain a positive correlation of education with the error and a strong one at that (0.89 with theoretical maximum 1.00). Combining, we get a clear message: education without ability is not really valued by the firms in the sample, maybe because it is then viewed as years wasted away from on-the-job training and experience.

Regarding potential experience, we also observe the anticipated correlation sign (negative). To see why we anticipated a negative correlation, in generic notation, let x denote potential experience, and, since it is a maximal estimation of actual experience x^*, their relation can be written as $x = x^* + \xi, \quad \xi \geq 0, \quad \text{Cov}(x, \xi) > 0$. Then the true regression relation is

$$y = \cdots + x^*\beta + \varepsilon^* = \cdots + (x - \xi)\beta + \varepsilon^* \Rightarrow y = \cdots + x\beta + \varepsilon, \quad \varepsilon = -\beta\xi + \varepsilon^*.$$

So $\text{Cov}(x, \varepsilon) = -\beta \, \text{Cov}(x, \xi)$. Given that $\text{Cov}(x, \xi) > 0$ and that the regression coefficient of potential experience is positive, we anticipate $\text{Cov}(\text{PEXP}, \varepsilon) < 0$, which is what we obtain. Moreover, note that the direct effect of potential experience reduces and becomes highly imprecise in the BTN copula model, compared to the NHN specification. Essentially it is overshadowed by the correlation with the error term, and this tells us that potential experience is not really a good proxy variable for actual experience.

Turning to the estimates related to the error term, we observe a negative estimate for the parameter ρ, indicating negative correlation between w and u, $\text{corr}(w, u) = -0.216$, which is consistent with the theoretical model. As we mention in Chap. 10, this is very close to the numerical limit of linear correlation that the BTN model can accommodate, and since the estimator pushed the estimate near the extreme, it is indication that the (w, u) are characterized by even stronger correlation.

In all, we see that the BTN model captures aspects of the data that the NHN model does not, both as regards regressor endogeneity but also the dependence between the two latent forces.

4.2.2 *Bargaining Space and Bargaining Performance*

The estimates in the BTN model of the primitive scale parameters are $\hat{\sigma}_w = 23.207$ and $\hat{\sigma}_u = 13.763$. Given these, the standard deviations of bargaining performances are $\mathrm{SD}(w) = 1.1117$ and $\mathrm{SD}(u) = 0.6593$. This means that the variable w, which represents bargaining performance of the worker side, has a higher mean than the employers' variable u but also a higher standard deviation, something that may reflect the higher number of unique workers than unique firms in the sample and hence more heterogeneity on the side of the workers.

Next, we focus on the individual metrics, which are the exponentiated ones, since the dependent variable is in logarithms. We present in Table 4.4 the median values of the series used to construct the metrics, because exponentiation magnifies the effects of estimation inaccuracy on averages.

To interpret the metrics, we remind again that in a bargaining model (and in contrast to the incomplete information setup), w represents a strength of the seller (not a weakness of the buyer), while u a strength of the buyer (not a weakness of the seller).

We first delineate the *bargaining space*, relative to the balanced wage (where the latent forces offset each other). Seeing how these magnitudes evolve over time provides a quantified sense of the bargaining space, as in Table 4.5.

The upper frontier gives us the wage (measured in terms of the balanced wage) if it were the case that $u = 0$ (no prudential discount from the part of the firm and/or zero bargaining power). The lower frontier gives the wage if $w = 0$ (no self-evaluation premium from the part of the worker and/or zero bargaining power).

The main information we get from Table 4.5 is that the median wage was above the balanced wage in all 3 years, showing a favorable bargaining outcome for

Table 4.4 Basic measures. Sample medians of the conditional expectation series

	1993	1994	1995	Full sample
$E(\exp\{w_i\}\|\varepsilon_i)$	2.867	2.383	2.624	2.605
$E(\exp\{-w_i\}\|\varepsilon_i)$	0.458	0.537	0.495	0.499
$E(\exp\{u_i\}\|\varepsilon_i)$	2.103	2.314	2.182	2.190
$E(\exp\{-u_i\}\|\varepsilon_i)$	0.591	0.547	0.574	0.573
$E(\exp\{w_i - u_i\}\|\varepsilon_i)$	1.400	1.076	1.243	1.230
$E(\exp\{u_i - w_i\}\|\varepsilon_i)$	0.794	1.029	0.892	0.901

Table 4.5 Relative wage frontiers. Sample medians

	1993	1994	1995
Upper frontier: $M_5 + 1 = E(\exp\{w_i\}\|\varepsilon_i)$	2.867	2.383	2.624
Actual wage: $M_3 = E(\exp\{w_i - u_i\}\|\varepsilon_i)$	1.400	1.076	1.243
Balanced wage	1	1	1
Lower frontier: $1 - M_6 = E(\exp\{-u_i\}\|\varepsilon_i)$	0.591	0.546	0.574
Volatility: $M_4 = M_5 + M_6$	2.276	1.836	2.050

Table 4.6 Bargaining performance. Sample medians

	1993	1994	1995
$\{M_{1i}\}$	0.542	0.463	0.505
$\{M_{2i}\}$	1.103	1.314	1.182
$\{M_{7i}\}$	0.400	0.076	0.243
$\{M_{10i}\}$	0.206	−0.029	0.108

workers. But also that the upper wage frontier fell visibly in the second year of the survey, and with it the observed mark-up that is incorporated in the actual wage (as well as the volatility metric M_4). On the other hand, the lower wage frontier remained stable, indicating again the relatively higher stability in the attitudes of the firms. The finding that the bargaining performance of workers impacts the outcome more strongly is in line with the conclusion of Abowd et al. (1999) in a study of the same topic using a different model that "person effects" appear to matter more than "firm effects."

We turn now to the metrics that reflect in economic terms how the two sides of the transaction may have experienced these bargaining outcomes, year after year. In Table 4.6 we show how much of the wage is due to the bargaining performance of the workers (M_1), while M_2 measures how much the bargaining performance of the employers managed to pull the wage down and away from its relative upper frontier (in terms of actual wage). We learn that the bargaining performance of the workers amounts to around half of the wage (M_1) and that the wage would be more than double if it weren't for the bargaining performance of the employers (M_2). For example, the value 1.103 for year 1993 reads "the distance between the upper wage frontier and the actual wage was equal to 1.1 times actual wages." Per our earlier discussion, these could be also viewed as measures of surplus extraction (qualified for relative bargaining power).

Turning to the other two metrics, M_7 measures the net effect of bargaining performance in terms of the *balanced* wage, while M_{10} measures the net effect but in terms of the actual wage. The metric M_7 indicates that the median value of the net bargaining effect on the balanced wage was positive in all 3 years. An *apparently* contradictory result is obtained though for the year 1994, where the median value of the metric M_{10} is estimated as negative. While we expect M_7 and M_{10} to differ due to them having different bases of measurement (and the more so the more Y^B and Y differ), should their *signs* not be the same, given the way that they are defined? In such a case, their medians should have the same sign.

The answer is "not necessarily," and however counterintuitive, it has to do with the stochastic nature of the magnitudes here. Let's recall the definition of the two metrics:

$$M_7 = E\left(\frac{Y - Y^B}{Y^B}\right), \qquad M_{10} = E\left(\frac{Y - Y^B}{Y}\right).$$

If the ratios involved were deterministic numbers, then indeed, M_7 would always have the same sign as M_{10}. But these are random variables and we compute their expected values. Considering a specific observation i and using the computational formulas, we have

$$M_{7i} = E\left(e^{w_i - u_i}|\varepsilon_i\right) - 1, \qquad M_{10i} = 1 - E\left(e^{u_i - w_i}|\varepsilon_i\right).$$

Suppose that for this observation i, we have

$$M_{7i} > 0 \implies E\left(e^{w_i - u_i}|\varepsilon_i\right) > 1 \implies \frac{1}{E\left(e^{w_i - u_i}|\varepsilon_i\right)} < 1,$$

while at the same time we have

$$M_{10i} < 0 \implies E\left(e^{u_i - w_i}|\varepsilon_i\right) > 1 \implies E\left(\frac{1}{e^{w_i - u_i}}|\varepsilon_i\right) > 1.$$

In a stochastic framework, and due to Jensen's inequality and the fact that the reciprocal function $h(z) = 1/z$ is convex, we have

$$\frac{1}{E\left(e^{w_i - u_i}|\varepsilon_i\right)} \leq E\left(\frac{1}{e^{w_i - u_i}}|\varepsilon_i\right).$$

In our specific case, the right-hand side lies on the other side of unity, and we have

$$\frac{1}{E\left(e^{w_i - u_i}|\varepsilon_i\right)} < 1 < E\left(\frac{1}{e^{w_i - u_i}}|\varepsilon_i\right).$$

There is nothing illegal here, but an indication that Jensen's inequality holds with extreme prejudice. So certain observations that have a positive M_{7i} individual metric could at the same time have a negative M_{10i}, and if this happens with many observations, the median of the series $\{M_{10i}\}$ may move to the other side of zero.

Certainly, this is a counterintuitive phenomenon, delicate to communicate if the presentation of the results treats the estimated quantities as deterministic numbers. Moreover, it could happen due to the presence of outliers. And indeed if we truncate the two series above and below (discarding 58 observations out of 1,545), we obtain the medians $\text{med}(M_7) = 0.18$ and $\text{med}(M_{10}) = 0.06$.

4.2.3 *Stratification: Bargaining Matters yet Again*

We look here at the relation between the common-information expected productivity and the actual wage in more detail, by stratifying the sample in two, according to whether the total negotiation effect was positive ($\hat{\varepsilon} > 0$) or negative ($\hat{\varepsilon} < 0$). The results are presented in Table 4.7.

Table 4.7 Winners and losers from wage negotiations

USD PPP (levels)	$\widehat{\varepsilon}_i < 0$	$\widehat{\varepsilon}_i > 0$	Full sample
Median values	$n = 692$	$n = 853$	$n = 1{,}545$
$\widehat{E}(p_i \vert I_{c,i})$	110.08	44.43	66.89
Actual monthly earnings (AME) ω_i	53.94	127.53	82.63
$\%\Delta$	−50.1%	+187.0%	+23.5%

The conclusion from Table 4.7 is that the agreed wage depended strongly on the negotiation process and the private information sets of the parties, relative to the formal qualifications of the worker as well as the main financial and structural indicators for the firm. Actual monthly earnings differ a lot from the common-information expectation for the worker's productivity. This strong effect of atypical forces on wage determination is an indication that a structured payroll and remuneration policy was largely absent in the firms sampled.

4.2.4 Bargaining Power

In relation to the relative bargaining power η, with the BTN specification, we cannot disentangle it from the self-evaluation premium and the prudential discount variables (g, d). In Chap. 10, where we present various ways to model the dependence between the two latent forces, we also include a conditioning technique that allows us to estimate the distribution of η and individual estimates of its value for each observation (see specifically Sect. 10.4).

Here, what we can do is to compute an upper bound for its variability. One can verify that

$$\mathrm{Cov}\,(w, u) = -\mathrm{Var}(\eta)E(g)E(d), \quad E(g) = E(w)/E(\eta), \quad E(d) = E(u)/(1-E(\eta)).$$

Combining and rearranging, we obtain

$$\mathrm{Var}(\eta) = E(\eta)(1 - E(\eta)) \cdot \frac{-\,\mathrm{Cov}\,(w, u)}{E(w)E(u)} = E(\eta)(1 - E(\eta)) \cdot \frac{\left|\mathrm{Cov}\,(w, u)\right|}{E(w)E(u)}$$

The term $|\,\mathrm{Cov}\,(w, u)\,|/E(w)E(u)$ can be computed from the estimates and the related expressions for the covariance and the means of (w, u). Moreover, the term $E(\eta)(1 - E(\eta))$ has maximum value $1/4$ since $E(\eta)$ ranges in $(0, 1)$. It follows that

$$\max \mathrm{Var}(\eta) = \frac{1}{4}\frac{\left|\mathrm{Cov}\,(w, u)\right|}{E(w)E(u)}, \quad \max \mathrm{SD}(\eta) = \frac{1}{2}\left(\frac{\left|\mathrm{Cov}\,(w, u)\right|}{E(w)E(u)}\right)^{1/2} \tag{4.8}$$

Using the estimates, we obtain $\mathrm{SD}(\eta) \leq 0.1941$, for a variable that ranges in $(0, 1)$. A reasonable way to model such a random variable is to assume that it follows the Beta distribution (we exploit this in Sect. 10.4). In the Beta distribution family, the Uniform $(0, 1)$ has standard deviation $1/\sqrt{12} \approx 0.2887$. So in our sample, relative bargaining power has standard deviation no bigger than $2/3$ of the variability it would have if it was distributed as a Uniform. This indicates a degree of concentration and consequently of unequal distribution of relative bargaining power among workers.

We conclude with these remarks the empirical application of the 2TSF Nash bargaining model to wage determination. We believe that we have shown how much our understanding of the phenomenon under study can be enriched by the capabilities that this model offers.

4.3 No-Nash Bargaining

As we have mentioned, the Nash bargaining model assumes that the bargaining parties respect the fundamentals of expected value theory and of Pareto optimality, and they are fully aware and accepting of their interrelation. This has a specific consequence for the statistical model: the creation of intra-error dependence between the latent forces w and u through the relative bargaining power η. In turn, this requires the use of special statistical specifications that incorporate such dependence. This operates as an indirect specification test: if the premises of the Nash bargaining model do not hold, it is likely that the data will not work well with the corresponding specifications. In such a case, we have to accept that the bargaining that took place and is reflected in the data is not of the Nash-bargaining variety and the two parties are not interconnected as the model assumes. The critical component that characterizes the statistical distribution is bargaining power and its modeling. Since the presence of bargaining power in a bargaining model is a *sine qua non*, we need to model it in a way that maintains its relative nature but at the same time does not create statistical dependence between the latent forces (w, u).

Let then b^w denote some absolute measure of the bargaining power of a "seller," and let b^u represent the corresponding measure of bargaining power of the "buyer." These are random variables, and because they reflect *absolute*, as opposed to relative measures, they can be assumed statistically independent.

Then we can define the latent forces, the bargaining performance variables (w, u) as follows, using familiar notation also from the previous part of the chapter:

$$w \equiv \frac{b^w}{E(b^u)} \cdot g, \qquad u \equiv \frac{b^u}{E(b^w)} \cdot d. \tag{4.9}$$

Note what we have done here: we divided the absolute bargaining power of the seller, b^w, with the average bargaining power of *buyers*, $E(b^u)$, and vice versa for the buyer. This is in place of the "relative bargaining power" of the Nash bargaining approach. It still represents a notion of "relative strength" in bargaining power, but a more general one, an index of how powerful is the specific seller compared to the "average" buyer, and how strong is the specific buyer compared to the "average" seller (average in terms of bargaining power). This way of modeling relative bargaining power does not create statistical dependence since we scale absolute bargaining power by an expected value, which is a constant, and so statistically independent from everything.

Moreover, the variables (g, d) have the same interpretation as before and are statistically independent (since they are the output of disjoint information sets). It follows that (w, u) remain independent. In every other respect, the no-Nash bargaining model is the same as its Nash counterpart. But for the no-Nash variant, we should use the benchmark 2TSF models of Chap. 7.

4.4 Application: No-Nash Bargaining on US Military Aid, 1992–2011

We present in this section an application of the 2TSF model to examine the bargaining effects in relation to the provision of military aid from the United States (US) to recipient countries. The original study is Wang (2021), where readers are referred for a fuller discussion of the issue and a review of the related literature. Here, we take the opportunity to expand on certain aspects of the original application.[5]

Wang (2021) points out that the prevailing conceptual model of US military aid in the relevant literature is a two-stage unilateral decision-making process: the US decides which countries will receive military aid, and then it decides how to allocate aid between them. But as he writes (p. 9), "Although such a unilateral decision-making paradigm is quite common and helpful in modeling charitable behavior, it is largely at odds with empirical observations that point to U.S. foreign aid as the outcome of transactions with aid-recipient countries rather than unilateral behaviour of giving." The author sides with another strand of the literature that envisions the situation as an "aid-for-policy" deal, where the donor provides aid (military in this instance) to obtain favorable policy concessions from the recipient.

If this is a bargaining situation, what could these "favorable policy concessions" be? Note that since the United States is paying, it is the "buyer," and the recipient country is the "seller." What does it sell? Its geopolitical alliance to the United States, its "friendliness" to or its importance for US economic interests, its criticality in the geopolitical security, and balance-of-power nexus... in such a situation, all structural conditions hold for the 2TSF Nash bargaining model to be applicable: there is a set of common information, and most certainly both United States and each

[5] We would like to thank the author, professor Yu Wang, for finding the time to discuss with us his paper, clarifying a few points and contributing some ideas for the empirical expansion.

recipient country hold closely private ("classified") information, which will affect their back-and-forth communications on deciding the level of military aid. There are certainly targets pursued by both sides, as regards this level. There is uncertainty as to the actual "worth" of what the United States is buying by providing military aid. And we expect that the recipient country tends to mark up the worth of what the donor is buying (honestly or for purely bargaining reasons), while the donor tends to downplay that value, for analogous reasons. Finally, bargaining power is expected to play a role.[6]

So everything is in place for the 2TSF model to be applied, with one exception: the buyer, the United States, is always the same. In such a situation, if we only had a cross section, we could not include in the regressor matrix covariates characterizing the buyer, because they would be invariant across the sample. But the available sample ranges across countries *and* years, and we will exploit this.

4.4.1 Determinants of US Military Aid

Wang (2021) examines US military aid in the post Cold War era, and his yearly data sample covers the period 1992–2011 (end points included), so we have a span of 20 years. He initially constructs a data set that includes 185 possible recipients of US military aid, so in all 3,700 potential data points. Of these, only 2,482 country-year points received military aid. Further, in order to include covariates as explanatory variables, the effective sample is restricted to 1,789 observations, which is 72% of the full sample. In this reduced sample, the minimum data points per year are 68, while the maximum is 109. So there is enough variation throughout. Moreover, the total value of military aid captured in the effective sample is 70% of all military aid in these 20 years, so again, it is adequate to obtain reliable inference.

The author includes only characteristics of the recipient countries, based on his review of which determinants of US military aid have been proposed and examined in the literature, dividing them into socioeconomic and political-strategic determinants. The socioeconomic regressors included are the level of *economic* aid to the recipient (there are studies that indicate a strong connection between economic and military aid); gross domestic product (GDP) as a representative of the economic size of the recipient; population, for which there exists an argument in the literature that it may *negatively* affect the size of military aid (since populous countries may rely on larger local armies); and US exports to the recipient which

[6] Here we consider just the financial level of US military aid, although we acknowledge that the functional composition of this aid (cutting-edge technology or not, what areas of military equipment covers, whether it includes training or not, etc.) is also an important issue, which nevertheless is indirectly captured up to a point by its monetary value.

indicates how large are the economic stakes for the US economy related to the recipient country.[7] Monetary values are expressed in constant 2005 USD.

The political-strategic group of explanatory variables includes voting consistency with the United States in the United Nations General Assembly (political alignment, if not alliance), US military presence in the recipient country measured in military headcount, an index related to "respect of human rights," another index for the "level of democracy," and geographical location grouped in five major geopolitical areas (Africa, America, Asia, Europe, and the Pacific). There is also a binary variable that takes the value "1" from the year 2002 on, to capture the possible effect on military aid due to the "9/11" terrorist attacks on US soil in September 2001 and the declared global "war on terrorism" by the United States that followed.

The time dimension allows us to expand the data set of the author and include political and economic variables that characterize the donor, in order to fully match the conceptual framework of the 2TSF bargaining model. The political variables are three binary ones: over the period examined, there were two political parties that almost exhausted the political system in the United States: the Democratic and Republican parties. We define three binary variables each indicating who controlled the presidency, the senate, and the house of representatives, respectively. We assigned the value "0" to Democrats and the value "1" to Republicans, so the estimated slope coefficient in a regression setting will indicate whether Republican control contributed positively or negatively to military aid, compared to Democratic control. The US economic variables we use are three: the yearly unemployment and inflation rates, as indicators of possibly politically sensitive hardship in the US economy that could pressure politicians into holding back "spending abroad," and the yearly growth rate of the US economy, as an indicator of a climate of economic health that could, when positive, make decisions on spending abroad easier. These three variables were included as percentiles so, for example, an inflation rate of 0.021 appears in the data set as 2.1.

4.4.2 *Data, Specification, and Estimation*

There is a clear positive trend in the number of recipient countries as time passes and of the total value of military aid. Also, in the full sample, one can see a positive time trend, from 2002 onward, in the monetary value of military aid. But in the effective sample of $n = 1{,}789$ the trend does not appear. So we can pool the data without a time trend. We note that there is a spike in military aid in 2003, due to a large jump in its value toward European and Asian countries (for Europe, this is mainly related to Poland; for Asia it is mainly related to Iraq).

---

[7] The original study also included GDP per capita, but since these variables enter in logarithms, it created a severe collinearity issue with the GDP and the population covariates, so we excluded it. Moreover we included a constant term while the original study did not.

We attempted first to estimate the Nash bargaining model. We tried three different specifications for the composed error term, with variants (for all of them see Chap. 10): the Truncated Bivariate Normal, the η-conditioned Normal-Exponential model, and a copula model to account for the dependence between w and u using four different copula densities (Gaussian, Frank, Clayton, Farlie-Gumbel-Morgenstern). In all cases, estimation failed and for different starting values of the parameters. Therefore, we accepted that the Nash bargaining model was not suitable for the specific sample, and we reverted to the non-Nash bargaining variant, which allows us to implement the benchmark Normal-Exponential specification, which was the one also used in the original study. The results are presented in Table 4.8.

We do not show them here, but we report that the estimates for the slope coefficients of the regressors common with the original study (that did not include US-related covariates) are generally very close. Of them, we single out the negative effect that a larger population of the recipient country has on military aid, as has been argued in the literature, and the strong positive effect of aligned UN-voting. The US political variables provide interesting insights: in general they confirm the

Table 4.8 US military aid 1992–2011. Normal-Exponential 2TSF model

Variable	Units	Estimate	Std error	Elasticity	% change
const		0.1014	0.7019		
Economic aid (ln)	ln 2005 USD	0.5121	0.0200	0.51	
Population (ln)	ln 2005 USD	−0.8204	0.0540	−0.82	
GDP (ln)	ln 2005 USD	0.5486	0.0631	0.55	
US Exports (ln)	ln 2005 USD	0.1973	0.0439	0.20	
UN voting	[0,1]	2.1719	0.4058		21.7
Human Rights	5−point scale	0.0906	0.0416		9.1
Democracy	21−point scale	−0.0045	0.0079		−0.4
US milit manpower (ln)	ln headcount	0.2375	0.0247	0.24	
War on Terror (2002 on)	Binary	0.4103	0.1228		50.7
Rg2: America	Binary	−0.2689	0.1342		−23.6
Rg3: Asia	Binary	0.977	0.1022		165.6
Rg4: Europe	Binary	0.6506	0.1609		91.7
Rg5: Pacific	Binary	0.3995	0.3747		49.1
US President (Dem/Rep)	Binary	0.4629	0.1139		58.9
US Senate (Dem/Rep)	Binary	−0.6044	0.1293		−45.4
US House (Dem/Rep)	Binary	0.8112	0.1079		125.1
US unemployment	percentiles	−0.0446	0.0378		−4.5
US growth	percentiles	0.1389	0.0307		13.9
US inflation	percentiles	−0.2231	0.0459		−22.3
σ_v		0.3613	0.1039		
σ_w		1.2335	0.0615		
σ_u		1.2665	0.0478		

Dependent Variable: US military aid per country/year, in 2005 USD (ln)
Standard errors based on outer products

casual impression that Republican politicians are "hawkish," since military aid tends to increase when they have political control, even though it appears that there is a "balancing act" from a Republican-controlled US senate. The counterfactual where we move from a Democrat trifecta to a Republican one has military aid almost doubling.

Economic variables also appear to matter, although less: the domestic economic situation tends to affect military aid in the same direction, but the indicator for internal hardship is inflation and not unemployment (which does not appear to affect military aid), something that supports the notion that politicians in USA are sensitive to citizens/voters viewed as consumers, not as workers. The last two columns provide directly interpretable measures. For variables in logs, the estimate is a direct elasticity, so for example, a country with GDP higher by 10% than another country receives 5.5% more military aid on average. For the U.N. voting index, which reflects the degree of alignment of the recipient country with how the US votes, and ranges in the interval [0, 1], we give the percentage increase in military aid for a 0.1 increase in the index. So going from, say an alignment index of 0.5 to an alignment index of 0.6 will increase military aid by 21.7%. For the binary variables we computed percentage changes as we move from zero to one.[8] We note that the baseline military aid is for African countries so the percentage increase in military aid per region is in relation to the African-aid level. For the economic US variables, the indicated percentage change in military aid is for a one-percentile change in a variable.

The sigmas are the scale parameters of the zero-mean Normal noise σ_v and of the two Exponentials that represent the bargaining performance of each side. The value for σ_w relates to the recipients, and the value for σ_u relates to the United States. We see that they are essentially identical. These are average values over the whole sample, so we will stratify the sample, specifically per geopolitical region and for the periods before and after "9/11," to detect variations along these dimensions. Moreover, what must be pointed out is that both σ_w and σ_u are higher than unity. This, as we will discuss in Chap. 7, implies that the exponentiated measures over the sample (which we want, since the dependent variable is in logarithms) are not defined if they contain $E(e^w)$ or $E(e^u)$. Moreover, as regards the individual metrics, the expected value of the conditional expectations series does not exist. But, because $\sigma_w^{-1} + \sigma_u^{-1} > 1$, the conditional expectations *themselves* do exist and can be computed, and we can use their *median* value as a representative metric. This is what we do in Table 4.9. We do not show metrics for the Pacific region because the observations are too few.

We observe a remarkable stability across time periods and regions as regards the volatility of military aid (M_4) with one exception: the European region after the end of the Cold War. This accords with a highly uncertain period for Europe when

[8] We computed these by using the exact formula $\exp\{\beta\} - 1$ and not the usual approximation which is just $\beta \approx \exp\{\beta\} - 1$, because this approximation is good only for much smaller values of β than the estimates obtained.

Table 4.9 Bargaining metrics of US military aid before and after "9/11." Median values of individual measures

Metric	Period	Geopolitical region			
		Africa	America	Asia	Europe
M_1-contribution of recipient bargaining	1992–2001	0.44	0.42	0.41	0.53
Performance in military aid	2002–2011	0.44	0.45	0.47	0.42
M_4-volatility	1992–2001	2.63	2.61	2.61	3.19
	2002–2011	2.62	2.64	2.67	2.66
M_7-net effect	1992–2001	0.03	−0.09	−0.17	0.39
	2002–2011	0.00	0.06	0.13	−0.11
M_{10}-net gain	1992–2001	−0.07	−0.22	−0.34	0.20
	2002–2011	−0.10	−0.05	0.01	−0.25
$M_{10}^{(+)}$ Recipient gain	1992–2001	0.61	0.71	0.45	0.64
	2002–2011	0.55	0.49	0.65	0.42
$M_{10}^{(-)}$ US gain	1992–2001	0.87	1.92	3.19	2.87
	2002–2011	1.28	0.94	1.04	0.94

the countries of Eastern Europe were in turmoil and political and socioeconomic transition, and with the Western Europe allies of US in serious angst.

From the point of view of recipient countries, the relative contribution of their bargaining performance to the received military aid (M_1) is also generally stable (40–45% of military aid received can be attributed to how they bargained over it), with the same temporal/geopolitical exception: Europe immediately after the Cold War, where this contribution was 53%. In all, this is structural indication that bargaining does play a role in determining US military aid, irrespective of whether it ends up being balanced or not.

The net effect of the bargaining process (M_7, percentiles above or below the balanced outcome) reveals how the focus of the US moved from Europe-in-transition before "9/11" to Asia after "9/11." Asian countries enjoyed increased political leverage in this second period, while European countries lost much of what they had before.

Finally, we turn to the M_{10} metric, the net gain, for which we also present its decomposition in recipient gains and US gains. A first conclusion is that the full M_{10} series hides large variations, comparing the two rows that relate to it, with the last four that show measures for the "positive only" and "negative only" M_{10} metrics. The median value of the full series was almost everywhere negative (indicating a net gain for the buyer/US), with the exception of Europe before "9/11" and Asia after, where the essentially zero value must not draw attention away from the fact that it came from a −0.34 value in the previous era. At the same time Europe's move from 0.20 to −0.25 is impressive.

The recipient gains metric reflects another effect of shifting US geopolitical concerns toward Asia after "9/11," since we see that Asia is the only region where recipient gains, when materialized, increased after "9/11" compared to before. Here too we observe how Asia and Europe switched places.

As regards US gains, when they materialize, they tend to be much higher than recipient gains. Also, and with the exception of Africa, in the other three regions the US gain fell substantially in the period after "9/11." This is an indication that the US drove a softer bargain in this period, which also aligns with expectations.

We conclude here this second empirical presentation of the 2TSF bargaining model. Either in its Nash or no-Nash variety, the model provides additional insights on how bilateral transactions are determined, quantifying the unobservable bargaining performance of the parties involved.

Chapter 5
A Jamboree of Latent Forces

In this chapter, we present four different situations (and their associated empirical applications) where two-tier stochastic frontier analysis can be applied usefully: the case of the "omitted variable/latent force" management in production which faces off inefficiency, idiosyncratic forces in secondary education, social pressure on businesses to become "greener" and operate broadly in a sustainable way, and finally, national psychological profiles as regards their self-perceived level of happiness. Readers can march through the chapter in one take, or make stops along the way, or choose the reading order based on their own interests.

5.1 The 2TSF Management Model of Production

As mentioned in Chap. 1, "organization" was one of the four "agents of production" in Alfred Marshall's *Principles of Economics*, alongside land, labor, and capital. But once researchers started in earnest to model and estimate production, it disappeared from the list of production inputs, for a very mundane reason: there were no quantitative data on organization/management.

There have been various attempts over the years to handle this lack-of-data problem for the management factor. In various instances, researchers attempted to construct a "measure of management," collecting data on managerial aspects of businesses through surveys. This culminated in the World Management Survey research program that collected systematic data on management practices from many countries all over the world and constructed a "management index." One can visit https://worldmanagementsurvey.org/ to learn more about this still active theoretical and applied program.

A. Papadopoulos, C. F. Parmeter, *Two-Tier Stochastic Frontier Analysis for the Social Sciences*, https://doi.org/10.1007/978-3-031-81513-3_5

The 2TSF approach treats unobservables as unobservables, and we will present next a 2TSF production model to measure the effects of management. The presentation is based on Papadopoulos (2021b).[1]

5.1.1 A Production Function with Management

We consider the following maximum production function, which is an abstract mathematical expression that links productive resources that are used to produce some output:

$$Q_{max} = A \cdot G(\boldsymbol{x}) \cdot \exp\{h(m)\}, \quad m \geq 0,\ h(0) = 0,\ h'(m) > 0,\ h''(m) < 0. \tag{5.1}$$

Here, A is a technology index, $G(\boldsymbol{x})$ is the function representing the *maximum* output that can be produced given the inputs $\boldsymbol{x}$, and $\exp\{h(m)\}$ is the term representing the effects of management on the output, with m representing the (unknown) "quantity/intensity" of management and $h(\cdot)$ being a transformation function that translates the effects of m in output terms.

Note that A is treated as a constant, and so it represents some "average" technology index over the whole sample. If we were to omit management from the specification, then some measure of "average management effect" would be incorporated and estimated in the technology index A. But is that enough?

We argue that it is not. While "hard" technology exhibits more homogeneity across firms of a sector in a cross-sectional setting (because it is partly based on natural laws and is embodied in machines and equipment), management practices, which we can view as a "soft" technology (the "way of doing things"), are consistently observed to vary wildly across firms even in the *same* sector, perhaps because management is where non-artistic people get to express their creativity (hence also the view of management as a "recipe"). So modeling management as a *variable*, and not being content with capturing some average level of it together with other unmodeled factors, appears to be a desirable enhancement of the standard production model.

We also note that the management factor enters the production function in such a way that for $m = 0$ management becomes "neutral" ($\exp\{h(0)\} = e^0 = 1$). Management can only increase the output that production inputs would have produced with "zero management" (i.e., with only a minimum level of coordination but not real monitoring and purposeful steering). This implies a distinction between *leadership* and *management*: While leadership can certainly lead a business to fail and stop operations, management is the net that keeps everything together, given the leadership's goals and the constraints from the socioeconomic environment.

---

[1] A comprehensive review of the theoretical methods and the empirical literature that has tried to estimate the effects of management on production can be found in Papadopoulos (2021a).

An empirically testable theoretical consequence of the above formulation is that management is a *strategic complement* to the conventional inputs: as the scale of operations expands (i.e., as the size of the firm increases), and naturally we expect production inputs like labor and capital to be employed in higher quantities, the 2TSF model presumes that management will also expand, in terms of resources dedicated to management activities. This is in line with real-world experience, and we will test it in the empirical application of the model.

5.1.2 The 2TSF Stochastic Specification

To turn equation (5.1) into a stochastic frontier model and obtain an expression for actual output, we introduce a random disturbance v and an "inefficiency" component u

$$Q = A \cdot G(\mathbf{x}) \cdot \exp\{h(m)\} \exp\{v - u\}, \quad E(v) = 0, u \geq 0. \tag{5.2}$$

Real-world experience has shown that businesses do not succeed in being fully efficient and producing the maximum possible level of output given the inputs they employ. Learning curves for new technologies, products and procedures, incompleteness of managerial control (especially on labor/employees), and external negative factors like regulations or destructive competition, all lead to a loss of efficiency, and the term u captures this negative effect on the output.

Setting for compactness $w \equiv h(m)$ and taking logarithms, we have

$$\ln Q = \ln A + \ln G(x) + \varepsilon, \quad \varepsilon = v + w - u. \tag{5.3}$$

This has the 2TSF structure with a transformed dependent variable. Note that the two opposing forces on output, management and inefficiency, are very different in nature, but what we will estimate is not their "magnitude," but their effects on the output, measured in output units or in relative terms.

In the empirical application, we will estimate this model with the dependent variable in logarithms and use maximum likelihood. We will adopt the Normal-Exponential distributional specification for the composite error term (see Chap. 7), with one additional twist: a copula to deal with endogeneity of regressors (Sect. 14.1.4).

The consensus in the literature is that in production models, the regressors (production inputs) are "endogenous," namely statistically correlated with the unobserved error term, which leads to estimates that are "inconsistent," i.e., permanently biased irrespective of the statistical power gained by increasing the sample size. By including management as an explicit unobservable in the composite error term of the model, we make the existence of endogeneity even more probable, since we cannot argue that management is not operationally related to the production inputs (and operational links create statistical dependence). To handle this technical problem, we will use the Gaussian Copula method that can account for endogeneity without the need to use additional variables as instruments.

5.1.3 Metrics of Interest

Since the model will be estimated with the dependent variable transformed, we need to use the metrics in exponentiated form (see Chap. 2). If we view management as another production factor, we can measure how much of output can be attributed to management, in percentage terms. This approach and measurement can be compared and analyzed together with the costs of the management system of a company, in a cost-benefit analysis. The contribution of management to production as an input corresponds to the metric M_1:

$$M_1 = 1 - E(e^{-w}).$$

But perhaps it is more useful to view management as a structured attempt to increase the overall efficiency of the firm and so as a shift-factor of the output relation. This approach corresponds to the metric M_5

$$M_5 = E(e^{w}) - 1,$$

or its "gross" variant which is simply $M_5 + 1 = E(e^{w})$. This measures how much management tends to increase output, *in the absence of inefficiency*. But inefficiency exists, and irrespective of whether it can be attributed causally to management or external factors, we hold management *accountable* for inefficiency. So it is natural to pit the positive shift-effect of management against the negative drag-down effect of inefficiency. This corresponds to the M_7 metric

$$M_7 = E(e^{w-u}) - 1,$$

or its gross variant $M_7 + 1 = E(e^{w-u})$. If the latter exceeds unity, it means that management succeeds in increasing output above its maximum unmanaged level. If it is lower than unity, management offsets inefficiency only partly, and output is below its maximum unmanaged level.

These are the most pertinent metrics of Chap. 2 for the specific topic, but of course users can explore the rest.

5.2 Application: Measuring Management Across the Globe

For our empirical application, we use a subset of the data used in Bloom et al. (2012), specifically the 1,888 observations from the year 2006.[2] The data are from various countries and industries. The regression specification is

[2] The full data set is "Manufacturing: 2004–2010 combined survey data (AMP)," freely available at http://worldmanagementsurvey.org/survey-data/download-data/download-survey-data/.

Table 5.1 Sales data for 2006 from various countries, $n = 1{,}888$

	OLS	2TSF	2TSF-Cop
Constant	4.087	4.305	5.593
	[0.140]	[0.098]	[0.219]
lppent	0.374	0.325	0.103
	[0.022]	[0.012]	[0.028]
lemp	0.617	0.678	0.824
	[0.031]	[0.019]	[0.038]
σ_ε	0.765	0.758	0.815
σ_v	–	0.230	0.242
		[0.037]	[0.040]
σ_w	–	0.461	0.474
		[0.021]	[0.028]
σ_u	–	0.556	0.617
		[0.023]	[0.025]
$\max\lvert\widehat{\rho}(K,\varepsilon)\rvert$	–	–	0.351
			[0.045]
$\max\lvert\widehat{\rho}(L,\varepsilon)\rvert$	–	–	0.149
			[0.044]
$\widehat{\gamma_1}(\varepsilon)$	– 0.320	– 0.380	– 0.591
$\widehat{\gamma_2}(\varepsilon)$	5.076	5.347	5.033

Standard errors in brackets (robust-HC2 for OLS). Description of variables: *lnS* logarithm of sales, *lppent* logarithm of capital, *lemp* logarithm of labor (number of workers)

$$\ln S = \beta_0 + \beta_1 \ln K + \beta_2 \ln L + \varepsilon, \qquad \varepsilon = v + w - u\,, \tag{5.4}$$

where the dependent variable is sales, in logarithms. Table 5.1 contains the estimation results using OLS, the 2TSF Normal-Exponential likelihood without a copula, and the same with the Gaussian copula attached.

While the results from OLS and the basic 2TSF model are broadly similar, the estimates from the Copula 2TSF model differ: first note that the correlation coefficients, the two ρ parameters, although not very strong, are estimated with high statistical confidence. These are the parameters that characterize the copula density that we have added to the log-likelihood. This result lends support to the conjecture that there exists endogeneity, questioning the validity of the other two models. Adopting this model, we observe that the slope coefficient for capital is reduced while for labor it has increased (last column compared to the other two). We remind the reader that the slope coefficients represent here the elasticity of sales with respect to each regressor, all else being constant, *including the error term.* So the use of the copula model, by disentangling this *ceteris paribus* effect from other avenues through which the inputs affect sales, allows us to see how much stronger is the effect of the labor input, *given* the management factor (which lives in the error term and is considered fixed under the ceteris paribus condition). At the same time, by detecting a degree of dependence between the inputs and the composite error

Table 5.2 Management metrics

	Sample level	Individual series	
Variable	Mean	Mean	Median
M_1	0.322	0.321	0.245
M_5+1	1.901	1.987	1.440
M_7+1	1.176	1.239	0.948

term, the model reveals the anticipated connection between the management factor and the inputs.

Turning to the scale parameters that relate to noise (σ_v), management (σ_w), and external inefficiency (σ_u), we observe that the inefficiency scale parameter σ_u is estimated to be higher than the management scale parameter σ_w and that the residuals in all three models exhibit negative skewness ($\widehat{\gamma}_1$) and positive excess kurtosis ($\widehat{\gamma}_2$). But in order to assess their effects on output, we must go back to the original units of measurement.

Table 5.2 contains the values of the three metrics mentioned earlier, both at the sample level, where the mean value is computed based on the estimated parameters, and also the sample means of the corresponding individual metrics that form a series. For example, if we look at M_7, then we report

$$E(M_{7i}+1) = \frac{1}{n}\sum_{i=1}^{n} E(\exp\{w_i - u_i\}\,\varepsilon_i).$$

But we also include the median values of the individual series, to assess the reliability of the empirical means of these series.

Note that the empirical mean of the individual metrics is very close to the sample-level mean (first two columns). In theory, they should be identical, since the expected value of the conditional expectation (second column) is the unconditional expected value (first column); the fact that they are actually almost equal is a (positive) diagnostic for the fit of the model with the data.

The management contribution metric M_1 is on average 32% of output, which appears a reasonable value. Put another way, output would be 32% lower if management was not present. The mean value of the management shifter M_5+1 is close to 2 (meaning that management, absent external inefficiency, tends to double unmanaged output). We observe though that its median value is visibly smaller, indicating a long right tail in the empirical distribution: there are firms with exceptionally productive management.

For the metric $M_7+1 = E(\exp\{w-u\})$, which represents the net multiplier effect of management (and also the relative strength of w over u in original measurement units), we see that, even though at the logarithmic level $E(w-u) < 0$, M_7+1 is estimated as being higher than unity, implying that on average management succeeds in offsetting externally imposed inefficiency, lifting output above the

Table 5.3 Deciles of the empirical distributions of the individual metrics

Theor. min	*0.00*	*1.00*	*0.00*
Cum. Prob.	M_1	$M_5 + 1$	$M_7 + 1$
0.1	0.21	1.37	0.37
0.2	0.21	1.37	0.57
0.3	0.22	1.37	0.73
0.4	0.23	1.40	0.84
0.5	0.25	1.44	0.95
0.6	0.28	1.52	1.06
0.7	0.33	1.67	1.20
0.8	0.42	1.94	1.42
0.9	0.57	2.63	1.93
Theor. max	*1.00*	∞	∞

maximum unmanaged level. But the median of the series of the individual metric is below unity.

This is a good example to illustrate two issues: the first one is to alert us to the fact that, when using expected values and not deterministic numbers, the qualitative relation at the estimated level ($E(w - u) < 0$) does not necessarily carry over to the exponentiated level (since here $E(e^{w-u}) > 1$).[3] So, when we estimate with the dependent variable in logarithms, we should always go back and compute our metrics exponentiated and not rely on the relation between the direct estimates obtained.

At the same time, this example also alerts us to the danger that, when using expected values of $E(e^w)$, we may get high values due to outliers in the data which do not necessarily represent the real-world situation. This is why it is advisable, when assessing the individual metrics, to look also at quantile measures like the median, which here, indeed, is below unity and much lower than the expected value, restoring the qualitative result at the estimation level (that external inefficiency is stronger than the management effect).

More generally, looking in greater detail at the quantiles of the empirical distributions provides additional insights. We present in Table 5.3 the empirical deciles of the series of individual metrics.

What can we learn from the empirical deciles? We observe that the individual metrics do not get close to the theoretical minimum quantile value; there are no firms that have close-to-minimum values as regards management. We also learn that for 70% of firms, the individual metric for the management contribution M_1 ranges in the narrow interval $[0.21, 0.33]$. We verify that the management shifter $M_5 + 1$ does have a long right tail, while below the median, firms are squeezed in the *very* narrow $[1.37, 1.44]$ interval. Finally, as regards the management net multiplier effect $M_7 + 1$, for roughly half of the firms, it is below unity and then for the next 40% of

[3] With deterministic numbers, we always have that $w - u < 0 \implies \exp\{w - u\} < 1$.

Table 5.4 Association between management and conventional inputs

Variables	Kendall's tau	Spearman's rho
$E(w\|\varepsilon)$, Capital	0.253	0.369
$E(w\|\varepsilon)$, Labor	0.130	0.194

firms rises gradually up to almost doubling the maximum unmanaged output. The distribution of this metric is much less concentrated than the other two.

We close this empirical application by examining the theoretical implication that between management and the conventional inputs, there is strategic complementarity, as we discussed earlier. The closest we get to the management variable itself is $E(w|\varepsilon)$ where $w = h(m)$ is assumed to be a nonlinear function of management. For this reason, we present in Table 5.4 two well-known nonparametric measures of association, Kendall's tau and Spearman's rho, which are invariant to nonlinear transformations, while the, perhaps more familiar, Pearson's correlation coefficient is not.

In all cases these association measures are positive (and statistically significant), supporting the assumption that management is a strategic complement to the input factors along the expansion path.

With this we conclude this empirical illustration. Papadopoulos (2021b) contains more theoretical elaborations on the matter and the full empirical application, but we believe we have presented enough results to show how much the 2TSF model can further our knowledge for an important unobservable like management.

5.3 Idiosyncratic Forces in Secondary Education

Public school systems shoulder one of the most critical tasks in securing the survival of a society: providing knowledge to the new members of the human race and also forcing them to learn to coexist with complete strangers, outside the narrow confines of their family. Being public, they are in principle accountable to any citizen, and they feel the pressure. No wonder they are obsessed with data collection and statistics to monitor performance.

Still, there are idiosyncratic factors possibly at play in relation to performance in school: family situation, neighborhood situation, status among peers in school, and even the psychological make-up of each pupil individually. In many instances, some of these factors may be available as data; in others they are not. We will examine one such data set related to pupils, where we do not know any of these attributes, and so a question arises: does the individuality of pupils, this unobserved heterogeneity, pierce through the educational system? It is important to stress that we are talking about persons in their teenage years whose individual identity is still very much a work in progress.

5.4 Application: Race and Income in Boston's Public Schools

In this application, we look at data from public schools in Boston, Massachusetts (United States), and we focus on three things: on the existence (or not) of "peer effects" related to a pupil's race and also to their income level and on the estimation of unobservable positive and negative forces on test scores.[4]

The full data set has 16,587 observations with data on pupils from 32 public schools in the Boston area, grades 6–11 (secondary education), at the beginning of the millennium. Among the personal attributes of the pupils is a three-level family income band, based on the "lunch status" of each student (what kind of lunch each was entitled to at school, which is based on the family's income level), as well as their race (Asian, Black, Hispanic, White).[5]

To measure peer effects in relation to race and income we proceeded as follows, based on the full sample: for race, we determined the prevailing/majority race in each school, and we created a "matching dummy" that takes the value "1" for each student that belongs to the majority race of its school and zero otherwise.

For income, we found the prevailing income band in each school and then created a set of three binary dummies that together indicated whether each pupil belonged to the prevailing income band, or to an income band higher than the prevailing, or to one that is lower.

These variables have the purpose of examining whether a racial and/or an income socio-psychological "peer effect" exists on test scores: if the majority of the students in your school belong (or not) to the same racial category as you, does this have an effect on your test scores? Also, does it matter whether pupils see around them in school pupils richer or poorer than them? We note that since the three income levels here were determined based on lunch status, the distinction would be pretty obvious to the pupils, not to mention that teenagers are much more forthcoming and vocal about their family income than adults usually are.

The actual data set used had only 3,194 observations, due to lack/nonexistence of data on test scores. In this smaller sample, the data came mainly from 14 schools, with very few observations (just 75) coming from another 14. Also, the majority of pupils were 11th graders (16–17 years old). The allocation is shown in Table 5.5.

The contingency Table 5.6 shows what was the prevailing race/income on these 28 schools (as said before, these characterizations of each school were based on the full data set).

From Table 5.6 we see that the sample consists mostly of schools where the majority race was "Black" and the prevailing income band was "Low."

---

[4] We would like to thank Roland B. Stark, www.integrativestatistics.com, for providing this data set, completely anonymized, and with observations from schools with fewer than 200 students removed.

[5] We note that "race" is determined based on whatever statistical/institutional/self-declaration criteria were used by the entities that recorded the data.

Table 5.5 Distribution of pupils in schools and grades

School group	Grade-11	Other grades	Total
Main 14 schools	2,330	789	3,119
Other 14 schools	0	75	75
Total for 28 schools	2,330	864	3,194
Main 14 schools	72.9%	24.7%	97.7%
Other 14 schools	0.0%	2.3%	2.3%
Total for 28 schools	72.9%	27.1%	100.0%

Table 5.6 Distribution of schools per majority race and income band

	Maj. Income band			
Majority race	Low	Middle	High	Total
Asian	0	0	0	0
Black	20	0	1	21
Hispanic	5	0	0	5
White	0	0	2	2
Total	25	0	3	28

Table 5.7 Distribution of pupils per race and sex

	Counts		
Race	Female	Male	Total
Asian	195	147	342
Black	838	809	1,647
Hispanic	355	313	668
White	272	265	537
Total	1,660	1,534	3,194

	Joint rel. frequency			Marginal	
Race	Female	Male	Total	Female	Male
Asian	6.1%	4.6%	10.7%	57.0%	43.0%
Black	26.2%	25.3%	51.6%	50.9%	49.1%
Hispanic	11.1%	9.8%	20.9%	53.1%	46.9%
White	8.5%	8.3%	16.8%	50.7%	49.3%
Total	52.0%	48.0%	100.0%		

Turning to the observations, we first present in Table 5.7 the allocation of pupils per race and sex. We observe that the sex ratio is reversed compared to the usual one, and we have more females than males.[6]

Next, we present in Table 5.8 the distribution of the 3,194 pupils in relation to their participation in majority groups in their school. We see that Asian pupils never

[6] In 2001, the sex ratio in the United States at ages 15–20 was approximately 105 males for 100 females, see e.g. https://ourworldindata.org/gender-ratio.

Table 5.8 Distribution of pupils per majority race and income band

	Belongs in the majority				
Actual race	Race only	Income only	Race & Income	None	Total
Asian	0	174	0	168	342
Black	743	63	769	72	1,647
Hispanic	12	352	38	266	668
White	49	83	250	155	537
Total	804	672	1,057	661	3,194
	Row distributions				
Asian	0.0%	50.9%	0.0%	49.1%	100.0%
Black	45.1%	3.8%	46.7%	4.4%	100.0%
Hispanic	1.8%	52.7%	5.7%	39.8%	100.0%
White	9.1%	15.5%	46.6%	28.9%	100.0%
Marginal	25.2%	21.0%	33.1%	20.7%	100.0%

Table 5.9 Distribution of pupils above or below the majority income band

	Relation to majority income band			
Actual race	Below	On	Above	Total
Asian	108	174	60	342
Black	144	832	671	1,647
Hispanic	40	390	238	668
White	53	333	151	537
Total	345	1,729	1,120	3,194
	Row distributions			
Asian	31.6%	50.9%	17.5%	100.0%
Black	8.7%	50.5%	40.7%	100.0%
Hispanic	6.0%	58.4%	35.6%	100.0%
White	9.9%	62.0%	28.1%	100.0%
Marginal	10.8%	54.1%	35.1%	100.0%

belong to the racial majority of a school (and Hispanic pupils almost never). On the other hand almost all Black students belong either to the race majority only or to both the race and income majorities. Asian and Hispanic pupils are much more likely to belong to the income majority only than either Black or White students.

A final descriptive statistic of interest, before proceeding with the estimation results, is the distribution of pupils above or below the prevailing income band in their school. This is shown in Table 5.9.

An interesting observation from Table 5.9 is that the vast majority of Black, Hispanic, and White pupils that are *not* aligned with the majority income band in their school are above it, while for Asian pupils the opposite holds.

We turn now to the estimation of a Normal-Exponential 2TSF model, using maximum likelihood. Related to the composite error term, we will estimate the

Table 5.10 List of regressors

Name	Kind	Comments
Constant term	Baseline	Black belonging to school's income majority
D_{Sex}	binary	1 = male
$D_{Hispanic}$	binary	1 = Hispanic
D_{White}	binary	1 = White
D_{Asian}	binary	1 = Asian
$D_{Billingual}$	binary	1 = bilingual
Income band	categorical	{1,2,3}
Special needs	categorical	{1,2,3,4}
$D_{MajRace}$	binary	1 = belongs in the school's majority race
$D_{Rel.Income-0}$	binary	1 = Income below the school majority
$D_{Rel.Income-2}$	binary	1 = Income above the school majority

three unknown scale parameters, σ_v, σ_w and σ_u, that here coincide with the standard deviation of the three components of the 2TSF error term (at the logarithmic level).

As dependent variable we used test scores from the Massachusetts Comprehensive Assessment System (MCAS) in Mathematics and in English Language Arts (ELA). Specifically, we took the natural logarithm of their average from Spring Semester 2001. In this "semi-log" specification, the estimated coefficients of binary and discrete/categorical random variables are directly interpretable as the proportional change on the dependent variable in its original measurement units, for a one unit/level change in the regressor (so an estimate, say, of 0.05, is interpreted as a tendency for the actual test scores to increase by 5% on average). The regressors are listed in Table 5.10.

Table 5.11 presents the estimation results, and there are many interesting conclusions: first, note that the unknown coefficients are estimated with very high precision. Second, as a general observation, the effects of these regressors on test scores are small, the highest being a tendency to increase test scores by 7.8% if the pupil is Asian. So these effects exist but are small. This speaks to a situation where the educational system manages fairly homogeneously the pupils irrespective of personal characteristics represented by our regressors. Being male, bilingual or with special needs, tends to reduce test scores (although again, not by much). As regards race, all races tend to perform better than Black pupils.

Turning to peer effects, the one statistically and essentially immaterial variable here is the participation in the race majority in one's school, which was one of our exploration targets: it does not appear to affect test scores. On the contrary, the results related to the income level are more nuanced: the income level itself tends to increase somewhat test scores (by 2.9%), but being *below* the majority income level in one's school tends *also* to increase test scores (by 4.6%), while being *above* the prevailing income level tends to decrease them: it would appear that educational attainment is indeed a way to counterbalance a disadvantaged family income status, even if only as a psychological need (or in order to have something to shout back

Table 5.11 2TSF Normal-Exponential model for educational outcomes in Boston secondary schools

Regressor	Estimate	Standard error	p-value
Constant	5.373	0.006	0.000
D_{Sex}	−0.009	0.002	0.000
$D_{Hispanic}$	0.011	0.005	0.034
D_{White}	0.065	0.005	0.000
D_{Asian}	0.078	0.007	0.000
$D_{Bilingual}$	−0.032	0.003	0.000
Income band	0.029	0.002	0.000
Special needs	−0.017	0.001	0.000
$D_{Inracemajority}$	0.005	0.005	0.295
$D_{BelowIncomemajority}$	0.046	0.004	0.000
$D_{AboveIncomemajority}$	−0.050	0.004	0.000
σ_v	0.032	0.003	0.000
σ_w	0.033	0.002	0.000
σ_u	0.025	0.002	0.000

Dependent variable: logarithm of average of test scores
Maximum likelihood with robust standard errors

at), but also that belonging to the comparatively affluent pupils of the school makes one's grades go down. If this was not about young persons, we would dare say that it is amusing how the well-known cliches surrounding wealth and education are born out by our sample (who needs the latter if one has the former?).

The estimation results on the sigma parameters are also interesting: all three exist, being also very precisely estimated, but are quite small, much smaller than applied experience. This indicates again that "homogenizing" is one of the core characteristics of the formal educational outcomes of secondary education.

Moreover, with such small latent effects, further exploration through the M-metrics from Chap. 2 will not really be interesting or revealing. So we conclude this empirical application here.

5.5 Environmental Policy and Social Pressure: Do Businesses Become Greener?

Environmental issues are all the rage these days—perhaps because the environment itself appears to have burst into a fit of rage that does not seem like it will abate any time soon. One of the ways societies try to cope with this is through environmental policy aimed at businesses. Environmental policy takes many forms and uses various instruments; some are financial incentives like investment subsidies to make business operations "greener"; others are disincentives if a company does not conform, like environmental taxes according to the principle "polluter pays."

And there is also environmental regulation: limits on pollutants and obligatory upgrading of technology, with the associated fines/penalties if these regulations are not observed. While this looks like the usual "carrot and stick" approach to policy making, as with any policy, the forever question is: what are its actual effects? Did it promote a more environmentally sustainable economic activity? Was it a waste of public resources, affecting nothing really? Or, even, did it make matters worse?

Suppose now that we can construct a measure related to business activity that indicates the degree to which a company walks the path of environmental sustainability (the higher the better). Moreover, suppose we also have measures of environmental policy: for example, the rates of, or the amounts paid by businesses due to, environmental taxation. We could then form a regression relation between them and estimate the effects of the latter on the former, first and foremost as regards the direction of influence, and then also its quantitative impact.

But there is another force that has risen on these matters as of late: social pressure on businesses from consumers and civic society to become more "sustainable." This has broadened the scope from a strict focus on activities that degrade the environment directly in an ecological sense to a more general demand that businesses rise to higher "quality standards" in all aspects of their operations and socioeconomic conduct.

This pressure would exert an increasing influence on the aforementioned measure of sustainable activity. On the other hand, we can be certain that the tendency of most businesses is to somehow evade all these, since, to begin with, they eat up financial resources and they lower profits. And so, the tug-of-war of 2TSF analysis emerges: beyond the effects of official environmental policy and regulations, there are these latent forces that may boost or dilute how much do businesses change their ways, in the broader natural and social environment.

We measure all these in the empirical application that follows, based on a data set from China covering the period 2010–2020.

5.6 Application: Going Green in China 2010–2020

We will base our application on Lei et al. (2024). They applied a 2TSF model to a data sample of Chinese manufacturing companies ("listed on the A-share market, excluding ST and *ST companies," as they write). The start year 2010, was chosen because it was the end of the "11th Five Year Plan for the Construction of National Environmental Protection Regulations and Environmental Economic Policies" of the Chinese government. The sample extended up to and including the year 2020. An important policy change during the period was the transition in 2018 from a "pollution fee" to an "environmental tax." The authors did not use their sample as a panel, but pooled the data, and we will do the same, but not before detecting certain issues that led us to adjust both the sample and our regression specification compared to theirs.

Dependent Variable One of the main contributions of Lei et al. (2024) is the computation for their sample of a "high-quality development" (HQD) index for each company and year. This is a weighted average of 21 indicators grouped into six broad categories: financial health, innovation capability, market presence, environmental and social responsibilities, human resources, and the effectiveness of corporate governance. They then use this dependent variable in logarithms. Examining in more depth the logarithmic series, out of the 15,863 observations that they ended up using, there were 2,330 observations where log(HQD) had the value 0. This implies that the original HQD index took the value "1." Inspecting, we discovered that this was not the case; the observations for which log(HQD)=0 were observations for which the HQD series had missing values. So the value "0" in the log(HQD) series was artificial—and led to a serious distortion in the sample: it falsely created a subgroup of observations with very high log(HQD) value of the dependent variable, since in all other observations with actual values for the index, the maximum HQD was just 0.61 leading to a corresponding logarithmic value of -0.49. This distortion showed in their estimation of the 2TSF latent forces, where they estimated the mean value of the positive force w as very high (in part, this was the dutiful attempt by the model to reflect the existence of this subgroup of observations with a very high value of the dependent variable, which was not supported by the levels of the control variables).

So for our work, we dropped these observations and kept a sample of size $n = 13{,}533$. Further, the empirical frequency of this cleaned sample is shown in Fig. 5.1.

It is clear from this figure that the empirical distribution of the dependent variable is still bimodal, even after cleaning out the artificial values. Initial estimation

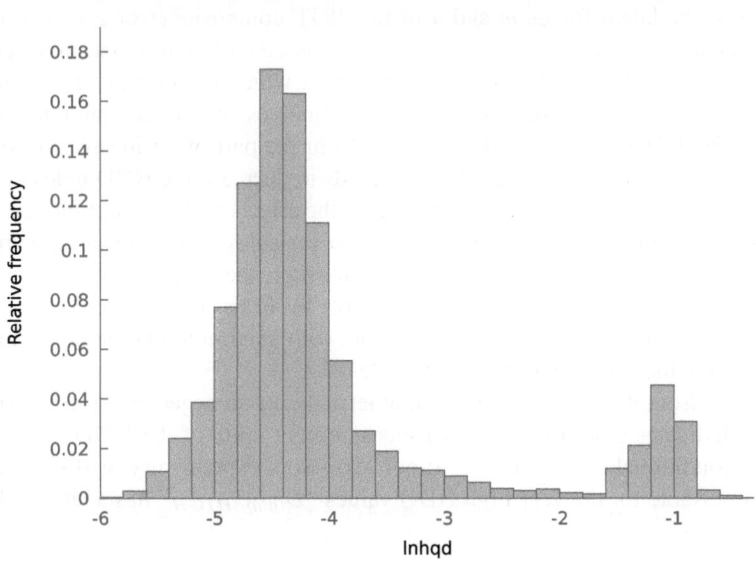

Fig. 5.1 Relative frequency of dependent variable $\ln hqd$

attempts indicated that this bimodality cannot be explained by the regressors used in the specification. It is an instance of heterogeneity that is beyond the explanatory abilities of the specific data sample. If the sample was used as a panel, this should be cauterized by the individual effects variable (see Chap. 9). But because we are pooling the data, what we did was to determine a threshold of $\ln hqd = -1.6$ and create a dummy variable taking the value 1 for those observations with higher (closer to zero) dependent variable.[7] There were 1,553 such observations (out of the 13,533). These high values related to 453 unique companies out of 2,313 that exist in the sample. Out of the 453, only 9 had all their observations in this high-HQD subgroup.

Control Variables Except for the constant term, we used the same nine control variables that were used in the original study. Some pertain to the firm and some to its socioeconomic environment: firm size in terms of total assets (lnsize), debt-to-asset ratio (lnlev), board size (lnboard), cash ratio (lncash), separation of ownership and control (dual), GDP growth rate (gdp), financial level of the city (fin), the number of Internet users (Internet), and the number of mobile phone users (mobile) at the city level.

Main Explanatory Variables Following Lei et al. (2024), we included in the specification two quantitative measures of environmental policy: the variable with the label "market" that records pollution fees and environmental taxes and the variable "command" that proxies for the intensity of environmental regulation by recording the number of "local administrative penalty cases," namely, those cases where companies were found in breach of environmental regulations.

Once environmental policy variables become observables and are included as regressors, the latent forces w and u of the 2STF composite error $\varepsilon = v + w - u$ do not relate any more to them, but, as we discussed earlier, to other, unobservable forces that could impact the dependent variable, which in our case are reasonably interpreted as the larger social pressure on businesses (the w variable that tends to increase the HQD index) and the attempts from the part of businesses to avoid all these additional costs (the u variable that tends to decrease the HQD index).

So the model will give a direct estimate for the effects of the environmental policy through the estimated coefficients on the two variables that represent government environmental policy; and it will estimate through the moments of w and u the unobservable social pressure and the attempts by firms to evade it. We apply the benchmark 2TSF Normal-Exponential model with maximum likelihood estimation, and we show the estimation results in Table 5.12.

Starting from the value of the constant term, being so negative at the logarithmic level, it indicates that at the original measurement units of the HQD index, there is no unconditional base value for it; our regressors explain very well its size. The dummy variable for the very high HDQ values "$D_{lnhqdHIGH}$" has a large value. In

[7] This threshold corresponds to the point where the tail of the main body of the distribution ends.

Table 5.12 Maximum likelihood estimation of the 2TSF Normal-Exponential model. Period: 2010–2020 (pooled sample). $n = 13{,}533$. Dependent variable is logarithm of HQD index. Robust standard errors

Variable	Coefficient	Std. Error	p-value
Constant	−9.7080	0.0956	0.0000
$D_{lnhqdHIGH}$	3.1234	0.0097	0.0000
lnsize	0.2348	0.0042	0.0000
lnlev	−0.0026	0.0086	0.7585
lnboard	0.0432	0.0184	0.0190
lncash	0.0674	0.0045	0.0000
dual	0.0031	0.0004	0.0000
gdp	0.0212	0.0064	0.0009
fin	−0.0106	0.0041	0.0103
internet	−0.00002	0.0001	0.8585
mobile	0.0324	0.0059	0.0000
market	0.0869	0.0299	0.0036
command	−0.0031	0.0007	0.0000
σ_v	0.1583	0.0072	0.0000
σ_w	0.3308	0.0057	0.0000
σ_u	0.2203	0.0057	0.0000

fact, for the related observations, it should be added to the constant term to form the constant term for them.

Of the control variables that appear to affect negatively the HQD index, neither the debt-to-assets balance sheet ratio (lnlev) nor the number of Internet users (Internet) have large coefficients, and in any case, they are very imprecisely estimated, with very high p-values. The level of local economic development (fin) does have a (small) negative effect on the HQD index, which would require further digging into the particulars to be understood.

Among the controls that affect positively the HQD index, the size of the firm stands out (lnsize); it supports a general impression that such comprehensive upgrading of operations and socioeconomic conduct is undertaken by large firms, because it requires massive financial resources not usually available to smaller firms. This is supported by the positive effect of cash resources (lncash). The size of the board (lnboard) also appears to have a discernible positive effect, aligning with the thought that a larger board of directors leads to more collective decisions that eventually take into consideration broader aspects and consequences of the company's decision-making. It is also comforting that the growth rate of GDP (gdp) has a positive effect on the index, indicating that growth does not necessarily come at the expense of sustainability.

Turning to the environmental policy variables, we see that the monetary burden on firms through environmental fees and taxes (the "market" variable) has a visible and *positive* effect on the HQD index, while the "penalty" variable (command) has a much smaller but still existing and *negative* effect. We interpret this as follows:

Table 5.13 2TSF metrics

	Sample level	Individual series	
Variable	Mean	Mean	Median
M_5	0.494	0.539	0.266
M_6	0.181	0.181	0.136
M_7	0.225	0.262	0.076

firms have accepted that society's demand through the government for greener and higher-quality operations is a legitimate one, and they eventually respond positively to *proportional financial* pressure; on the other hand, everybody gets mad when fined; and many would think of how to offset that fine, and in a vindictive sort of way, like pulling resources away from the very thing they were fined for (which is a fully rational response, by the way, so that the specific area of activities costs as much as they have decided that it should cost).

Latent Forces and 2TSF Metrics Already from the estimation results, we see that the latent forces of social pressure w and evasion tendencies u are present and influence the HQD index. Because the dependent variable was in logarithms, we revert back in its original units, and we present estimates for the three exponentiated metrics that are the most pertinent here, namely, the counterfactual effects of the latent forces on the index if the other was absent (M_5 for w and M_6 for u), as well as their combined multiplicative effect (M_7). In Table 5.13, we present the sample-level means based on the sigma estimates, as well as the empirical mean and median of the corresponding individual series.

As mentioned earlier, in theory the sample-level means should coincide with the empirical means of the series, the latter being conditional expectations. This is what we observe for the M_6 series, but not for the others. This is due to (anticipated) outliers in the M_5 series. Qualitatively, the social pressure force appears stronger. The combined multiplicative effect of the two latent forces on the HQD index (M_7) is to increase it (in mean-value terms) between 22.5% and 26.2%.

What is interesting is to decompose the M_7 metric into its positive and negative values and do so per year. We show this in Table 5.14.

The last column of Table 5.14 shows the proportion of times social pressure was estimated as stronger than the tendency to evade. Roughly, this happens for 2/3 of cases (firms). We can detect a slow downward trend, which possibly reverts in the last years. A quantitative assessment of this same aspect is shown in Table 5.15.

Table 5.15 shows percentage net markups and markdowns on the HQD index. We see that, per year, the effect of social pressure when it prevails is on average always stronger than the strength of the evasion tendency when the latter prevails. The average value of the M_7 metric was positive in all years, which indicates that in China throughout the 2010s, there was a well-rooted pressure on businesses to treat seriously the whole matter of sustainable operations and presence and to act on it.

Table 5.14 Observations for which M_7 was positive, per year

Year	$M_7 > 0$	Sample	%
2010	636	916	69.4
2011	719	1,087	66.1
2012	725	1,082	67.0
2013	732	1,116	65.6
2014	818	1,187	68.9
2015	765	1,193	64.1
2016	890	1,456	61.1
2017	1,104	1,870	59.0
2018	1,068	1,877	56.9
2019	532	804	66.1
2020	587	945	62.1

Table 5.15 Average values of positive and negative M_7 metric, per year

Year	$M_7 < 0$	$M_7 > 0$	Overall
2010	−0.1204	0.6240	0.3964
2011	−0.1212	0.5348	0.3127
2012	−0.1367	0.4447	0.2529
2013	−0.1501	0.4780	0.2619
2014	−0.1631	0.5304	0.3148
2015	−0.1436	0.4914	0.2636
2016	−0.1652	0.4384	0.2038
2017	−0.1706	0.4840	0.2159
2018	−0.1911	0.4660	0.1827
2019	−0.1720	0.6150	0.3487
2020	−0.2365	0.6092	0.2888

5.7 2TSF Analysis of Subjective Welfare

"Quality of life" becomes an issue when it is absent or when *we think* it is absent. The rise of the individual *en masse* in the western civilization during the last centuries has had very many consequences; one of them is the accompanying rise of subjectivity, and the "right" to decide for yourself how much "quality" characterizes your life. And while health and wealth remain the two prominent indicators for a *realistic* assessment of the quality *surrounding* one's life, a person can have both and be miserable—but can we have none and feel happy? Not to mention that even western medicine has come to accept the existence of psychosomatic interactions, with causality going both ways. Various studies have documented that there is such a hard-wired character trait as "pessimism" and "optimism," and it does affect the *perceived* quality of life even for people with such serious illnesses as cancer.

Pessimism and optimism are clearly antagonistic; but the words convey an end result. The *Oxford Advanced Learner's Dictionary of Current English* (some old edition) describes optimism as the *tendency to look upon the bright side of things* and an optimist as a *person who is always hopeful*; and it describes pessimism as the *tendency to believe that the worst thing is most likely to happen*. It appears that there is a scale from pessimism to optimism, and somewhere in the middle, there is "realism." If this is so, pessimism and optimism emerge as eventual psychological biases that affect our perception, our experiences, and ultimately our own assessment for the quality of life that we have. But how do we move up and down this scale? It does not appear very convincing to say that "realism is the absence of either pessimism or optimism." It is more likely that realism is the point where two tendencies balance, and they balance long-term since we are talking about character traits here. In statistics, "long-term" translates to "distribution," since the moment we assert that a phenomenon follows/is characterized by a specific distribution, we are asserting a long-term (or at least a mid-term) piece of information that we are to take for granted and to use subsequently in exploring the phenomenon, until further (and unanticipated) notice.

From this angle, realism is a symmetric distribution, while pessimism and optimism are skewed distributions, having opposite skewness. And skewed distributions come about when we setup a face-off between two opposing forces of unequal strength. Both tendencies exist in all of us; we look at things both from a bright side and a dark side. We are characterized as pessimists if the dark side proves strong enough to dim the brightness sufficiently more often (or even "always"). We are characterized as optimists when we perceive darkness as just the absence of light, dispersing when the latter lights up in our eyes. It is evident that, if we want to go quantitative on this, the two-tier stochastic frontier model was made for this purpose.

This is exactly what Groot and van den Brink (2007) did on a sample from the United Kingdom, where they correlated self-reported quality of life with income, family status, and other characteristics including health/disease conditions (with a focus on heart and blood disease). The 2TSF composite error term was there to capture deviations from the estimated realistic assessment based on these control variables. And deviations there were, supporting the presumption that these psychological tendencies exist. Three years later Poggi (2010) returned to the labor market, the empirical origin of 2TSF analysis, but now in order to measure "job satisfaction" and how it is affected by people's aspirations and the biases these may create.

Both these studies pioneered the application of 2TSF analysis in the field of psychology, being microeconometric works based on samples comprised of individual data. We dared to go macro, global really, and in the next section, we offer a 2TSF take on a worldwide sample that deals with national averages of subjective happiness, to see whether latent forces still operate at this aggregate plane, transitioning thus from psychology to sociology.

For operational convenience we will carry with us the labels "pessimism" and "optimism," invoking Groot and van den Brink (2007) that usefully distinguish between these concepts as referring to the current evaluation of the *future* state of the world, from their use referring to the current evaluation of the *current* state of the world. It is in this second meaning that we will use them, tracing universal latent forces affecting subjective happiness around the world.

5.8 Application: World Happiness Report 2020

On July 19, 2011, the UN General Assembly adopted resolution 65/309 "*Happiness: Towards a Holistic Definition of Development*," which was sponsored by the Kingdom of Bhutan. In it, the General Assembly recognized that "*the gross domestic product indicator by nature was not designed to and does not adequately reflect the happiness and well-being of people in a country*" and invited member states to "*pursue the elaboration of additional measures that better capture the importance of the pursuit of happiness and well-being in development with a view to guiding their public policies*".

As is perhaps more usual than hoped with such UN calls, countries did not heed the UN advice—but pollsters and scientists did. Since the first report in 2012, and with the exception of 2014, a World Happiness Report (WHR) is published yearly, tracking, measuring, explaining, and visualizing in straightforward ways the level of happiness in most countries. "Straightforward" does not mean "naive"; the report's findings are supported by proper statistical work, freely available to the world[8]. Here we take a 2TSF look on WHR 2020 and ask: can whole nations be pessimistic or optimistic?

The *measurement* approach of WHR is bottom-up, with any explanations to come top-down: people around the globe answer the "Cantril ladder" question that asks respondents to value their lives today on a scale of 0 to 10, with 0 representing the worst possible life, and 10 representing the best possible life. It is called a "ladder" because the question is framed visually as follows:

1. Imagine a ladder with steps numbered from zero at the bottom to 10 at the top.
2. The top of the ladder represents the best possible life for you and the bottom of the ladder represents the worst possible life for you.
3. On which step of the ladder would you say you personally feel you stand at this time?

It is the "Cantril" ladder because it was conceived by American psychologist Albert Hadley Cantril, Jr., in his 1965 book "*The Pattern of Human Concerns*" (under the name "Self-Anchoring Striving Scale"—no wonder "ladder" won the popularity contest). So individuals, without explanations or other elaborations, speak their subjective mind. This is aggregated into an average score per country,

[8] For readers interested in finding out more, the WHR website https://worldhappiness.report/ has everything.

and then scientists take over and bring all sorts of descriptive and explanatory tools to visualize the state of happiness around the world, its evolution, with what it correlates and what can explain it. We will focus on the first step undertaken in WHR 2020 Helliwell et al. (2020), which is to show the six variables that appear to explain these national averages of happiness. It is useful (and illuminating, for the more "soft" variables) to reprint here how data are collected on them (p. 17 of WHR 2020):

1. **Gross domestic product per capita** (GDP, in logarithms) is in terms of purchasing power parity (PPP) adjusted to constant 2011 international dollars, taken from the World Development Indicators (WDI) released by the World Bank on November 28, 2019.
2. The time series of **healthy life expectancy at birth** are constructed based on data from the World Health Organization (WHO) Global Health Observatory data repository, with data available for 2005, 2010, 2015, and 2016, measured in years.
3. **Social support** is the national average of the binary responses (0 =no, 1 =yes) to the question, "If you were in trouble, do you have relatives or friends you can count on to help you whenever you need them, or not?"
4. **Freedom to make life choices** is the national average of binary responses to the question, "Are you satisfied or dissatisfied with your freedom to choose what you do with your life?"
5. **Generosity** is the residual from regressing the national average of responses to the question "Have you donated money to a charity in the past month?" on GDP per capita.
6. **Perceptions of corruption** are the average of binary answers to two questions: "Is corruption widespread throughout the government or not?" and "Is corruption widespread within businesses or not?" Where data for government corruption are missing, the perception of business corruption is used as the overall corruption-perception measure.[9]

Data Set and Descriptives The original data set used for the first results of WHR 2020 (shown in their Table 5.16, 1st column, p. 16) had 1,848 country/year observations from 166 countries and for the period 2005–2019. Discarding observations with some missing values, they ended up with 1,627 observations from 156 countries, still covering the same period (this is "unbalanced" panel data in that not all countries are present in all years). They then pooled the sample and applied OLS, using "year fixed effects" (for all years and so eliminating a universal regression constant term). These year fixed effects are just a dummy variable for each year, representing essentially an intercept that may vary per year.[10] First we replicated (successfully) their results in order to make sure that we have handled properly the

[9] We suspect that many readers may wonder why an index of income/wealth inequality is not included in this list. It is, in extensions of this benchmark specification, and readers are advised to read the WHR 2020 document on the matter.

Table 5.16 World Happiness data. Observations per year and geopolitical region

	Geopolitical region					
Year	Africa	America	Asia	Europe	Pacific	Total
2006	19	19	20	16	1	75
2007	20	19	27	27	2	95
2008	25	22	32	21	2	102
2009	22	20	35	29	0	106
2010	21	21	34	34	2	112
2011	38	23	33	36	2	132
2012	31	21	28	39	2	121
2013	31	23	31	37	2	124
2014	36	23	29	39	2	129
2015	34	21	31	40	2	128
2016	36	21	26	40	2	125
2017	39	23	28	40	2	132
2018	38	19	30	38	2	127
Total	390	275	384	436	23	1,508

sample and have understood the specification. Then we decided to restrict further the sample by a) dropping the single observation that remained from year 2005 and b) dropping all observations for year 2019 because for them the GDP variable was a forecast (since official data were not yet available at the time of the writing and publication of the report). With these, we arrived at a sample of $n = 1{,}508$ observations across 156 countries for the period 2006–2018. In Table 5.16 we present the decomposition of the sample per year and geopolitical region, using the same grouping that we used in the US military aid application of Chap. 4.

The sample is acceptably balanced as regards representation of the world of humans, and we always have to keep in mind that the composition of such samples depends on data availability rather than any biased decisions by researchers. Before going forward with our estimation, we show in Table 5.17 the evolution of the World Happiness score, Cantril's ladder, per year.

The picture emerging from Table 5.17 is not good. While the average and the maximum happiness score remain relatively stable, the minimum score has a visible downward trend, and this is also reflected in the widening distance between minimum and maximum.[11]

From a technical point of view (always handy to handle bad news), the data is adequately stable across time, and so we can safely treat it as a single pooled

[10] This is a simple and economical way to exploit somewhat the panel structure even when we pool the data, since these year fixed effects represent unobserved heterogeneity across the time dimension. See Chap. 9 for more on this.

[11] One can only wonder whether the COVID-19 pandemic brought people closer in that respect in a morbid kind of way, but this is left for future studies.

Table 5.17 World Happiness scores (Cantril ladder), per year

Year	min	average	max	min-max range
2006	3.20	5.20	7.67	4.47
2007	3.28	5.42	7.83	4.55
2008	2.81	5.42	7.97	5.16
2009	3.41	5.46	7.68	4.27
2010	3.23	5.50	7.77	4.54
2011	2.94	5.42	7.79	4.85
2012	3.16	5.44	7.78	4.62
2013	2.69	5.39	7.59	4.90
2014	2.84	5.39	7.51	4.67
2015	2.70	5.40	7.60	4.90
2016	2.69	5.40	7.66	4.97
2017	2.66	5.46	7.79	5.13
2018	2.69	5.50	7.86	5.17

sample. Before presenting and discussing estimation results, we should explain what the two-tier stochastic frontier approach is looking for here. Remember that our dependent variable, Cantril's ladder, is a subjective perception of happiness/well-being, and as such we are looking for factors that push it up or down the ladder beyond what "rational, realistic" assessments the individuals themselves may internally undertake in order to answer the happiness question. Of course, the data we have on the dependent variable are *national averages*, and one could expect that such idiosyncratic traits would dissipate in this aggregate level. Indeed, this is why a finding that 2TSF latent forces are present here will be a surprising but also an intriguing result, pointing toward *universal* tendencies regardless of culture, religion, history, or socioeconomic situation.

Estimation We implemented joint GMM estimation using the benchmark Normal-Exponential 2TSF specification (for both see Chap. 7). In the joint GMM approach, we estimate in one fell swoop, both the regression coefficients and the parameters of the error distribution. It also allows inclusion of more moment conditions than the minimum required, which is usually good for the efficiency of the estimator. This we did, by including one overidentifying orthogonality condition in the form of the fifth cumulant equation of the composite error term distribution. In Table 5.18, we present the WHR 2020 estimation results mentioned earlier (that used the larger sample), the WHR 2020 specification on our smaller sample (pooled OLS with year fixed effects), and the results from joint GMM estimations that uses the same regression equation.

The results in Table 5.18 are practically identical across the two samples and/or the two estimation methods, which is a welcome result. We note that the standard errors of WHR are visibly higher, which has to do with the clustering and not with the sample (applying clustering to pooled OLS would increase the standard errors

Table 5.18 World Happiness data. Regression estimation results. Dependent variable: Cantril ladder. Robust standard errors in parentheses (for WHR, clustered by country). Year fixed effects included, not shown

	WHR 2020	OLS pooled	2TSF joint GMM
logGDP per capita	0.31	0.325	0.324
	(0.066)	(0.028)	(0.028)
Social Support	2.362	2.304	2.338
	(0.363)	(0.181)	(0.173)
Healthy life expectancy at birth	0.036	0.035	0.035
	(0.01)	(0.004)	(0.004)
Freedom to make life choices	1.199	1.176	1.171
	(0.298)	(0.138)	(0.135)
Generosity	0.661	0.688	0.685
	(0.275)	(0.114)	(0.112)
Perceptions of corruption	– 0.646	– 0.618	– 0.613
	(0.297)	(0.112)	(0.109)
$n =$	1,627	1,508	1,508

Table 5.19 World Happiness data. Estimates of composite errors parameters. Robust standard errors in parentheses

	WHR 2020	OLS pooled	2TSF joint GMM
σ_ε	0.567	0.568	0.563
σ_v	–	–	0.443
			(0.035)
σ_w	–	–	0.193
			(0.065)
σ_u	–	–	0.289
			(0.034)

in the second column). Still, all coefficients are estimated accurately. We leave for readers the task to translate these coefficient estimates into marginal effects on Cantril's ladder, hinting that the WHR 2020 document contains focused analysis on the matter. It is time to turn to matters of 2TSF analysis.

2TSF Estimates and Inference After estimation we conducted the Doornik-Hansen specification test for the 2TSF model (see Sect. 7.3.2). The null hypothesis is that the Normal-Exponential specification is an acceptable fit for the model. The statistic of the test is distributed under the null as a Chi-square with two degrees of freedom. The computed value was 1.956 which gives a p-value of 0.376, which in turns implies that if we decide to reject the specification, we will do it with statistical confidence just 62.4%, which is considered low. So we keep it and we show in Table 5.19 the estimated parameters related to the error term.

Table 5.20 Pearson's correlation and Kendall's τ concordance coefficients between latent forces and regressors

	$E(w\|\varepsilon)$		$E(u\|\varepsilon)$	
Variable	Pearson	Kendall	Pearson	Kendall
logGDP per capita	−0.0590	0.0269	−0.0367	−0.0269
Social Support	−0.0897	0.0525	−0.0398	−0.0525
Healthy life expectancy at birth	−0.0519	0.0439	−0.0339	−0.0439
Freedom to make life choices	−0.0333	0.0313	0.0214	−0.0313
Generosity	0.0106	0.0464	0.0272	−0.0464
Perceptions of corruption	0.0096	−0.0443	−0.0232	0.0443

From Table 5.19, we see that the 2TSF latent forces are present, and the negative force is stronger than the positive force: even at the level of national averages, beyond the six reasonable explanatory variables there is something persistently negative, across years and countries that permeates the subjective perception of life happiness and drags it somewhat below the level of "realistic" assessment. Since the dependent variable is in levels, these estimates translate directly as effects on its size: the positive force w tends to increase its value by 0.193 and the negative force to decrease it by 0.289. These may not amount to a whole step in Cantril's ladder, but they are not negligible either (exactly because we are looking at national averages). Next we provide evidence that the series of $E(w|\varepsilon)$ and $E(u|\varepsilon)$ are not correlated with the regressors, so they do not indirectly reflect the latter. Table 5.20 holds Pearson's correlation and Kendall's tau concordance coefficients between them.

The magnitudes in Table 5.20 are totally negligible. So w and u are their own masters, bringing into the picture of World Happiness other factors than those represented by the regressors. Since this is a global result, it is tempting to turn philosophical or even metaphysical about it. So we will not, but rather, we will call it "universal pessimism" and stop there, leaving to other social scientists and philosophers the burden of coming up with explanations and rationalizations. What we will do is to dig a little deeper by looking at one particular 2TSF metric, M_7, the net effect of w minus u on the realistic happiness level. We will separate, as suggested in Chap. 2, its positive from its negative values. Table 5.21 shows the frequency of obtaining a positive M_7 per year. The relative frequency remains rather stable at approximately 1/3.

Next, we examine the same information per geopolitical region, in Table 5.22. We admit that we are a bit mystified here. Traditionally, it is the North-South divide that creates expectations of some degree of clustering (which would also be interesting to explore). But here, it appears that as regards the frequency of a positive M_7 metric, regions tend to cluster per *Longitude*. Imagining the planet vertically, we have Europe and Africa at one broad Longitude region, Asia and Pacific in another, and the Americas in a third one. This is one more intellectual challenge.

We close the application of 2TSF analysis on World Happiness scores by looking at the averages of positive and negative M_7 predictions per region, presented in Table 5.23.

Table 5.21 Observations for which M_7 was positive, per year

Year	$M_7 > 0$	Total obs.	%
2006	19	75	25.3
2007	30	95	31.6
2008	27	102	26.5
2009	31	106	29.2
2010	31	112	27.7
2011	42	132	31.8
2012	40	121	33.1
2013	42	124	33.9
2014	42	129	32.6
2015	43	128	33.6
2016	42	125	33.6
2017	46	132	34.8
2018	41	127	32.3

Table 5.22 Observations for which M_7 was positive, per geopolitical region

Region	$M_7 > 0$	Sample	%
Africa	118	390	30.3
America	174	275	63.3
Asia	60	384	15.6
Europe	122	436	28.0
Pacific	2	23	8.7

Table 5.23 Averages of positive and negative M_7 subsamples, per region.

Region	$M_7 < 0$	$M_7 > 0$	Overall	Distance
Africa	−0.049	0.025	−0.027	0.074
America	−0.025	0.022	0.005	0.047
Asia	−0.045	0.027	−0.034	0.072
Europe	−0.028	0.009	−0.017	0.037
Pacific	−0.006	0.002	−0.005	0.008
Sample means	−0.038	0.02	−0.02	0.058

From Table 5.23 we learn that what differentiates people in the Americas is not a stronger optimistic tendency, but a weaker pessimistic one. Also, we learn that Africa appears now rather closer to Asia, and what stands out in these two regions is the larger distance between the negative and positive averages, indicating a more extreme lived experience, compared, for example, to Europeans.

And with this, we proclaim our Jamboree of latent forces concluded.

Chapter 6
Binary Responses and Treatment Effects

6.1 Latent Forces in a Latent Regression

Two-tier stochastic frontier analysis exists because unobservables exist. We model these unobservables as part of the "error term" in our regression relations, and we develop ways that will allow us to obtain some quantitative evidence about them.

In "conditional probability" models, there is no apparent structural error term. Among these models, binary response (BR) setups are very popular, because there are many real-world situations where our interest focuses on whether something happened or not and why; hence it can be represented by a binary {0, 1} classification.

It would appear then that 2TSF analysis is inapplicable in such a setting. But this is not the case. Binary response models arise from an underlying regression model with an unobservable dependent variable y^{ℓ} and a structural error term indeed. What BR models estimate are the parameters of this underlying regression, overcoming, by the use of a connected probability relation, the absence of data on the unobservable dependent variable. But if this is so, then we can conduct 2TSF analysis also in the context of BR models.

In Chap. 13, we develop the model and its tools in detail, and we call it the 2TSF Probit Exponential model. Here we show what additional insights it can provide with an application. The essence of the theoretical matter is this: specific BR models arise when we assume a particular statistical distribution for the error term of the aforementioned underlying regression (and not because we want to use a mathematically convenient function to model probabilities). We get the famous Logit model when we assume that this error term follows a zero-mean Logistic distribution. We get the probit model when we assume that this error term follows a zero-mean Normal distribution. In the same manner, we get the "2TSF Probit Exponential" model if we assume that the error term is composite, $\varepsilon = v + w - u$ and that it follows the 2TSF Normal-Exponential distribution (see Sect. 7.2.1). The main difference is that this is a non-zero mean and asymmetric distribution; therefore the

A. Papadopoulos, C. F. Parmeter, *Two-Tier Stochastic Frontier Analysis for the Social Sciences*, https://doi.org/10.1007/978-3-031-81513-3_6

fundamental probability relation that we estimate cannot be simplified as it does with symmetric distributions like the Logistic and the Normal. Also, there is a well-known identification issue as regards BR models that relates to the variance of the underlying error term. In standard BR models, it is of no interest. But in our setting, it implies the restriction

$$\mathrm{Var}(\varepsilon) = 1 \implies \mathrm{Var}(v) + \mathrm{Var}(w) + \mathrm{Var}(u) = 1.$$

The implication of this is that we are estimating essentially the relative variance of each latent force and so its relative strength rather than its absolute magnitude. But this does not impede useful inference.

Denoting F_ε the distribution function of this 2TSF distribution, our model is characterized by the otherwise familiar probability mass function

$$\Pr(y|\boldsymbol{x}) = \left[1 - F_\varepsilon(-\boldsymbol{x}'\boldsymbol{\beta}|\boldsymbol{x})\right]^y \cdot \left[F_\varepsilon(-\boldsymbol{x}'\boldsymbol{\beta}|\boldsymbol{x})\right]^{1-y}, \tag{6.1}$$

where y is a binary variable linked to the unobservable dependent variable of the underlying regression. Its likelihood for observation i is

$$\ell_i = y_i \cdot \ln\left[1 - F_\varepsilon(-\boldsymbol{x}_i'\boldsymbol{\beta}|\boldsymbol{x}_i)\right] + (1 - y_i) \cdot \ln F_\varepsilon(-\boldsymbol{x}_i'\boldsymbol{\beta}|\boldsymbol{x}_i),$$

which is estimated by maximum likelihood, as most BR models are.

The only special characteristic of the 2TSF-BR model is that it *should not* contain a constant term in the regressor vector, as we explain in Chap. 13. Armed with this model, we now go into delicate territory, to showcase its potential.

6.2 Application: Probability of Extramarital Affairs

The fundamental motivation for an extramarital affair from an economist's point of view is very simple and always present: people like variety in their lives; they get positive "utility" from it. So, in principle the tendency to become involved in an extramarital affair is always there, irrespective of the situation in the marriage or cultural norms and/or self-interest reasons that could eventually make a person avoiding having one.

Fair (1978) presented a theory of extramarital affairs based on the theory of utility maximization. Looking at the survey data that he used for his empirical application, he observed that between 1/5 and 1/3 of married people disclosed of having at least one extramarital affair during their marriage, therefore the phenomenon could not be labeled as negligible.

In true economics spirit, Fair (1978) considered as arguments in the utility function of the individual, the allocation of their time between time spent with spouse, time spent with "paramour," and time spent working and of the "goods

Table 6.1 Regressors in the extramarital affairs model

Description	Values	Mean
D_{Sex}	0 = female, 1 = male	0.476
Age	midpoints of 5-year intervals	32.5
No. of years married	midpoints of 2-year intervals	8.18
$D_{Children}$	0 = no, 1 = yes	0.715
How religious	1–5 (1 = anti, 5 = very)	3.12
Level of education	measured in years of schooling	16.2
Occupation	1–7, reverse Hollingshead classification	4.19
Marriage rating	1–5 (1 = very unhappy, 5 = very happy)	3.93

supplied" (whatever they may be) by the individual, as well as the spouse and the paramour in their respective time together with the individual. Utility maximization of course happens under an income constraint, and the situation gets all the standard treatment of an economics study: dry and strictly based on materialistic, rational considerations centered on self-interest.

Following acceptable academic norms of the day, the empirical part of the paper instead focused on explaining extramarital affairs as a function of certain basic individual, marital, and socioeconomic characteristics. The author used two surveys conducted by US magazines. Here we work with the data from *Psychology Today*. A (rather long) questionnaire on sexual matters was published in the July 1969 issue of the magazine, and readers were asked to mail-in their answers. From this questionnaire, Fair (1978) extracted a number of explanatory variables that are shown in Table 6.1. The total usable sample size was $n = 601$.

The proportions of the sexes are a bit below the actual allocation in United States at the time for males (88% of the sample was in the age interval $20 - 44$, and males were estimated as 49.39% of this segment of the population in 1969). The average age reflects a rather young but fully active sample of individuals especially considering that we are looking 55 years back. This is also supported by the average length of marriage which is above 8 years already, indicating marriage at around age 25 (currently it is above 30 for both men and women in the country). 71.5% of the individuals in the sample had children (currently this percentage is about 50%). Religiosity is somewhere between "slightly" (=3) and "somewhat" (=4). 16 years of schooling corresponds to college graduates. The "reverse" Hollingshead classification means that a higher number is assigned to unskilled/manual labor, while high-level executives, owners of large businesses, and "major professionals" are assigned lower numbers. The mean value of 4.19 is above the arithmetic average of the 1–7 scale. Finally, the marriage rating is close to the value 4 ("happier than average").

Fair (1978) used as dependent variable the number of times an individual had extramarital sexual intercourse during the last year, and due to many zeros, he applied a tobit model. For our purposes, we transformed this dependent variable into a "have/have not" binary, in order to be able to estimate the 2TSF Probit

Table 6.2 Men, women, and extramarital affairs: proportions

	Extramarital affair		
Sex	No	Yes	Total
Female	40.4%	12.0%	52.4%
Male	34.6%	13.0%	47.6%
Total	75.0%	25.0%	100%

Table 6.3 Extramarital affairs conditioned on sex

	Extramarital affair		
Sex	No	Yes	Total
Female	77.1%	22.9%	100%
Male	72.7%	27.3%	100%

Table 6.4 Extramarital affairs conditioned on having children

	Extramarital affair		
Sex	No	Yes	Total
No children	84.2%	15.8%	100%
Children	71.4%	28.6%	100%

Exponential model. 150 observations have a non-zero dependent variable, i.e., 25% of the persons in the sample had an extramarital affair.

But before proceeding with the model, let's discuss some global pre-conceptions: men are considered more prone to extramarital affairs than women. Table 6.2 show us what holds in the specific sample.

While both men and women appear to engage almost equally in extramarital affairs, Table 6.2 holds the *joint* empirical relative frequencies. The conditional such (with conditional distributions to be read per row) are shown in Table 6.3, and here a visible gap emerges, supporting the general belief.

Another popular belief is that children affect negatively the sexual life of the parents, which would imply that individuals with children have a greater tendency for extramarital affairs. Table 6.4 holds the related conditional probabilities. We see that this preconception also is born out by the specific sample.

It is time now to turn to our model. We present in Table 6.5 the estimated regression coefficients from a standard Probit model, and from the 2TSF Probit Exponential model, and also the estimates of the scale parameters of our two latent forces (they are only two because σ_v is computed and not estimated so as to satisfy the identification constraint discussed earlier).

In many respects, the two models agree: the signs of the coefficients are everywhere the same.[1] They both estimate accurately/inaccurately the same parameters,

[1] The direction of influence also agrees for all variables with what Fair (1978) found in his Tobit model, see his table 3, p. 59, first column.

Table 6.5 Probit and 2TSF Probit Exponential models: Estimation results. Robust standard errors in parentheses

Variable	Probit	2TSF Probit-Exp
Constant	0.7794	–
	(0.5390)	
D_{Sex}	0.1734	0.1111
	(0.1405)	(0.1081)
Age	– 0.0246	– 0.0172
	(0.0111)	(0.0077)
# of years married	0.0543	0.0401
	(0.0189)	(0.0140)
$D_{Children}$	0.2166	0.1953
	(0.1686)	(0.1448)
How religious	– 0.1854	– 0.1370
	(0.0531)	(0.0426)
Level of education	0.0112	0.0228
	(0.0291)	(0.0166)
Occupation	0.0136	0.0101
	(0.0414)	(0.0322)
Marriage rating	– 0.2717	– 0.2009
	(0.0533)	(0.0395)
σ_w	–	0.6697
		(0.0432)
σ_u	–	0.2219
	–	(0.0595)
Log-likelihood	– 305.19	– 305.17
Predictive Performance	76.54%	76.04%

and they have essentially the same log-likelihood value and the same predictive performance. We do not show it here, but the predicted probability $\Pr(y_i = 1|\boldsymbol{x}_i)$ series of the two models have correlation coefficient 99.4%.

But note that the standard errors of the 2TSF probit Exponential model are everywhere smaller than the standard probit, which is a clear performance improvement. More importantly, the scale parameters σ_w and σ_u of the two latent forces are estimated very accurately, while also the regressor coefficient values are closer to zero in the 2TSF probit Exponential model (smaller in absolute magnitude, with the exception of education). This combination of results tells us that the standard probit coefficient estimates are biased away from zero, i.e., overestimated in absolute value, and this happens because they "take on" the effects of the latent forces, and selectively, those regressors that have an increasing effect on the probability of an extramarital affair take on the effect of the positive latent force w, and those that have a decreasing effect take on the effects of the negative force u.

Looking at the coefficient estimates more specifically, the stronger influences come from sex, having children, religiosity, and happiness in marriage, although only the last two are estimated with conventional statistical confidence (above 95%), while the first two are estimated with confidence ∼80% (a confidence level which, we are in a position to reveal, is considered an acceptable level of certainty in empirical studies outside academia). All have the anticipated sign: being male increases the tendency for an extramarital affair, having children also, while higher religiosity and happiness in marriage both decrease such tendencies.

If we compute the series for $\Pr(y_i = 1|\boldsymbol{x}_i)$ and compute its mean, for the probit model, it is 0.2499, and for the 2TSF probit Exponential model, it is 0.2481. Both are spot-on the proportion of observations in the sample with an extramarital affair (and this is expected up to a point).[2]

But the models differ when we look at the probability for the "average individual" rather than the average probability. The probability of an extramarital affair computed at the mean values of the regressors is, per model,

$$\Pr(y = 1|\boldsymbol{x} = \bar{\boldsymbol{x}})|_{\text{Probit}} = 0.1991,$$

$$\Pr(y = 1|\boldsymbol{x} = \bar{\boldsymbol{x}})|_{\text{2TSF}} = 0.3120.$$

The 2TSF Probit Exponential model suggests a much higher probability of extramarital affairs for the average individual in the sample.

But the remarkable result here is that the latent force that favors an extramarital affair (w) is *three times stronger* than the other that pushes against it. This supports the economist's presumption that, beyond the explanatory variables that carry characteristics realistically expected to affect the decision to have an extramarital affair or not, there is some strong latent tendency toward an extramarital affair: the desire for variety.

***M*-Metrics** In the 2TSF probit Exponential model, the latent forces operate directly on the latent variable y^ℓ which by construction is in its original measurement units. But also in many cases y^ℓ and $\boldsymbol{x}'\boldsymbol{\beta}$ will range in the negative values. This renders some of the metrics developed in Chap. 2 not directly usable/interpretable here (although all can be computed), and we use the absolute value in those that we do compute. Moreover, we can only obtain sample-level metrics (as long as we do not have covariates that explain σ_w and σ_u; see Chaps. 8 and 13). We present them in Table 6.6.

The average value of the systematic part was -0.845. We see in Table 6.6 that the w positive force is three times stronger than u, as already mentioned. Volatility (M_4) is equal to 1.05, meaning that combined the latent forces have a lot of influence on the decision to have an extramarital effect or not. M_5 and M_6 measure the individual strength of w and u with respect to the systematic part $\boldsymbol{x}'\boldsymbol{\beta}$ (on average), and M_7

[2] In the logit model this holds as an exact mathematical result.

Table 6.6 Extramarital affairs: metrics of latent forces for the 2TSF probit Exponential model

Metric	Formula	Value
M_3	$\frac{E(w)}{E(u)}$	3.02
M_4	$\frac{E(w)+E(u)}{\lvert E(\mathbf{x}')\boldsymbol{\beta}\rvert}$	1.05
M_5	$\frac{E(w)}{\lvert E(\mathbf{x}')\boldsymbol{\beta}\rvert}$	0.79
M_6	$\frac{E(u)}{\lvert E(\mathbf{x}')\boldsymbol{\beta}\rvert}$	0.26
M_7	$\frac{E(w)-E(u)}{\lvert E(\mathbf{x}')\boldsymbol{\beta}\rvert}$	0.53

Table 6.7 Probit and 2TSF probit Exponential models: marginal effects on the probability, computed at the mean values of the variables

		2TSF
Variable	Probit	Probit-Exp
D_{Sex}	0.0525	0.0395
Age	−0.0074	−0.0061
# of years married	0.0164	0.0142
$D_{Children}$	0.0631	0.0694
How religious	−0.0561	−0.0487
Level of education	0.0034	0.0081
Occupation	0.0041	0.0036
Marriage rating	−0.0822	−0.0714
$E(w)$ $(+0.1)$	–	0.0100
$E(u)$ $(+0.1)$	–	−0.0614

tells us that their average net effect is equal to approximately half the systematic part.

Marginal Effects We will finish this application with the computation of the marginal effects of the regressors on the probability and, courtesy of the 2TSF approach, of the latent forces also. This is one of the most interesting insights a conditional probability model can provide. We present them in Table 6.7.

For the regressors, these are probability points (and not percentage changes of the probability), per "one absolute unit of change" $(+1)$. In the probit model, for example, being male adds 5.25 points to the probability of having an extramarital affair while having children adds 6.3 probability points. Being more religious (moving one level up) decreases this probability by 5.6 points, while being happier in the marriage (moving one level up) decreases it by 8.22 points. This is the strongest marginal effect according to the probit model, and it shows that marriage is like a general election: it is lost by the government, not won by the opposition.

We note that the estimates from the two models are very close, with those from the 2TSF probit Exponential being slightly lower, as should be expected from the coefficient estimates.

But the 2TSF model offers more, namely, the marginal effects on the probability of the latent forces, specifically, the effect due to an increase in their mean value. Here, the scale is different. Consider $E(w) = \sigma_w$. The computation of its marginal effect using the related formula from Chap. 13 gives the value 0.0975. But remember that here, σ_w measures the *relative* strength of w in the composite error term, so contemplating an increase of "one unit" has no meaning, since by necessity σ_w ranges in (0, 1). The marginal effect we show is for an increase of 0.1 (i.e., to go from 0.6697 to 0.7697 while the relative position of σ_u remains the same and what reduces is the contribution of the noise term through σ_v), so we present $0.0975 \times 0.1 = 0.01$. The same holds for σ_u: the computation of the formula gives the value -0.6142, and we present $-0.6142 \times 0.1 = -0.0614$, imagining an increase from 0.2219 to 0.3229.

The large difference in their effect possibly reflects the opposite large difference in their current level: the latent factor w that favors an extramarital affair is already high, and it appears that it has plateaued in its influence, hence the very low marginal effect. On the other hand, the latent factor u is currently at a low level and its marginal effect is much stronger. This indicates the existence of diminishing marginal effects, which is only reasonable.

In a full-blown empirical study, more avenues could be explored: we could compute odds ratios, we could examine subsamples per certain characteristics (e.g., the subsample of males, the subsample of very religious people, etc.) and do all the above analysis per subsample. But we believe that we have shown how the 2TSF approach and the 2TSF probit Exponential model deepens and enriches our understanding. So we will stop here, and we will move to a second empirical study with a binary response model, where the focus is on compounded but conflicting treatment effects and how they can be estimated separately by the two-tier stochastic frontier methodology.

6.3 The 2TSF Treatment of Treatment Effects

Coming from ancient Greece thinkers, the wisdom "there is no bad without some good" (similar to the saying "every cloud has a silver lining") conveys the complexity and the contradictory nature of the effects that any causal factor has. Of course, the reverse also holds: there is no good without some bad. We see this in economic activity, where increased production efficiency may also lead to increased unemployment or where production itself creates pollution of the environment. We see this in justice, where incarcerating an offender may impact very negatively the members of the offender's family, which may have had nothing to do with the offender's crimes. We see this in medicine, where every drug has "side effects";

this is a gentle way to say that the drug, the *treatment*, has more than one, and contradictory, effects.

This phenomenon is what we called "Rudrian (dualistic)" treatments in Chap. 1, and we see that essentially, all treatments are Rudrian; all treatments have desirable and undesirable effects. To be precise, these effects become *opposing* when what we actually examine is an aggregate that inherently incorporates both. So, a drug for stomach aches with a drowsiness side effect has desirable and undesirable consequences that do *not* oppose one another, if we examine separately the organ of the stomach and the organ of the brain. But if we want to measure a composite metric like "how do you feel after taking the drug?," the two effects antagonize each other.

These are the situations where we end up measuring the *net* effect of the treatment, without having recourse to data on the separate effects in order to assess them individually. And these are the situations where the 2TSF setup allows us to decompose these opposing treatment effects. The theoretical and statistical models are developed in Chap. 13; here we present an application from healthcare in the United States.

6.4 Application: Home Healthcare and US Hospital Readmissions

In the healthcare system of the United States, the government reimburses hospitals for treating patients under the federal Medicare program, using, among other things, indicators of performance in terms of medical and healthcare efficacy, which may adjust (usually downward) the final monetary amount paid. One of these performance indicators is the "rate of 30-day hospital readmissions." The rationale of using this metric appears realistic. If a patient is discharged from a hospital, and they are readmitted no later than 30 days after, this can indeed be an indication that the hospital did not do a particularly good job: maybe the medical and healthcare services offered were not of the quality required; maybe the decision itself to discharge was not thought out as carefully as it should have been.

Certainly, medicine and healthcare (i.e., "repair and maintenance" of the human body) are far from reaching the efficacy levels of machine repair and maintenance—the degree of uncertainty is still high. But we should expect the beneficial effects of healthcare offered in a hospital to last at least for 30 days, should we not? It sounds reasonable, and it lends validity to the use of the 30-day hospital readmission rate as a performance and reimbursement metric.

Enter the "Home Health Care (HHC) paradox": discharged patients that were assigned home health services upon discharge were found to have consistently higher rates of 30-day hospital readmissions, even after controlling for multiple health factors that could be the leading causes or indicators of readmission despite the provision of excellent healthcare. The HHC paradox was uncovered as researchers in healthcare attempted to measure the HHC impact on patients, anticipating a positive effect that would also materialize in the form of lower rates of hospital readmissions. But they always found the opposite, using different models and methodologies (see, e.g. Weinberger et al., 1996; Madigan, 2008; Madigan et al., 2012; Bradley et al., 2013; Leppin et al., 2014).

The paradox was so widely documented, that, if we insisted that there was a unique unidirectional effect in operation, we would be compelled to argue that across United States, the HHC medical and nursing professionals have become so low-level in terms of competence and professionalism that they were actually hurting the patients. But there was never any evidence, even an indication, that such a massive deterioration and failure has happened. So, in trying to explain the paradox, scholars and healthcare professionals invoked the dualistic principle: HHC does have a positive effect on the patient's *health status*. But, came the Rudrian part, the maintained links of the patient with the health system *increased the chances of a readmission*: on their own, patients would be more reluctant to be readmitted to the hospital, for psychological and/or financial reasons; on their own, patients would not have the expertise to assess accurately their health condition and whether it has worsened to such a degree that hospital readmission would be the sensible decision to make. On the other hand, when visited regularly by the HHC professionals, such assessment would be more accurate; also, an HHC professional most likely felt the burden of responsibility in not letting their patient have a serious health incident outside the hospital, and that could increase risk aversion and the frequency of decisions to readmit the patient in the hospital. When the health system sees you, it may invite you in more often, and it will not take "no" for an answer.

So the treatment of HHC had a desirable medical effect and an undesirable "administrative/professional/monitoring" effect, related to hospital readmission, and many quantitative studies consistently found that this administrative effect was stronger *on average*, leading to the HHC paradox. Note that the reasons why monitoring leads to more frequent hospitalization is *also* an indication of increased efficacy; the argument is not that HHC professionals are "over-reacting" but that they actually do their jobs, monitoring and looking at the future than just care for the patient in the present, something that results in higher readmission rates.

Papadopoulos and Stark (2021) revisited the issue and tried to answer it comprehensively, by studying in detail whether the binary logistic (logit) model is a suitable tool for such an analysis, in the sense of whether it would allow a "sign reversal" of an *average* treatment effect to pass through as we enhance the specification with control variables, but also, by estimating the effect of HHC on the 30-day readmission probability in seven acute-care hospitals of different types and from different US states, using large samples (for all seven together the observation count was 228,041). The logit model proved not to impose any artificial restrictions on the estimated coefficients, and it was then used to estimate these samples. And here too, in all cases, the average probability of hospital readmission ("RA risk") for those treated with Home Health Care had increased. The sign of the treatment coefficient was positive in all hospitals; the HHC paradox was present in all samples. The authors summed up their findings and conclusions as follows:

> But this means that the statement "HHC is beneficial to the health of a patient and it increases the probability of hospital readmission" *is not a contradiction in terms*. The direct clinical actions improve the health of the patient but the more frequent, expert and sensitive monitoring of the patient as well as a closer association of the patients to the healthcare and hospital system appear to induce and facilitate readmissions at a higher rate.
>
> In turn this makes the use of the 30-day readmission rate as a measure in hospital reimbursement schemes questionable: If interventions like HHC are beneficial to the patients and at the same time lead to an increase in readmission rates of the treated, by using the RA risk as a financial stick as is currently used, we penalize the hospitals that prescribe HHC to benefit their patients.

Here we take up the topic again, but now applying the 2TSF random treatment effects (RTE) model of Chap. 13, in an attempt to measure separately the two competing forces operating on the probability of 30-day hospital readmission, but also to dig deeper on the treatment effect per treated individual. In this model we start from the latent regression

$$y^{\ell} = \boldsymbol{x}'\boldsymbol{\beta} + \varepsilon_i, \qquad \varepsilon_i = v_i + T_i(w_i - u_i),$$

where the unobservable y^{ℓ} is a health risk index (the higher the worse), $\boldsymbol{x}$ are the controls, T is the binary treatment, the two latent forces w and u are the opposing effects of this Rudrian treatment (treated themselves as random variables, hence the name of the model), and v completes the specification of the composite error term. Embedded in the model is a different specification of the error term per patient category: those treated ($T = 1$) have a 2TSF error term $\varepsilon = v + w - u$ (with the Normal-Exponential specification), but those not treated have a usual random Normal error $\varepsilon = v$. From this underlying specification, we arrive at the conditional probability model

$$\Pr(y_i | \{\boldsymbol{x}_i, T_i\}) = \Phi\left((2y_i - 1)\left(\boldsymbol{x}_i'\boldsymbol{\beta}\right)\right) + T_i(\mathcal{D}_{1i} - \mathcal{D}_{2i}),$$

where $\mathcal{D}_{1i} - \mathcal{D}_{2i}$ is a complicated expression that includes the regressors and the dependent variable, still perfectly estimable by maximum likelihood. Its two components though are *not* to be interpreted as the separate marginal effects of w and u (see Chap. 13 as to why). They are to be seen jointly as determining the net effect of treatment, while the average marginal effects of w and u are more involved expressions. Compared to the 2TSF Probit Exponential model that we presented in the first part of this chapter, the 2TSF random treatment effect model has two differences: first, the restriction on the error variance applies to the variance of v, not of ε,

$$\text{Var}(v) = 1.$$

It implies that here, what the obtained estimates of the scale parameters of w and u actually represent is not their relative magnitude to the composite error term strength (σ_ε) as before, but "signal to noise" ratios, σ_w/σ_v and σ_u/σ_v. Since the denominator is the same, they are directly comparable though.

The second difference with the 2TSF Probit Exponential model is that this one does include a constant term in the regressor matrix as conventional BR models do. An implied difference with *them* is that the treatment variable is not included as a regressor in the specification alongside the control variables, but in the additive way shown in the estimating equation above.

The big advantage of this model compared to conventional conditional probability setups like the Logit and Probit models is that *it allows the individual treatment effects on the treated to have different signs*, something that conventional models do not allow, as explained in Chap. 13. As we will see, this ability makes serious inroads in understanding the HHC paradox and pinning down its source.

We use a part of the data sample of Papadopoulos and Stark (2021), specifically their "Hospital # 3," an urban acute-care private hospital in the state of California. The sample comes from the years 2015–2017 (but we use it pooled) and has 17,424 observations, of which 2,977 observations of patients treated with HHC (17% of the sample) and 14,447 observations for HHC-untreated patients. Apart from data on HHC treatment, it includes binaries for 29 medical conditions, number of prior admissions in the previous 12 months, and length of hospital stay in days (the last one we use in logarithms). Papadopoulos and Stark (2021) used the 29 binaries as individual regressors; we instead created a simple "health score" in [0, 1] by summing them and dividing by 29.

The dependent variable is coded as "1" when readmission in the 30-day window occurred, and "0" if it did not. So the positive force w is the undesirable, administrative/monitoring force, while the negative force u is the desirable healthcare effect. According to all findings, $w > u$ in an appropriate average sense. In Table 6.8 we present the estimated coefficients from a conventional probit model, where the treatment variable enters as a regressor, and from the 2TSF RTE model, where we do not estimate a coefficient for the treatment variable, but instead the two scale parameters σ_w and σ_u that characterize the w and u opposing treatment effects (being their expected values and standard deviations).

In conditional probability models, the regression coefficient estimates themselves are not the primary focus, but they serve here the purpose of assessing whether the two models diverge in their common elements or not. From Table 6.8 we see that the two models are very close in many aspects: coefficient estimates and standard errors, likelihood value, and predictive performance. This also validates the 2TSF RTE model: as regards the control variables, the two models agree, and they also fit the data equally well.

But as regards our main interest which is the binary treatment of HHC, things diverge: the Probit model estimates the coefficient on treatment as positive, small but accurately estimated, as Papadopoulos and Stark (2021) also found using the Logit model. Namely, the average treatment effect of HHC is to *increase* the probability of 30-day hospital readmission. This is the HHC paradox. But the 2TSF RTE model clearly estimates the healthcare effect that reduces the probability of readmission (u) as being stronger on average compared to the monitoring effect (w) that tends to increase this probability. In fact their ratio is $\sigma_u/\sigma_w = 1.64$ which means that u is on average 64% stronger than w. We must remember that w and u do not operate

Table 6.8 Probit and 2TSF random treatment effects models: estimation results. Robust standard errors in parentheses

Label	Variable	Probit	2TSF RTE
	Constant	–1.6910	–1.7206
		(0.0330)	(0.0359)
M12	Prior admissions	0.1873	0.2015
		(0.0102)	(0.0124)
LgLOS	Length of stay (log)	0.3759	0.3998
		(0.0485)	(0.0521)
DiagnScore	Diagnoses Index [0, 1]	2.4841	2.6447
		(0.3930)	(0.4162)
HHSOrig	Home Health Services (binary)	0.0849	–
		(0.0340)	–
σ_w	Mean monitoring effect	–	0.6040
		–	(0.1251)
σ_u	Mean health effect	–	0.9910
		–	(0.3469)
Log-likelihood		–5643.616	–5637.348
Predictive Performance		89.1%	89.1%

Table 6.9 Probit and 2TSF random treatment effect models: estimated average probabilities of readmission

Label	Measure	Probit	2TSF RTE
APU	Untreated	0.103	0.103
AP^{c}T	Treated before treatment (counterfactual)	0.127	0.129
APT	Treated after treatment	0.144	0.143
ATT	Average treatment effect on the treated	0.017	0.014

directly on the probability but on the latent health-risk index y^ℓ, and we learn that the average effect of the HHC treatment on the health-risk index is negative, which is beneficial. On its own, this is already a comforting result, because it gives the primary spot to the intended action and effect: provision of health services at home do ameliorate the health status of the patient.

Does this mean that the average treatment effect in the 2TSF RTE model is also negative and in sharp contrast with what we found using the Probit model? Table 6.9 contains some standard probability measures from the two models: average probability of readmission (PRA) of the untreated (APU), counterfactual PRA for the treated before treatment (AP^{c}T), average PRA for the treated after treatment (APT), and average treatment effect on the treated (ATT).

The two models appear here to agree. Eventually, both estimate the average treatment effect on the treated as *positive*. But this means that while the probit model

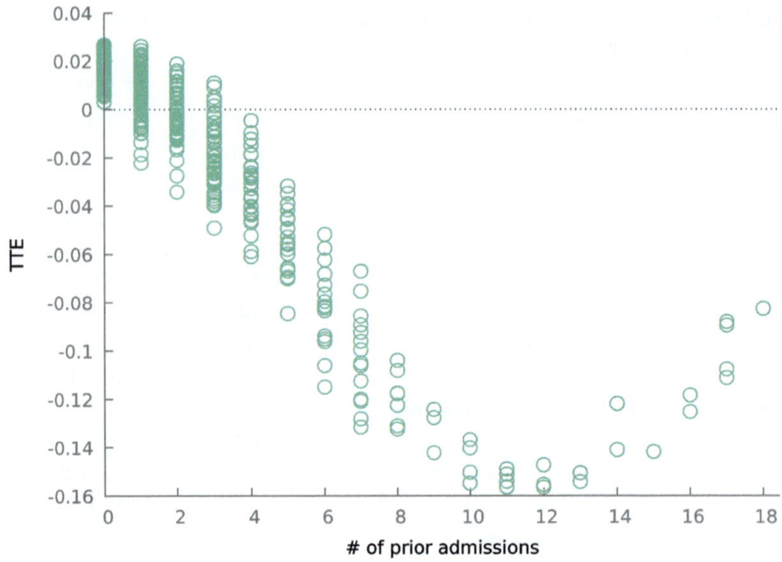

Fig. 6.1 Treatment effect on the treated against prior admissions. Scatterplot

can smile triumphantly with an "I told you so" wink, the 2TSF RTE model seems embarrassingly in contradiction with itself: previously it estimated the negative treatment effect u as stronger than the positive one w, but here it would appear that the opposite is implied... Not so fast. The influence of the treatment effects on the readmission probability is not a direct transmission of their influence on the underlying health index. It passes through the nonlinear stochastic transformation that leads to the conditional probability model, and it ends up depending in a complex way on the regressors also. So we should not necessarily expect the average result pertaining to the underlying latent regression to be qualitatively the same with the average effect at the level of the probability model. These are two distinct results, each conveying a different piece of information. So yes, the 2TSF RTE model tells us that HHC services improve the health status of the patient and at the same time increase the readmission probability of the treated patients. It is still a paradox, but now at least we have empirical support that the direct services offered are beneficial.

Moreover, at the level of the probability model where the HHC paradox emerges, the 2TSF can offer something that the probit model by design cannot. Namely, the possibility that, at the observation level, the overall treatment effect may be positive for some observations and negative for others. In Fig. 6.1, we present the scatterplot of the series for the treatment effect on the treated (TTE) from our model (whose mean value is ATT), against the "Prior hospital admissions during the last 12 months" regressor, for which applied experience indicates that it is among the strongest predictors of readmissions.

We could never have obtained Fig. 6.1 from a Probit or a Logit model, because in these models every single individual treatment effect on the treated has necessarily the sign of the treatment coefficient. But here we see both negative and positive net treatment effects, and with a very interesting and nonlinear pattern: *all* patients that received HHC and had four or more hospital admissions in the past 12 months experienced a reduction in the probability of readmission and of a large magnitude. The effect is maximized for patients with 11 prior admissions and then it weakens, remaining strongly negative (and hence desirable). On the other end, for patients with one to three prior admissions, the net treatment effect was sometimes positive and sometimes negative, and for patients with zero prior admissions, the effect was always positive. Still, its magnitude is considerably smaller in absolute terms than the negative effects observed, but it so happens that this last group, patients with zero prior admissions who received HHC, occupies 65% of the treated subsample, and another 19% is occupied by patients that had one prior admission, received HHC, and had a positive net treatment effect on the readmission probability. This is why the average treatment effect on the treated ends up being positive also in the 2TSF RTE model. And this is how one can reconcile the result $E(w) - E(u) < 0$ with ATT > 0: the fact that the negative effects are much larger in absolute value indicates that $E(w) - E(u) < 0$, while ATT weighs these influences by their frequency of appearance in the whole treated subsample.

Although we have not "solved" the Home Health Care paradox, we have offered important conclusions, segmentation, and re-focusing:

1. Home Health Care services benefit patients and the healthcare effect is stronger than the "monitoring" effect *on the perceived health-risk index*, on average.
2. Home Health Care *reduces* the 30-day readmission probability for patients with heavier history of hospital admissions.
3. Home Health Care *increases* the 30-day readmission probability for patients with a lighter record of past admissions, especially for those with zero prior admissions.
4. If the sample contains more patients with few/none past admissions, the overall average effect of HHC on the readmission probability may be positive.

The HHC paradox now resides only in this last subset of cases, and healthcare experts can think and elaborate much more efficiently as to why the "administrative/professional/monitoring" effect may be stronger for this subset of patients. We repeat that these conclusions could not be reached using a logit or a probit model due to their mathematical structure that imposes the restriction on the sign of individual treatment effects on the treated, to always be the same as the sign of the average treatment effect.

Marginal Effects In Table 6.10 we present marginal effects. We indicate in parentheses for each variable to what change of magnitude correspond these changes in probability points (we have transformed the effect for "length of stay" to correspond to days and not log-days). For the probit model, we present the results for the whole sample estimated at the sample means of the regressors, while for the 2TSF RTE model, we present separately the estimates for the untreated and for the treated sample. We see that the differences across rows are very small, except

Table 6.10 Probit and 2TSF random treatment effects models: marginal effects

	Probit	2TSF RTE	
Variable	Full sample	Untreated	Treated
Prior admissions (+1)	0.0330	0.0336	0.0314
Length of stay (+1 day)	0.0200	0.0225	0.0121
Diagnoses Index (+0.1)	0.0437	0.0441	0.0411
Home Health Services (binary)	0.0155	–	–
σ_w (+0.1)	–	–	0.0263
σ_u (+0.1)	–	–	−0.0117

for the length of stay variable, where the average length of stay for the treated group is implicitly revealed to be higher. The diagnoses Index appears to have the stronger effect on the probability of readmission. In the 2TSF RTE model, it is not surprising that the average partial effect of u is weaker in absolute terms than that of w, for the same proportional increase. Given the higher mean level of the former, this indicates that the treatment effects have a concave relation with the probability, which is plausible. We also note (not shown) that the marginal effect of σ_w is everywhere positive, and the marginal effect of σ_u is everywhere negative, so we have not encountered pathological cases.

Finally what is interesting is to examine the *series* of partial effects. This speaks to issues of "intensity/efficacy" of the marginal effects and what effect on the readmission probability such a change could have. But also, it will tell us whether the two opposing treatment effects move together or not and how strong the relation is, which are important pieces of the puzzle. To close this application, we present, in Fig. 6.2, scatterplots between the positive and the negative treatment effect (the latter expressed in absolute values), where we have also segmented the treated sample with few prior admissions (those categories that contributed to the HHC paradox) and many. The marginal effect series are those for an increase in the sigmas by 0.1.

These two figures provide new insights: the first, related to patients with few prior admissions (where we have encountered the HHC paradox), shows a clear positive monotonic relation between the marginal strength of the two effects: a more efficacious medical treatment (that would improve health more and reduce the readmission probability) goes hand-in-hand with a stronger monitoring effect that increases the probability of readmission. This connection between the two forces is even more clearly shown in the scatterplot for the patients with many prior admissions: the lower line shows a convex tendency to explode implying that as the medical marginal effect increases the monitoring marginal effect increases even more; but this does not last. Eventually we reach a point where the line regresses; in reality it maintains the same qualitatively positive monotonic relation at a higher mean level. It appears that the two forces are strongly "tied together," in their *marginal* effects on the probability of 30-day readmission, which makes operational interventions to improve the overall effect of the HHC treatment on the readmission probability more challenging.

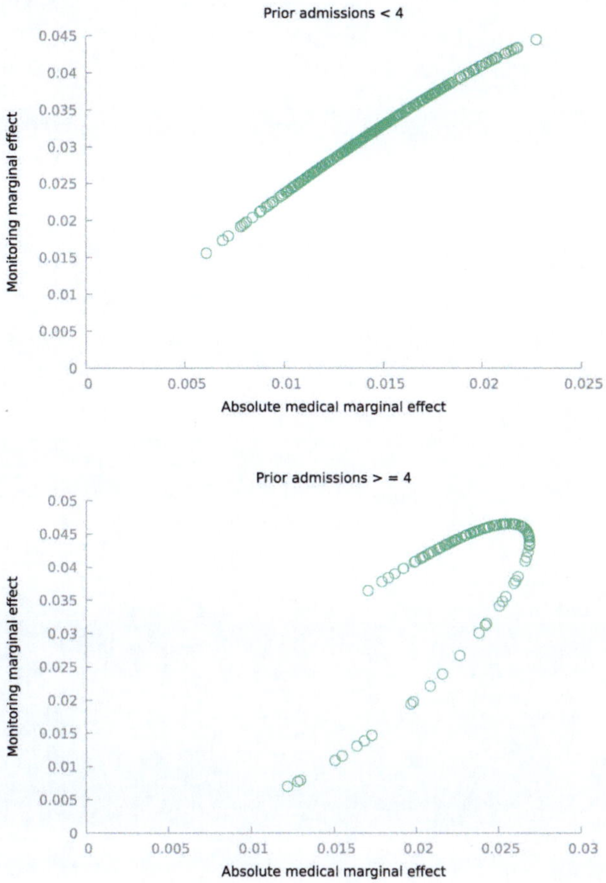

Fig. 6.2 Marginal Effects for a 0.1 increase in intensity. Positive ("monitoring") treatment effect against (absolute) negative (medical). Scatterplots for low and high number of prior admissions

Part III
Tools for Estimation and Inference

Chapter 7
Estimation Methods and Statistical Specifications

This chapter is one of the pivots of the book. We started with an introductory discussion about the general characteristics and the history of the 2TSF framework. We moved on to present metrics that meaningfully summarize the information on the effects of latent forces on the outcome. We then presented a number of applications, where the technical tools of the model were used without formal justification or systematic elaboration.

We now turn to present in a systematic way the two necessary ingredients for inference. In the first part of this chapter, we present the basic methods of estimation that can be used for the 2TSF model: maximum likelihood (ML), corrected ordinary least squares (COLS), and corrected nonlinear least squares (CNLS), as well as generalized method of moments (GMM). We present them here under the simplest stochastic assumptions: homogeneity of observations, intra-error independence, and regressor exogeneity. Later chapters examine how the estimation strategy must be adjusted when these assumptions are relaxed.

In the second part of the chapter, we present specific statistical specifications for the composite 2TSF error term, together with all the expressions and formulas one will need to run estimation, obtain estimates, and compute the metrics presented in Chap. 2.

In the finale we first offer informal forensic diagnostics for choosing the most appropriate distributional specification for the composite error term, which is needed irrespective of the estimation method used. We then present three formal statistical tests for that purpose. We close the chapter with an empirical example related to model selection.

Throughout we stay clear from mathematical computations or theoretical derivations. These have been incorporated and peer-reviewed in the original published studies that included these results (and any new ones have been audited and tested). Here we draw mostly from Kumbhakar and Parmeter (2009) and Papadopoulos

A. Papadopoulos, C. F. Parmeter, *Two-Tier Stochastic Frontier Analysis for the Social Sciences*, https://doi.org/10.1007/978-3-031-81513-3_7

(2018, 2021a).[1] The goal of this chapter is to be the benchmark reference for estimation methods and mathematical expressions needed to implement the 2TSF model, alongside practical advice and guidance.

7.1 Estimation Methods

Our goal is to estimate a regression model in conjunction with a cross-sectional sample of independent and identically distributed (iid) observations. *The crucial feature of the estimating equation is that the error term enters additively in it.* Apart from that, only the OLS estimator is paired with a regression equation that is "linear in the regression coefficients." The other three methods that we present can have a nonlinear structure (with respect also to the regression parameters). So in general, the estimating equation will be

$$y_i = g(\boldsymbol{x}_i; \boldsymbol{\beta}) + \varepsilon_i, \qquad i = 1, \dots, n. \tag{7.1}$$

The actual form of $g(\boldsymbol{x}_i; \boldsymbol{\beta})$ may depend on what is known about the relationship of the outcome with the regressors or on modeling conventions in the specific discipline where the model is applied. As we have mentioned earlier, y may be the logarithm of the dependent variable (but again, this does not necessarily imply that the regressors have to be in logarithms as well). So there is high flexibility as regards the functional form.

The error term is composite, $\varepsilon_i = v_i + w_i - u_i$, with $E(v_i) = 0$, $w_i, u_i \geq 0$, which is the 2TSF form. The maintained stochastic assumptions are that (v_i, w_i, u_i) are jointly independent and that the regressors are exogenous to the error term. This generic expression reflects slightly different properties per estimation method. In all cases, essentially it represents the existence of *some degree of*—statistical—independence between the regressors and the error term (which is needed), but usually not full statistical independence (which would be too restrictive, excluding, e.g., the existence of conditional heteroskedasticity). We will see in a little while why we need exogeneity of the regressors.[2] But not infrequently, especially with observational data, it is difficult to defend the existence of exogeneity. We briefly discuss different ways to handle regressor endogeneity in Sect. 14.1.4. Dependence between the components of the error term is treated in Chap. 10.

As mentioned, the maintained sampling assumption is that we have available a cross-sectional sample $\{y_i, \boldsymbol{x}_i\}$ of size n that is iid. This implies that it is also "ergodic," and ergodicity guarantees that sample averages converge to the

[1] The 2TSF Normal-Half Normal specification has been published in Papadopoulos (2015b), but the notation there was not standard. Here we follow the standard notational conventions in the literature.

[2] We will indicate what form of exogeneity is required for each estimation method.

corresponding population expected values, and we will assume that these expected values exist and are finite, or we will state the sufficient conditions for them to be so. Allowing for heterogeneity is treated in Chap. 8. Working with panel data sets, is presented in Chap. 9.

All four estimation methods that we will present must be combined with an assumption as to the distribution of the error term ε_i that is represented by the density $f_\varepsilon(\varepsilon)$. This relates to the 2TSF model and reflects our desire to decompose this composite error term. Otherwise, only ML estimation requires inherently a distributional assumption to provide estimates for the regression betas. But without a distributional assumption on the error term, we will not be able to estimate the effects of the latent forces. Such statistical specifications are presented in detail later in this chapter. For the purposes of this section, we can treat this aspect in abstract. The density of the composite error term is a convolution that is carried out more efficiently if we set, say, $z \equiv w - u \implies \varepsilon = v + z$, in which case we have

$$f_z(z) = \int_{S_u} f_w(z+u) f_u(u) du, \qquad f_\varepsilon(\varepsilon) = \int_{S_z} f_v(\varepsilon - z) f_z(z) dz.$$

From these general expressions, it is clear that the parameters characterizing the distribution of ε are transformations of the parameters of the distributions of v, w, and u, and we denote these primal distribution parameters summarily by $\boldsymbol{\theta}$. Distribution parameters are involved in determining the moments of a distribution, including the mean (expected value) and the variance. So since we will be estimating $\boldsymbol{\theta}$, we will be able to compute separately the moments that characterize the positive and the negative latent forces.

Under the stochastic assumptions made, all four estimators are (asymptotically) consistent and asymptotically Normal. Namely, writing the estimator as $\hat{\boldsymbol{q}} = (\hat{\boldsymbol{\beta}}, \hat{\boldsymbol{\theta}})$ (which depends on the sample size n) and denoting by $\boldsymbol{q}^o$ the true vector of unknown parameters, consistency implies

$$\Pr(|\hat{\boldsymbol{q}} - \boldsymbol{q}^o| > \epsilon) \to 0, \text{ as } n \to \infty, \forall \epsilon > 0,$$

and its shorthand is $\hat{\boldsymbol{q}} \to_p \boldsymbol{q}^o$ ("convergence in probability"). Asymptotic Normality implies that a properly centered and scaled function of the estimator converges to a multivariate zero-mean Normal random variable with finite variance-covariance matrix, namely,

$$\sqrt{n}(\hat{\boldsymbol{q}} - \boldsymbol{q}^o) \to_d \mathrm{N}(\mathbf{0}, \mathrm{Avar}),$$

which is "convergence in distribution." Note that consistency is a pre-requisite to obtain asymptotic Normality in a practically useful way. An inconsistent estimator may be asymptotically Normal, but the characterization of the distribution will include the unknown divergence from the true value at the limit.

These properties permit us to perform asymptotically valid statistical tests. MLE and GMM (with the optimal weight matrix) are also equivalent as regards asymptotic efficiency, while COLS is inferior in relation to efficiency. But there are reasons to have available different estimation methods in practice: sometimes the MLE will not converge to a solution, and then COLS is a simple alternative. NLS can be handy when MLE does not converge, and the estimating equation is nonlinear in the regression parameters. On the other hand, GMM can be used to test the distributional assumptions through its overidentifying restrictions.

7.1.1 *Maximum Likelihood*

To implement maximum likelihood estimation, we use the assumed $f_\varepsilon(\varepsilon)$ to form the log-likelihood of the sample

$$\mathcal{L} = \sum_{i=1}^{n} \ln f_\varepsilon(\varepsilon_i; \boldsymbol{\theta}) = \sum_{i=1}^{n} \ln f_\varepsilon(y_i - g(\boldsymbol{x}_i; \boldsymbol{\beta}); \boldsymbol{\theta}) \equiv \sum_{i=1}^{n} \ell_i(\boldsymbol{\beta}; \boldsymbol{\theta} | y_i, \boldsymbol{x}_i).$$

The last notation conveys the message that the log-likelihood is treated as a function of the unknown parameters, with the data being used as fixed values. Terminology-wise, sometimes ℓ_i is called the "likelihood contribution" (of observation i). Using the available sample $\{y_i, \boldsymbol{x}_i\}$, the sample log-likelihood is to be maximized with respect to the vector $\boldsymbol{q}$, given the data. This requires finding values for $\boldsymbol{q}$ that make the *score* of the log-likelihood equal to zero

$$\hat{\boldsymbol{q}} : \frac{\partial \mathcal{L}(\hat{\boldsymbol{q}})}{\partial \boldsymbol{q}} = \sum_{i=1}^{n} \frac{\partial \ell_i(\hat{\boldsymbol{q}} | y_i, \boldsymbol{x}_i)}{\partial \boldsymbol{q}} = \mathbf{0}. \tag{7.2}$$

$\partial \ell_i / \partial \boldsymbol{q}$ is the score vector for observation i. Computations and maximization are carried out by iterative software routines, which also compute second-order conditions in order to guarantee that the stationary point is indeed a maximum, at least a local one. For these routines to be initiated, they require the following ingredients:

- the data sample $\{y_i, \boldsymbol{x}_i\}$, $i = 1, \ldots, n$;
- the specification of the log-likelihood, usually the ℓ_i expression due to the iid sample;
- initial values for the vector $(\boldsymbol{\beta}, \boldsymbol{\theta})$ that is under estimation.

When coding the log-likelihood, sometimes there is the temptation to not include any normalizing constants that the density has, because they do not affect the MLE solution. We advise against this practice, for the following reason: if one wants to execute *non-nested* specification tests based on the likelihood, i.e., compare in some way two competing distributional specifications in terms of the achieved value of the log-likelihood, then the comparable values are those that include also the constants of the competing densities—and if they are different densities, these constants will have different values.[3] Only in likelihood ratio tests that use the same density (which are applied, e.g., in order to test different compositions of the regressor matrix) do these constants play no role. Even if we want to use the simple BIC/AIC information criteria to quantify the performance of a likelihood specification against another, they should take into account the normalizing constants of each density.

Initial values for the regression coefficients $\boldsymbol{\beta}$ can be obtained by a preliminary OLS regression. For the elements of the vector $\boldsymbol{\theta}$, we can set reasonable starting values using as guidance the estimated error variance from the OLS regression. Specifically, because we assume that (v, w, u) are independent, it follows that $\mathrm{Var}(\varepsilon) = \mathrm{Var}(v) + \mathrm{Var}(w) + \mathrm{Var}(u)$. These variances are functions of the distribution parameters that we want to estimate. For the statistical specifications that we will examine, the distribution parameters are scale factors that determine the standard deviation/variance. So after obtaining an estimate $\widehat{\mathrm{Var}}(\varepsilon)$ through OLS, we can "allocate" it over the three unknown parameters as their starting values.

Sometimes the MLE will not converge, and then a different set of starting values could be tried. But even if all goes smoothly with the first try, it is strongly advised to experiment with different sets of starting values, in order to guard against multiple local maxima of the log-likelihood. In case that different sets of starting values lead the MLE to converge to different estimates, one chooses the set for which the value of the log-likelihood is the greatest. This value (sometimes divided by sample size) is routinely printed out together with the estimation results.

Under ergodicity of the sample, and exogenous regressors, usually expressed as "mean-independence," namely, $E(\varepsilon_i|\boldsymbol{x}_i) = E(\varepsilon_i)$, the MLE will be consistent and asymptotically Normal, so we can compute standard errors based on the asymptotic variance-covariance matrix. This matrix can be obtained in three different ways. A consistent estimator of it is based on the following sample mean of second derivatives ("Hessian"),

$$\widehat{\mathrm{A}}\mathrm{var}_{\mathrm{H}}[\sqrt{n}(\hat{\boldsymbol{q}} - \boldsymbol{q})] = -\left(\frac{1}{n}\sum_{i=1}^{n} \frac{\partial^2 \hat{\ell}_i}{\partial \boldsymbol{q} \partial \boldsymbol{q}'}\right)^{-1}. \tag{7.3}$$

Under correct distributional specification the information matrix equality (IME) holds, and the above is asymptotically equivalent to another expression that is based on the outer products of the first derivatives of likelihood contributions

$$\widehat{\mathrm{A}}\mathrm{var}_{\mathrm{S}}[\sqrt{n}(\hat{\boldsymbol{q}} - \boldsymbol{q})] = \left(\frac{1}{n}\sum_{i=1}^{n} \frac{\partial \hat{\ell}_i}{\partial \boldsymbol{q}} \frac{\partial \hat{\ell}_i'}{\partial \boldsymbol{q}}\right)^{-1}. \tag{7.4}$$

There is some evidence that using the Hessian expression may perform better in smaller samples, but computing and inverting the Hessian is more demanding for

[3] It also facilitates replication studies.

the software. In case robustness is sought, one could use the "sandwich estimator" that accounts for the case where we have distributional misspecification and the IME no longer holds:

$$\widehat{\text{Avar}}_{\text{R}}[\sqrt{n}(\hat{\boldsymbol{q}}-\boldsymbol{q})] = n \cdot \left(\sum_{i=1}^{n} \frac{\partial^2 \hat{\ell}_i}{\partial \boldsymbol{q} \partial \boldsymbol{q}'}\right)^{-1} \cdot \left(\sum_{i=1}^{n} \frac{\partial \hat{\ell}_i}{\partial \boldsymbol{q}} \frac{\partial \hat{\ell}_i'}{\partial \boldsymbol{q}}\right) \cdot \left(\sum_{i=1}^{n} \frac{\partial^2 \hat{\ell}_i}{\partial \boldsymbol{q} \partial \boldsymbol{q}'}\right)^{-1}. \tag{7.5}$$

Statistical and econometric software allow the user to choose the variance-covariance estimator when running an MLE routine. Note that the above expressions estimate the asymptotic variance of the random variable $\sqrt{n}(\hat{\boldsymbol{q}}-\boldsymbol{q})$. So

$$\widehat{\text{V}}\text{ar}(\hat{\boldsymbol{q}}) = n^{-1}\widehat{\text{A}}\text{var}[\sqrt{n}(\hat{\boldsymbol{q}}-\boldsymbol{q})]. \tag{7.6}$$

Namely, to obtain the variance of $\hat{\boldsymbol{q}}$ itself, we must divide the asymptotic variance by n, and this is actually the magnitude that is printed out by software.

7.1.2 *Corrected OLS*

The COLS estimator uses a linear specification, $g(\boldsymbol{x}_i;\boldsymbol{\beta}) = \beta_0 + \boldsymbol{x}_i'\boldsymbol{\beta}$.[4] It is based on the following idea: under the stochastic assumptions made, OLS will be a consistent estimator for the slope regression coefficients, but not for the constant term, β_0.[5] This is because we do not assume that the composite 2TSF error term has mean zero; on the contrary, we expect that in general the mean will be non-zero:

$$E(\varepsilon) = E(v + w - u) = E(w) - E(u) \equiv \mu_\varepsilon(\boldsymbol{\theta}) \neq 0.$$

Then the linear regression specification can be rewritten:

$$y_i = (\beta_0 + \mu_\varepsilon(\boldsymbol{\theta})) + \boldsymbol{x}_i'\boldsymbol{\beta} + (\varepsilon_i - \mu_\varepsilon(\boldsymbol{\theta})) = \tilde{\beta}_0 + \boldsymbol{x}_i'\boldsymbol{\beta} + \tilde{\varepsilon}_i, \quad E(\tilde{\varepsilon}_i) = 0. \tag{7.7}$$

This shows that $\hat{\boldsymbol{\beta}}_{OLS} \rightarrow_p \boldsymbol{\beta}$ but $\hat{\beta}_{0,OLS} \rightarrow_p \beta_0 + E(\varepsilon)$.[6] Regardless, we have the result that the powers of the OLS residuals are consistent estimators of the *central* moments of the error term:

[4] The inclusion of a constant term is, as usual, strongly advised.

[5] In OLS estimation "regressor exogeneity" is represented either in its strong form, $E(\varepsilon_i|\boldsymbol{x}_i) = E(\varepsilon_i)$, or in its weak version, $E(\boldsymbol{x}_i\varepsilon_i) = \boldsymbol{0}$.

[6] When the error term is not zero-mean, the weak exogeneity condition for regression mutates into a zero-covariance one, $\text{Cov}(\boldsymbol{x}_i, \varepsilon_i) = \boldsymbol{0}$.

$$\frac{1}{n}\sum_{i=1}^{n}(\widehat{\varepsilon}_{i,OLS})^r \longrightarrow_p E[\varepsilon - \mu_\varepsilon(\boldsymbol{\theta})]^r \equiv \mu_r, \qquad r \geq 2. \tag{7.8}$$

If we now make a distributional assumption for ε, we can derive the theoretical expressions for its mean and central moments in terms of the parameters in $\boldsymbol{\theta}$.

We exploit two assumptions that characterize the statistical specifications in this chapter: that the noise component v is a zero-mean Normal random variable and that the distributions that govern (w, u) have a single parameter. In this framework, we will have to estimate three strictly positive distribution parameters, denoted σ_v, σ_w, and σ_u (note that these will not necessarily coincide with the standard deviation). So we need three moment conditions, to form a 3×3 system of (nonlinear) equations. It is convenient to use cumulants, κ_r, rather than central moments μ_r. The relations between the two are

$$\kappa_2 = \mu_2 = \text{Var}, \qquad \kappa_3 = \mu_3, \qquad \kappa_4 = \mu_4 - 3\kappa_2^2.$$

Using the properties of the cumulants for independent random variables and for multiplication by the negative sign, and the fact that, since we assume that v is distributed Normal, we have $\kappa_4(v) = 0$ as well as that all odd central moments of v are also zero, we have the population relations

$$\begin{aligned}
\kappa_2(\varepsilon) &= \kappa_2(v) + \kappa_2(w) + \kappa_2(u) &&\Longrightarrow \kappa_2(\varepsilon) - \varphi_2(\boldsymbol{\theta}) = 0,\\
\kappa_3(\varepsilon) &= \kappa_3(w) - \kappa_3(u) &&\Longrightarrow \kappa_3(\varepsilon) - \varphi_3(\boldsymbol{\theta}) = 0,\\
\kappa_4(\varepsilon) &= \kappa_4(w) + \kappa_4(u) &&\Longrightarrow \kappa_4(\varepsilon) - \varphi_4(\boldsymbol{\theta}) = 0.
\end{aligned}$$

The actual expressions for the terms $\varphi_r(\boldsymbol{\theta})$, $r = 2, 3, 4$ will depend on the statistical specification that we adopt. Applying method-of-moments we obtain estimates for $\boldsymbol{\theta}$ by solving the system of equations (viewed as deterministic)

$$\begin{aligned}
\frac{1}{n}\sum_{i=1}^{n}(\widehat{\varepsilon}_{i,OLS})^2 - \varphi_2(\boldsymbol{\theta}) &= 0\\
\frac{1}{n}\sum_{i=1}^{n}(\widehat{\varepsilon}_{i,OLS})^3 - \varphi_3(\boldsymbol{\theta}) &= 0\\
\frac{1}{n}\sum_{i=1}^{n}(\widehat{\varepsilon}_{i,OLS})^4 - 3\left(\frac{1}{n}\sum_{i=1}^{n}(\widehat{\varepsilon}_{i,OLS})^2\right)^2 - \varphi_4(\boldsymbol{\theta}) &= 0.
\end{aligned}$$

In practice, the second and third equations include only σ_w and σ_u, because the higher cumulants of v are zero since it is assumed Normal:

$$\varphi_2(\boldsymbol{\theta}) = \varphi_2(\sigma_v, \sigma_w, \sigma_u), \quad \varphi_3(\boldsymbol{\theta}) = \varphi_3(\sigma_w, \sigma_u), \quad \varphi_4(\boldsymbol{\theta}) = \varphi_4(\sigma_w, \sigma_u).$$

So, if we are interested only in the point estimates, we can solve a 2×2 system and substitute the results into the first equation to obtain σ_v. We note that the system is nonlinear in the parameters. This means that it may be the case that it has no solution or that some estimated elements of $\boldsymbol{\theta}$ have the "wrong sign." In our case, they should be strictly positive to be meaningful scale parameters. If we estimate them as negative, this is forensic evidence that the distributional assumption that we have made may be wrong, and we should change it.

After finding the solution $\hat{\boldsymbol{\theta}}$, we can compute $\mu_\varepsilon(\hat{\boldsymbol{\theta}})$ also and *correct* the OLS estimator for β_0 by

$$\hat{\beta}_{0,COLS} = \hat{\beta}_{0,OLS} - \mu_\varepsilon(\widehat{\boldsymbol{\theta}}).$$

With these we obtain consistent estimates for all regression coefficients. In turn, we should correct the OLS residuals accordingly and obtain a consistent predictor series for the true error term. This will allow us to compute reliably the individual metrics related to the unobservables.

7.1.3 *Corrected Nonlinear Least Squares*

Suppose now that in the model that we want to estimate, the regression function $g(\boldsymbol{x}_i; \boldsymbol{\beta})$ is not linear. There are two cases to consider here.

7.1.3.1 Theory Provides for a Constant Term

If our theoretical model leads to an estimating equation that includes an additive constant term,

$$y_i = \beta_0 + g(\boldsymbol{x}_i; \boldsymbol{\beta}) + \varepsilon_i,$$

it can be rewritten as

$$y_i = (\beta_0 + \mu_\varepsilon(\boldsymbol{\theta})) + g(\boldsymbol{x}_i; \boldsymbol{\beta}) + (\varepsilon_i - \mu_\varepsilon(\boldsymbol{\theta})), \tag{7.9}$$

and the procedure we have described for COLS holds exactly for CNLS also. We note that the presence of an additive constant term will force the NLS residuals to sum to zero, as is the case with the OLS residuals. The NLS estimator is consistent under the assumption that $\text{Cov}[\partial g(\boldsymbol{x}_i; \boldsymbol{\beta})/\partial \boldsymbol{\beta}, \varepsilon_i] = \mathbf{0}$. This will hold if it is the case that $E(\boldsymbol{x}_i|\varepsilon_i) = 0$; otherwise, it is an assumption that is made directly.

7.1.3.2 Theory Does Not Provide a Constant Term

Suppose now that the theoretical model does *not* lead to a constant term. Then an estimating equation that does contain an additive constant can be interpreted as

$$y_i = \mu_\varepsilon(\boldsymbol{\theta}) + g(\boldsymbol{x}_i; \boldsymbol{\beta}) + (\varepsilon_i - \mu_\varepsilon(\boldsymbol{\theta})). \tag{7.10}$$

In this case, through the constant term, we estimate directly the mean value $\mu_\varepsilon(\boldsymbol{\theta})$ of the composite error term. This provides an alternative moment condition to estimate the unknown distribution parameters, compared to COLS:

$$\mu_\varepsilon(\boldsymbol{\theta}) = \varphi_1(\sigma_w, \sigma_u).$$

This relation contains only σ_w and σ_u since the scale parameter of the noise component v does not contribute to the mean value of ε. One could use this equation instead of say the $\varphi_4(\sigma_w, \sigma_u)$ relation. The justification could be that lower moments usually are estimated with lower bias in finite samples or that the expression for the mean will be characterized by "less" nonlinearity (or even be linear) which also reduces the consequences of estimation bias.

Another difference is that here the "corrected residuals" are directly obtained by $\widehat{\varepsilon}_{\text{CNLS}} = \widehat{\varepsilon}_{\text{NLS}} + \mu_\varepsilon(\hat{\boldsymbol{\theta}})$, where $\widehat{\varepsilon}_{\text{NLS}}$ are the NLS residuals produced by the software. Namely, there is no theoretical constant term to correct for first.

It should be realized that in both cases, the NLS estimator will estimate the exact same equation. What changes is how *we* view and interpret the estimated constant term, either as an estimate of $\beta_0 + \mu_\varepsilon(\boldsymbol{\theta})$ or as an estimate of $\mu_\varepsilon(\boldsymbol{\theta})$ only. Namely, the issue here is whether we believe that the actual relation we describe includes an autonomous constant term β_0, or not. This must be decided based on out-of-sample information as regards the relation between the outcome and the available regressors.

7.1.4 Generalized Method of Moments

7.1.4.1 Combined COLS/GMM Estimation

As its name suggests, the GMM estimator generalizes the venerable "method-of-moments," in two ways. The first generalization is that, because GMM is a statistical method, instead of just "solving" a system of equations, it estimates it and provides standard errors of these estimates. It can therefore be combined with the COLS (or CNLS) methodology described earlier, where after we run OLS and obtain the residuals, we run GMM on the cumulant system of equations to obtain estimates for the distribution parameters but also their standard errors. A second generalization is that GMM can estimate an "overidentified" system, namely, a system where the number of equations is greater than the number of unknowns, something that usually lowers the variance of the estimator. In our case where the additional equations will be expressions for higher-order cumulants, overidentification works indirectly also as a specification test for the assumed distribution, since now we require that the data align with additional characteristics of this distribution.

Combined COLS/GMM estimation requires that after GMM we should still correct the estimated constant term $\widehat{\beta}_0$ and recompute the residuals, in order to calculate our 2TSF metrics. Its methodology is a truncated version of the next section, using only the orthogonality conditions that relate to the distribution moments.

7.1.4.2 Full GMM Estimation

We can also implement full GMM estimation, where we estimate jointly the regression coefficients and the distribution parameters. An advantage of joint GMM estimation is that we do not need to "correct" the obtained residuals, because the constant term of the regression (which is the reason for this needed correction in COLS and CNLS) is estimated properly from the beginning. GMM estimation can be used with a linear or a nonlinear regression function, but we will restrict our presentation here to the linear case.

The identifying restrictions at the population level in the case of linear regression and a non-zero mean error term are summarized by $\mathrm{Cov}(\boldsymbol{x}_i; \varepsilon_i) = 0$. It is convenient here to define the centered error:

$$\tilde{\varepsilon} \equiv \varepsilon - \mu_\varepsilon(\boldsymbol{\theta}) = y - \boldsymbol{x}'\boldsymbol{\beta} - (\beta_0 + \mu_\varepsilon(\boldsymbol{\theta})).$$

Then the population moment conditions to estimate the regression coefficients, including the constant term, are, in matrix notation, $E(\mathrm{X}'\tilde{\boldsymbol{\varepsilon}}) = \mathbf{0}$. To estimate the distribution parameters, we will use as in the case of COLS the cumulant relations $\kappa_2, \kappa_3, \kappa_4$ plus κ_5, to have an overidentifying restriction (remember that $\kappa_5(v) = 0$)

$$\kappa_5 = \mu_5 - 10\kappa_3\kappa_2, \qquad \kappa_5(\varepsilon) = \kappa_5(w) - \kappa_5(u) \implies \kappa_5(\varepsilon) - \varphi_5(\boldsymbol{\theta}) = 0.$$

Together with $E(\mathrm{X}'\tilde{\boldsymbol{\varepsilon}}) = \mathbf{0}$ the four cumulant relations formulate the full vector of orthogonality conditions, say $E[\boldsymbol{h}(\boldsymbol{q})] = \mathbf{0}$ for GMM estimation. To obtain its finite-sample incarnation, we proceed as follows: For $k = 1, \ldots, K + 1$ regressors (including the constant term), define

$$h_{k,i} = x_{ki}\tilde{\varepsilon}_i, \qquad h_{\beta,i} = (h_{1i}, \ldots, h_{K+1,i})'.$$

For the four moment conditions related to the distribution parameters, we have

$$h_{K+2,i} = \tilde{\varepsilon}_i^2 - \varphi_2(\boldsymbol{\theta}),$$
$$h_{K+3,i} = \tilde{\varepsilon}_i^3 - \varphi_3(\boldsymbol{\theta}),$$

$$h_{K+4,i} = \tilde{\varepsilon}_i^4 - 3\left(\frac{1}{n}\sum_{j=1}^{n}\tilde{\varepsilon}_j^2\right)\tilde{\varepsilon}_i^2 - \varphi_4(\boldsymbol{\theta}),$$

$$h_{K+5,i} = \tilde{\varepsilon}_i^5 - 10\left(\frac{1}{n}\sum_{j=1}^{n}\tilde{\varepsilon}_j^2\right)\tilde{\varepsilon}_i^3 - \varphi_5(\boldsymbol{\theta}).$$

Setting $\mathbf{h}_i = (h'_{\beta,i}, h_{K+2,i}, h_{K+3,i}, h_{K+4,i}, h_{K+5,i})'$, we have

$$\boldsymbol{h}(\boldsymbol{q}) = \frac{1}{n}\sum_{i=1}^{n}\mathbf{h}_i.$$

Note that the vector $\boldsymbol{\theta}$ exists also in $\tilde{\varepsilon}$ through $\mu_{\varepsilon}(\boldsymbol{\theta})$. As with the case of COLS and CNLS, the exact expression for the terms $\mu_{\varepsilon}(\boldsymbol{\theta})$, $\varphi_r(\boldsymbol{\theta})$, $r = 2, 3, 4, 5$ will depend on the statistical specification that we will adopt. Different software packages have different requirements as to how exactly one should input the above expressions in order to use the GMM estimation routine.[7]

Given the above notation, the GMM estimator is defined as

$$\hat{\boldsymbol{q}} = \arg\min\left\{n \cdot \boldsymbol{h}(\hat{\boldsymbol{q}})'\,\mathrm{W}\,\boldsymbol{h}(\hat{\boldsymbol{q}})\right\},$$

where W is a weight matrix. The optimal weight matrix that minimizes the asymptotic variance of the estimator can be consistently estimated by

$$\widehat{\mathrm{W}}^* = \left(\frac{1}{n}\sum_{i=1}^{n}\widehat{\mathbf{h}}_i\widehat{\mathbf{h}}_i'\right)^{-1}.$$

Nevertheless, evidence exists that in finite samples the asymptotic properties of GMM as regards efficiency do not materialize quickly and the "naive" estimator that uses the Identity matrix as the weight matrix, $\mathrm{W} = I_{K+5}$, often performs better than the GMM estimator with the estimated optimal weight matrix $\widehat{\mathrm{W}}^*$.

In practice, GMM estimation with overidentifying restrictions is implemented iteratively by software. In the first step, the objective function is minimized by using a weight matrix other than the optimal, with the only property that it is, or converges to, a positive definite matrix. Often, the Identity matrix is called for duty. Using the estimated parameters from this first step, we then proceed to compute $\widehat{\mathrm{W}}^*$ and re-minimize the objective function using it as the weight matrix. With the new set of estimates we re-compute $\widehat{\mathrm{W}}^*$ and so on, until convergence of the obtained estimates. The steps to convergence are carried out automatically by software.

[7] We provide `gretl` code for this in Chap. 11.

While the overidentified GMM estimator could be implemented as a single-step procedure, iteration is advised in order to achieve efficiency and sometimes convergence itself. And there are cases where iteration can safeguard against the opposite situation: a pseudo-result, where the one-step estimator appears to converge and produces estimation results, but if iterated, it fails to converge, revealing that the one-step results are not to be trusted.

We note that we multiplied the objective function with the sample size n, so that it has directly an asymptotic distribution. This will be needed in order to test the overidentifying restrictions, which in our framework corresponds to testing the specification and is presented in Sect. 7.3 later in this chapter.

7.2 Statistical Specifications

In this part of the chapter, we present all the necessary formulas related to different distributional specifications for the composite error term $\varepsilon = v + w - u$. These cover the alternative estimation methods we discussed in the first part of the chapter and are used to compute the various metrics we discussed in Chap. 2. We remind that the symbol "f" will denote the density of a distribution, while "F" will denote the (cumulative) distribution function, indexed appropriately. There is one exception: we will use the established symbols ϕ for the Standard Normal density and Φ for the Standard Normal distribution function, namely,

$$\phi(t) = \frac{1}{\sqrt{2\pi}} \exp\left\{\frac{-t^2}{2}\right\}, \qquad \Phi(t) = \int_{-\infty}^{t} \phi(t)\, dt.$$

While strictly speaking, there is not a closed-form mathematical expression for Φ, all software we are aware of offer it as a "special function" and compute it to adequate precision.

In all cases we assume

$$v \sim \mathrm{N}(0, \sigma_v), \quad f_v(v) = \frac{1}{\sigma_v\sqrt{2\pi}} \exp\left\{\frac{-v^2}{2\sigma_v^2}\right\}, \quad E(v) = 0, \quad \mathrm{Var}(v) = \sigma_v^2,$$

(v, w, u) are jointly statistically independent.

We start with the Normal-Exponential specification which is the industry standard in 2TSF studies (in contrast to the single-tier stochastic frontier model where the Normal-Half Normal specification is the dominant one).

7.2.1 Normal-Exponential (NE)

In this specification we assume that the two one-sided components each follow an Exponential distribution with scale parameter, namely,

$$w \sim \mathrm{Exp}(\sigma_w), \qquad u \sim \mathrm{Exp}(\sigma_u), \qquad f_t(t) = \frac{1}{\sigma_t} \exp\{-t/\sigma_t\}, \quad t = w, u.$$

To write compactly the composite error density, we define the symbols

$$a_1 = \frac{\sigma_v^2}{2\sigma_u^2} + \frac{\varepsilon}{\sigma_u}, \quad b_1 = -\left(\frac{\varepsilon}{\sigma_v} + \frac{\sigma_v}{\sigma_u}\right), \quad a_2 = \frac{\sigma_v^2}{2\sigma_w^2} - \frac{\varepsilon}{\sigma_w},$$
$$b_2 = \frac{\varepsilon}{\sigma_v} - \frac{\sigma_v}{\sigma_w}.$$

With these we have

$$f_\varepsilon(\varepsilon) = \frac{\exp\{a_1\}\,\Phi(b_1) \;+\; \exp\{a_2\}\,\Phi(b_2)}{\sigma_w + \sigma_u}. \tag{7.11}$$

The composite error distribution function is

$$F_\varepsilon(\varepsilon) = \Phi\left(\frac{\varepsilon}{\sigma_v}\right) + \frac{\sigma_u}{\sigma_w + \sigma_u}\exp\{a_1\}\,\Phi(b_1) \;-\; \frac{\sigma_w}{\sigma_w + \sigma_u}\exp\{a_2\}\,\Phi(b_2). \tag{7.12}$$

The distribution function can be used in order to estimate probabilities related to the composite error term, but also in a copula model when one wants to account for possible regressor endogeneity without using instrumental variables (see Sect. 14.1.4).

Maximum Likelihood Estimation The likelihood contribution of observation i and the log-likelihood of the sample are

$$\ell_i = -\ln(\sigma_w + \sigma_u) + \ln\Big(\exp\{a_{1i}\}\,\Phi(b_{1i}) \;+\; \exp\{a_{2i}\}\,\Phi(b_{2i})\Big), \tag{7.13}$$

$$\mathcal{L} = -n \cdot \ln(\sigma_w + \sigma_u) + \sum_{i=1}^{n} \ln\Big(\exp\{a_{1i}\}\,\Phi(b_{1i}) \;+\; \exp\{a_{2i}\}\,\Phi(b_{2i})\Big). \tag{7.14}$$

COLS, CNLS, and GMM Estimation For these estimation methods, we need the expressions of various moments/cumulants of the composite error term. We have

$$\varphi_1 = \mu_\varepsilon = \sigma_w - \sigma_u \tag{7.15}$$

$$\varphi_2 = \kappa_2(\varepsilon) = \sigma_v^2 + \sigma_w^2 + \sigma_u^2 \tag{7.16}$$

$$\varphi_3 = \kappa_3(\varepsilon) = \kappa_3(w) - \kappa_3(u) = 2\sigma_w^3 - 2\sigma_u^3 \tag{7.17}$$

$$\varphi_4 = \kappa_4(\varepsilon) = \kappa_4(w) + \kappa_4(u) = 6\sigma_w^4 + 6\sigma_u^4 \tag{7.18}$$

$$\varphi_5 = \kappa_5(\varepsilon) = \kappa_5(w) - \kappa_5(u) = 24\sigma_w^5 - 24\sigma_u^5. \tag{7.19}$$

For COLS and CNLS estimation, the above will be equated to the corresponding sample moments of the residuals from estimation, as described earlier (and we stress again that for further inference one should re-compute the residuals after correcting for the constant term using the sigma estimates from the above system). For GMM estimation, these expressions will be used to form the elements of the $\mathbf{h}_i$ vector and eventually of $\boldsymbol{h}(\boldsymbol{q})$. There is no need to provide mathematical expressions for the weight matrix, since it will be computed automatically by software. For all estimation methods, this holds true for the variance-covariance matrix as well.

Sample-Level Metrics The expected values needed for the sample-level metrics of Chap. 2 are

$$E(w) = \sigma_w \qquad E(u) = \sigma_u$$

$$E\left(e^w\right) = \frac{1}{1-\sigma_w} \qquad E\left(e^u\right) = \frac{1}{1-\sigma_u}$$

$$E\left(e^{-w}\right) = \frac{1}{1+\sigma_w} \qquad E\left(e^{-u}\right) = \frac{1}{1+\sigma_u}$$

$$E\left(e^{w-u}\right) = \frac{1}{(1-\sigma_w)(1+\sigma_u)} \qquad E\left(e^{u-w}\right) = \frac{1}{(1+\sigma_w)(1-\sigma_u)}.$$

All the above must be positive and are not defined if the estimated sigmas are such that they become negative. In practice for some of them, this requires that either $\sigma_w < 1$ or $\sigma_u < 1$. If this is not the case, the corresponding expected values do not formally exist (they are not finite), so they cannot be used. In practice, the related formulas would turn negative if the scale parameter involved is greater than unity, which is meaningless. Essentially their derivation would have stopped abruptly because we would encounter an infinite integral in the intermediate computations (since expected values are definite integrals). From the point of view of statistics, if, say, $\sigma_w > 1$, it would mean that the distribution of the random variable e^w has "heavy/fat tails" and its moments do not exist. In our specific case, if w is Exponential with scale parameter σ_w, then e^w follows a Pareto distribution with shape parameter $1/\sigma_w$. Suppose $\sigma_w > 1$. Then the Pareto random variable e^w has shape parameter *smaller* than unity, and so the mean $E(e^w)$ does not exist.

Individual Metrics To express the conditional expected values that are used in computing individual metrics, we first define

$$\lambda = \sigma_w^{-1} + \sigma_u^{-1}, \quad \chi_{1i} = \Phi(b_{2i}) + \exp\{a_{1i} - a_{2i}\}\Phi(b_{1i}),$$
$$\chi_{2i} = \exp\{a_{2i} - a_{1i}\}\chi_{1i}.$$

Conditional expectations are obtained by the basic formula

$$E(g(t)|\varepsilon) = \int_{S_{t|\varepsilon}} g(t) \cdot f_t(t|\varepsilon)\,dt, \qquad t = w, u.$$

Here $g(t)$ is any function of t and $S_{t|\varepsilon}$ is the support of t conditional on ε. For reference purposes, we present next the conditional densities $f(t|\varepsilon)$, $t = w, u$:

$$f(w_i|\varepsilon_i) = \frac{\lambda \exp\{-\lambda w_i\}\Phi(w_i/\sigma_v + b_{1i})}{\chi_{2i}}, \tag{7.20}$$

$$f(u_i|\varepsilon_i) = \frac{\lambda \exp\{-\lambda u_i\}\Phi(u_i/\sigma_v + b_{2i})}{\chi_{1i}}. \tag{7.21}$$

Carrying out the integrations, we have

$$E(w_i|\varepsilon_i) = \frac{1}{\lambda} + \frac{\sigma_v[\phi(b_{2i}) + b_{2i}\Phi(b_{2i})]}{\chi_{1i}}, \tag{7.22}$$

$$E(e^{w_i}|\varepsilon_i) = \frac{\lambda}{(\lambda - 1)\chi_{2i}}\left[\Phi(b_{1i}) + \exp\left\{\frac{1}{2}\left[(b_{2i} + \sigma_v)^2 - b_{1i}^2\right]\right\}\Phi(b_{2i} + \sigma_v)\right], \tag{7.23}$$

$$E(e^{-w_i}|\varepsilon_i) = \frac{\lambda}{(1 + \lambda)\chi_{2i}}\left[\Phi(b_{1i}) + \exp\left\{a_{2i} - a_{1i} - b_{2i}\sigma_v + 0.5\sigma_v^2\right\}\Phi(b_{2i} - \sigma_v)\right], \tag{7.24}$$

and

$$E(u_i|\varepsilon_i) = \frac{1}{\lambda} + \frac{\sigma_v[\phi(b_{1i}) + b_{1i}\Phi(b_{1i})]}{\chi_{2i}}, \tag{7.25}$$

$$E(e^{u_i}|\varepsilon_i) = \frac{\lambda}{(\lambda - 1)\chi_{1i}}\left[\Phi(b_{2i}) + \exp\left\{\frac{1}{2}\left[(b_{1i} + \sigma_v)^2 - b_{2i}^2\right]\right\}\Phi(b_{1i} + \sigma_v)\right], \tag{7.26}$$

$$E\left(e^{-u_i}\,|\varepsilon_i\right) = \frac{\lambda}{(1+\lambda)\,\chi_{1i}}\left[\Phi\left(b_{2i}\right)+\exp\left\{a_{1i}-a_{2i}-b_{1i}\sigma_v+0.5\sigma_v^2\right\}\Phi\left(b_{1i}-\sigma_v\right)\right]. \tag{7.27}$$

$E(e^w|\varepsilon)$ and $E(e^u|\varepsilon)$ are defined only if $\lambda > 1$, as should be obvious from the corresponding expressions above, since if we had $\lambda < 1$ they would turn negative, which is meaningless for these non-negative variables, or become infinity (for $\lambda = 1$). This restriction on the scale parameters is milder than the one encountered in relation to the unconditional expected values. It suffices that at least one of σ_w or σ_u is smaller than unity to have $\lambda > 1$, and then, the other parameter could be higher than unity. The case, say, $\sigma_u < 1 < \sigma_w$, is where the conditional expectation $E(e^w|\varepsilon)$ exists while the corresponding unconditional one, $E(e^w)$, does not.

Moreover, even if both are greater than unity (and so *both* unconditional expected values do not exist), we will still have $\lambda > 1$ if $1 < \sigma_w, \sigma_u < 2$ (because then, each reciprocal will be higher than $1/2$ and so their sum will be higher than unity), and the conditional expectations will be defined. Empirical practice indicates that, when the dependent variable is in logarithms (which is the case where we need the exponentiated conditional expectations), neither σ_w nor σ_u are estimated as being higher than 2 (the authors certainly have never encountered such a case), so we expect that the condition $\lambda > 1$ will hold and it will be legitimate to use all of the above expressions.

Finally, we turn to the difference $z = w - u$. Since here w and u are assumed Exponential and independent, their difference is an asymmetric Laplace distribution, with density

$$f_z(z) = \frac{1}{\sigma_w+\sigma_u}\cdot\begin{cases} e^{z/\sigma_u} & z \le 0 \\ e^{-z/\sigma_w} & z > 0. \end{cases} \tag{7.28}$$

Using this and the previous results, we obtain

$$E\left(e^{w_i-u_i}\,|\varepsilon_i\right) = \frac{\exp\left\{(1+\sigma_u)\left(a_{1i}+\frac{\sigma_v^2}{2\sigma_u}\right)\right\}\Phi\left(b_{1i}-\sigma_v\right)+\exp\left\{(1-\sigma_w)\left(a_{2i}-\frac{\sigma_v^2}{2\sigma_w}\right)\right\}\Phi\left(b_{2i}+\sigma_v\right)}{\exp\{a_{1i}\}\Phi\left(b_{1i}\right)+\exp\left\{a_{2i}\right\}\Phi\left(b_{2i}\right)}, \tag{7.29}$$

and

$$E\left(e^{u_i-w_i}\,|\varepsilon_i\right) = \frac{\exp\left\{(1-\sigma_u)\left(a_{1i}-\frac{\sigma_v^2}{2\sigma_u}\right)\right\}\Phi\left(b_{1i}+\sigma_v\right)+\exp\left\{(1+\sigma_w)\left(a_{2i}+\frac{\sigma_v^2}{2\sigma_w}\right)\right\}\Phi\left(b_{2i}-\sigma_v\right)}{\exp\{a_{1i}\}\Phi\left(b_{1i}\right)+\exp\{a_{2i}\}\Phi\left(b_{2i}\right)}. \tag{7.30}$$

We remind the reader that even though (w, u) are statistically independent, conditional on ε, they stop being so; therefore, the expected value $E\left(e^{w-u}|\varepsilon\right)$ is not equal to the product of the two separate conditional expectations. It is important for applied researchers to bear this in mind when engaging in empirical work.

7.2.2 *Normal-Half Normal (NHN)*

In this specification we assume that the two non-negative components $w \geq 0,\ u \geq 0$ each follow a Half Normal distribution, namely,

$$w \sim \mathrm{HN}(\sigma_w), \qquad u \sim \mathrm{HN}(\sigma_u), \qquad f_t(t) = \frac{2}{\sigma_t\sqrt{2\pi}}\exp\left\{\frac{-t^2}{2\sigma_t^2}\right\}, \quad t = w, u.$$

To write compactly the composite error density, we need to define certain composite coefficients and shorthands (necessarily re-using some symbols that were used earlier in the chapter with different content):

$$\theta_1 \equiv \frac{\sigma_w}{\sigma_v}, \quad \theta_2 \equiv \frac{\sigma_u}{\sigma_v}, \quad s \equiv \sqrt{\sigma_v^2 + \sigma_w^2 + \sigma_u^2} = \sigma_v\sqrt{1+\theta_1^2+\theta_2^2},$$

$$\omega_1 \equiv \frac{s\sqrt{1+\theta_2^2}}{\theta_1}, \quad \omega_2 \equiv \frac{s\sqrt{1+\theta_1^2}}{\theta_2}, \quad \lambda_1 \equiv \frac{\theta_2}{\theta_1}\sqrt{1+\theta_1^2+\theta_2^2},$$

$$\lambda_2 \equiv \frac{\theta_1}{\theta_2}\sqrt{1+\theta_1^2+\theta_2^2}.$$

Also, let $\Phi_2(x_1, x_2; \rho)$ denote the bivariate Standard Normal distribution function of random variables (x_1, x_2) with correlation coefficient ρ. Then the density of the composite error is

$$f_\varepsilon(\varepsilon) = \frac{4}{s}\phi\left(\varepsilon/s\right)\cdot \mathcal{D}, \tag{7.31}$$

$$\mathcal{D} = \left[\Phi_2\left(\frac{\varepsilon}{\omega_1}, 0; \rho = \frac{\lambda_1}{\sqrt{1+\lambda_1^2}}\right) - \Phi_2\left(\frac{\varepsilon}{\omega_2}, 0; \rho = \frac{-\lambda_2}{\sqrt{1+\lambda_2^2}}\right)\right].$$

Due to the presence of $\mathcal{D}$, the distribution function of the composite error has no closed-form expression, and using the NHN error term in 2TSF models that require the use of the CDF would add an additional layer of computations (by numerical integration or simulation). But in a model that requires only the density, the $\Phi_2(x_1, x_2; \rho)$ bivariate integral is routinely computed by software packages, so we can proceed with maximum likelihood estimation.

Maximum Likelihood Estimation The likelihood contribution of observation i and the log-likelihood of the sample are

$$\ell_i = \ln\left(\frac{2\sqrt{2}}{\sqrt{\pi}}\right) - \ln s - \frac{\varepsilon_i^2}{2s^2} + \ln \mathcal{D}_i, \tag{7.32}$$

$$\mathcal{L} = n\ln\left(\frac{2\sqrt{2}}{\sqrt{\pi}}\right) - n\ln s - \frac{1}{2s^2}\sum_{i=1}^{n}\varepsilon_i^2 + \sum_{i=1}^{n}\ln \mathcal{D}_i. \tag{7.33}$$

COLS, CNLS, and GMM Estimation The various moments/cumulants of the composite error term here are

$$\varphi_1 = \mu_\varepsilon = (\sigma_w - \sigma_u)\sqrt{\frac{2}{\pi}} \tag{7.34}$$

$$\varphi_2 = \kappa_2(\varepsilon) = \sigma_v^2 + (\sigma_w^2 + \sigma_u^2)\left(1 - \frac{2}{\pi}\right) \tag{7.35}$$

$$\varphi_3 = \kappa_3(\varepsilon) = \kappa_3(w) - \kappa_3(u) = (\sigma_w^3 - \sigma_u^3)\frac{(4-\pi)\sqrt{2}}{\pi^{3/2}} \tag{7.36}$$

$$\varphi_4 = \kappa_4(\varepsilon) = \kappa_4(w) + \kappa_4(u) = (\sigma_w^4 + \sigma_u^4)\frac{8(\pi - 3)}{\pi^2} \tag{7.37}$$

$$\varphi_5 = \kappa_5(\varepsilon) = \kappa_5(w) - \kappa_5(u) = (\sigma_w^5 - \sigma_u^5)\frac{\sqrt{2}}{\pi^{5/2}}(96 - 40\pi + 3\pi^2). \tag{7.38}$$

We note that the fifth cumulant of the Half Normal distribution is a negative number.

Sample-Level Metrics The expected values needed for the sample-level metrics of Chap. 2 are here

$$E(w) = \sigma_w\sqrt{\frac{2}{\pi}} \qquad E(u) = \sigma_u\sqrt{\frac{2}{\pi}}$$

$$E\left(e^{w}\right) = \frac{\Phi(\sigma_w)}{\phi(\sigma_w)}\sqrt{\frac{2}{\pi}} \qquad E\left(e^{u}\right) = \frac{\Phi(\sigma_u)}{\phi(\sigma_u)}\sqrt{\frac{2}{\pi}}$$

$$E\left(e^{-w}\right) = \frac{\Phi(-\sigma_w)}{\phi(\sigma_w)}\sqrt{\frac{2}{\pi}} \qquad E\left(e^{-u}\right) = \frac{\Phi(-\sigma_u)}{\phi(\sigma_u)}\sqrt{\frac{2}{\pi}}$$

$$E\left(e^{w-u}\right) = \frac{\Phi(\sigma_w)}{\phi(\sigma_w)}\frac{\Phi(-\sigma_u)}{\phi(\sigma_u)}\frac{2}{\pi} \qquad E\left(e^{u-w}\right) = \frac{\Phi(\sigma_u)}{\phi(\sigma_u)}\frac{\Phi(-\sigma_w)}{\phi(\sigma_w)}\frac{2}{\pi}.$$

Individual Metrics To express the conditional expected values for the individual metrics, we first introduce expressions that involve the Skew Normal distribution function

$$F_{1i} \equiv F_1(\varepsilon_i; 0, \omega_1, -\lambda_1) = \int_{-\infty}^{\varepsilon_i} \frac{2}{\omega_1}\phi(t/\omega_1)\,\Phi(-\lambda_1 t/\omega_1)\,dt$$

$$= 2\Phi_2\left(\frac{\varepsilon}{\omega_1}, 0; \rho = \frac{\lambda_1}{\sqrt{1+\lambda_1^2}}\right)$$

$$F_{2i} \equiv F_2(\varepsilon_i; 0, \omega_2, \lambda_2) = \int_{-\infty}^{\varepsilon_i} \frac{2}{\omega_2}\phi(t/\omega_2)\,\Phi(\lambda_2 t/\omega_2)\,dt$$

$$= 2\Phi_2\left(\frac{\varepsilon}{\omega_2}, 0; \rho = \frac{-\lambda_2}{\sqrt{1+\lambda_2^2}}\right),$$

while the corresponding densities are

$$f_{1i} = \frac{2}{\omega_1}\phi(\varepsilon_i/\omega_1)\,\Phi(-\lambda_1\varepsilon_i/\omega_1), \qquad f_{2i} = \frac{2}{\omega_2}\phi(\varepsilon_i/\omega_2)\,\Phi(\lambda_2\varepsilon_i/\omega_2).$$

We compact further by setting

$$\psi_{1i} \equiv \frac{f_{1i}}{F_{1i} - F_{2i}}, \qquad \psi_{2i} \equiv \frac{f_{2i}}{F_{1i} - F_{2i}}, \qquad \psi_i = \psi_{1i} - \psi_{2i} = \frac{f_{1i} - f_{2i}}{F_{1i} - F_{2i}},$$

and

$$s_1 = \sqrt{\sigma_w^2 + \sigma_v^2}, \qquad s_2 = \sqrt{\sigma_v^2 + \sigma_u^2}, \qquad \omega_w \equiv \frac{\sigma_w s_2}{s}, \qquad \omega_u \equiv \frac{\sigma_u s_1}{s}.$$

With these, the conditional densities are

$$f_{w|\varepsilon}(w_i|\varepsilon_i) = (F_{1i} - F_{2i})^{-1}\frac{2}{\omega_w}\phi\left(\frac{w_i}{\omega_w} - \frac{\varepsilon_i}{\omega_1}\right)\Phi\left(\lambda_1\frac{(w_i - \varepsilon_i)}{\omega_1}\right), \tag{7.39}$$

$$f_{u|\varepsilon}(u_i|\varepsilon_i) = (F_{1i} - F_{2i})^{-1}\frac{2}{\omega_u}\phi\left(\frac{u_i}{\omega_u} + \frac{\varepsilon_i}{\omega_2}\right)\Phi\left(\lambda_2\frac{(u_i + \varepsilon_i)}{\omega_2}\right). \tag{7.40}$$

The conditional expected values that we need are then

$$E(w_i|\varepsilon_i) = s^2\psi_{1i} + \frac{\sigma_w^2}{s^2}\left(\varepsilon_i - s^2\psi_i\right), \tag{7.41}$$

$$E\left(e^{w_i}|\varepsilon_i\right) = 2(F_{1i} - F_{2i})^{-1}\exp\left\{\frac{1}{2}\omega_w^2 + \frac{\omega_w}{\omega_1}\varepsilon_i\right\}\left[\Phi\left(-\frac{\left(\varepsilon_i - \sigma_w^2\right)}{\omega_2}\right)\right. \tag{7.42}$$

$$\left. - \Phi_2\left(-\frac{\left(\varepsilon_i - \sigma_w^2\right)}{\omega_2}, -\left(\omega_w + \frac{\varepsilon_i}{\omega_1}\right); \rho = \frac{-\sigma_w\sigma_u}{s_1 s_2}\right)\right],$$

$$E\left(e^{-w_i}|\varepsilon_i\right) = 2(F_{1i} - F_{2i})^{-1}\exp\left\{\frac{1}{2}\omega_w^2 - \frac{\omega_w}{\omega_1}\varepsilon_i\right\}\left[\Phi\left(-\frac{\left(\varepsilon_i + \sigma_w^2\right)}{\omega_2}\right)\right. \tag{7.43}$$

$$\left. - \Phi_2\left(-\frac{\left(\varepsilon_i + \sigma_w^2\right)}{\omega_2}, \omega_w - \frac{\varepsilon_i}{\omega_1}; \rho = \frac{-\sigma_w\sigma_u}{s_1 s_2}\right)\right],$$

and

$$E\left(u_i|\varepsilon_i\right) = s^2\psi_{2i} - \frac{\sigma_u^2}{s^2}\left(\varepsilon_i - s^2\psi_i\right), \tag{7.44}$$

$$E\left(e^{u_i}|\varepsilon_i\right) = 2(F_{1i} - F_{2i})^{-1}\exp\left\{\frac{1}{2}\omega_u^2 - \frac{\omega_u}{\omega_2}\varepsilon_i\right\}\left[\Phi\left(\frac{\varepsilon_i + \sigma_u^2}{\omega_1}\right)\right. \tag{7.45}$$

$$\left. - \Phi_2\left(\frac{\varepsilon_i + \sigma_u^2}{\omega_1}, \frac{\varepsilon_i}{\omega_2} - \omega_u; \rho = \frac{-\sigma_w\sigma_u}{s_1 s_2}\right)\right].$$

$$E\left(e^{-u_i}|\varepsilon_i\right) = 2(F_{1i} - F_{2i})^{-1}\exp\left\{\frac{1}{2}\omega_u^2 + \frac{\omega_u}{\omega_2}\varepsilon_i\right\}\left[\Phi\left(\frac{\varepsilon_i - \sigma_u^2}{\omega_1}\right)\right. \tag{7.46}$$

$$\left. - \Phi_2\left(\frac{\varepsilon_i - \sigma_u^2}{\omega_1}, \frac{\varepsilon_i}{\omega_2} + \omega_u; \rho = \frac{-\sigma_w\sigma_u}{s_1 s_2}\right)\right].$$

Finally, we turn to the difference $z = w - u$. The marginal distribution for z is

$$f_z(z) = \frac{4}{s_h}\phi\left(z/s_h\right) \times \begin{cases} \Phi\left(\frac{\sigma_w}{\sigma_u}\left(z/s_h\right)\right) & z \le 0 \\ \Phi\left(-\frac{\sigma_u}{\sigma_w}\left(z/s_h\right)\right) & z > 0 \end{cases} \qquad s_h = \sqrt{\sigma_w^2 + \sigma_u^2} \tag{7.47}$$

With this and previous results, we obtain

$$
\begin{aligned}
E\left(\exp\left\{w_i - u_i\right\}|\varepsilon_i\right) = \exp\left\{\frac{\left(\sigma_w^2+\sigma_u^2\right)}{s^2}\left(\frac{\sigma_v^2}{2}+\varepsilon\right)\right\} \\
\times \mathcal{D}_i^{-1}\cdot\left[\Phi_2\left(\frac{\varepsilon_i+\sigma_v^2}{\omega_1}, 0; \rho=\frac{\lambda_1}{\sqrt{1+\lambda_1^2}}\right)\right. \\
\left. -\Phi_2\left(\frac{\varepsilon_i+\sigma_v^2}{\omega_2}, 0; \rho=\frac{-\lambda_2}{\sqrt{1+\lambda_2^2}}\right)\right],
\end{aligned}
\tag{7.48}
$$

and

$$
\begin{aligned}
E\left(\exp\left\{u_i - w_i\right\}|\varepsilon_i\right) = \exp\left\{\frac{\left(\sigma_w^2+\sigma_u^2\right)}{s^2}\left(\frac{\sigma_v^2}{2}-\varepsilon\right)\right\} \\
\times \mathcal{D}_i^{-1}\cdot\left[\Phi_2\left(\frac{\varepsilon_i-\sigma_v^2}{\omega_1}, 0; \rho=\frac{\lambda_1}{\sqrt{1+\lambda_1^2}}\right)\right. \\
\left. -\Phi_2\left(\frac{\varepsilon_i-\sigma_v^2}{\omega_2}, 0; \rho=\frac{-\lambda_2}{\sqrt{1+\lambda_2^2}}\right)\right].
\end{aligned}
\tag{7.49}
$$

In Chap. 14, we discuss briefly, with pointers to the literature, additional statistical specifications. But the NE and NHN models are currently the two main ones.

7.3 Model Selection

By "model selection" in this section, we refer neither to the selection of the theoretical model (which will provide the core explanatory variables for our regressor matrix) nor to the selection of the auxiliary control variables that are used in econometric models, to remove various sources of unobserved heterogeneity that are not of direct interest. Our focus is the determination of the distributional specification on the composite error term that matches better the data at hand. Forensic diagnostics depend also on the estimation method used.

As a general first taste, we can run an OLS regression and examine the residuals: if they are characterized by positive excess kurtosis, and possibly also by skewness, we have indicative support that a 2TSF model can match the data.

7.3.1 Forensic Diagnostics

Maximum Likelihood When we estimate by MLE parameters that should be positive, we impose this restriction on the estimation algorithm. This is because the search of the parameter space conducted by the MLE algorithm may be a bit wild before it converges to some extremum, which may not be unique, in which case a local extremum may reside in a forbidden subset of the parameter space. Restricting the scale parameters to be positive as they should keeps the algorithm away from regions where a mathematical solution may exist but is nonsensical and useless in terms of statistical interpretation. If, then, a parameter is estimated as being virtually zero, this means that the algorithm tries stubbornly to go to the "other side of zero," indicating that the only solution lies in the forbidden subspace, and this is evidence that the specified distribution and likelihood is not a good match for the data at hand.

Suppose now that what is estimated as "virtually zero" is σ_v, the scale parameter of the noise term. One could think that in this specific case there is no need to re-specify the likelihood, since what we are mainly interested in is the two opposing latent forces. But this is wrong: the asymptotic theory for the maximum likelihood estimator when a parameter is estimated at the boundary of the feasible space becomes much more complicated which translates into complicated testing procedures. So even if the estimates themselves may not differ much, the standard errors which are obtained are reliable only in a specification where the noise term has been set a priori equal to zero and eliminated from the likelihood before estimation. We will present such distributional specifications in the "Extensions" part of the book, Chap. 12.

Information criteria like the AIC and BIC are also useful indicators to choose between distributions that perform comparably. Care must be taken in how each software computes them exactly and how one should interpret larger or smaller values.

COLS-CNLS-GMM These methods are based on solving a system of nonlinear equations. If we obtain no solution, or "wrong signs" (e.g., a negative value for a scale parameter), it is forensic evidence that the specification is not well suited for the data at hand. The same comments and suggestions apply here too if a scale parameter is estimated as virtually zero.

Suppose now that we obtain multiple admissible solutions. Then a way to choose among the candidate vectors of estimates is to compute the sample-likelihood values at each vector given the data and choose the one with the higher value. Likelihood values can be computed even if we did not use maximum likelihood estimation, because in all cases we need to make a distributional assumption, which implies a specific likelihood.

7.3.2 *Formal Statistical Tests*

Doornik-Hansen Test

In order to formally test the assumed specification for the composite error term, we can use the residuals and the estimated parameters to test whether the assumed distribution function of the error term $F_\varepsilon(\varepsilon)$ behaves like a Uniform(0,1) when viewed as a random variable. But this is equivalent to test whether $\Phi^{-1}(F_\varepsilon(\varepsilon))$ is distributed as Standard Normal. Among the many Normality tests that exist, we suggest the Doornik and Hansen (2008) test using $\Phi^{-1}(F_\varepsilon(\hat{\varepsilon}))$ as the series to be tested. The test statistic is rather complicated, and it is based on the sample skewness and kurtosis of this series, using transformations that almost eliminate their correlation (correlation that exists initially due to them being estimated quantities). It is asymptotically distributed as χ^2_2. The test is usually part of the menu of specification tests offered by most software; hence there will be no need to be coded.

If the assumed distribution function of the error term is not available in closed-form, it can be numerically evaluated by Gauss-Laguerre quadrature after the following transformation:

$$F_\varepsilon(\varepsilon_i) = \int_{-\infty}^{\varepsilon_i} f_\varepsilon(t)dt = 1 - \int_{\varepsilon_i}^{\infty} f_\varepsilon(t)dt = 1 - \int_0^{\infty} e^{-t} \cdot [e^t f_\varepsilon(t + \varepsilon_i)]dt. \tag{7.50}$$

In practice, we just need to choose the number m of node-weight (k_j, ξ_j) pairs for Gauss-Laguerre quadrature. The software will provide these pairs, and it will compute for each observation i

$$F_\varepsilon(\widehat{\varepsilon}_i) \approx 1 - \sum_{j=1}^{m} \xi_j e^{k_j} f_\varepsilon(\kappa_j + \widehat{\varepsilon}_i). \tag{7.51}$$

The resulting series will be the input for the Doornik-Hansen specification test.

GMM: Testing the Specification Through the Overidentifying Restrictions

Hansen's test of overidentifying restrictions is based on the fact that under correct specification:

$$J = n \cdot \boldsymbol{h}(\hat{\boldsymbol{q}})' \widehat{\mathrm{W}}^* \boldsymbol{h}(\hat{\boldsymbol{q}}) \rightarrow_d \chi^2_1.$$

The asymptotic chi-square distribution has one degree of freedom because the orthogonality conditions are $K + 5$, while the parameters under estimation are $K + 4$. In practice, after estimation we compute the value of the objective function (including the multiplication by the sample size n), and we compare it with the tail value of a chi-square distribution with one degree of freedom at the desired confidence level. If the value we obtain is higher than the critical value, the test rejects the overidentifying restrictions, which amounts to rejecting the distributional assumptions we have made with the statistical confidence that we chose.

Non-nested Likelihood Ratio Test to Select a Distributional Specification
Vuong (1989) has developed simple likelihood ratio tests for "model selection" between two competing distributional specifications. We present here just the steps to execute this test, which belongs to the sub-category of "strictly non-nested models" (Section 5 of Vuong's paper). This test can be used in order to choose between the Normal-Exponential and the Normal-Half Normal specifications, for example.

We want to choose between distributional specifications A and B. We formulate a test where the null hypothesis is that the two are equivalent.

Step 1. Estimate using specification A, and compute the series of log-likelihood contributions $\{\ell_i^A\}$, $i = 1, \ldots, n$.

Step 2. Estimate using specification B, and compute the series of log-likelihood contributions $\{\ell_i^B\}$.
Most, if not all, software generate and save these series automatically. Otherwise, we can compute them using the expressions provided earlier, using the residuals and the estimated parameters.

Step 3. Compute the likelihood ratio (LR) sum

$$LR_n = \sum_{i=1}^{n} (\ell_i^A - \ell_i^B).$$

Note that LR_n can be positive or negative.

Step 4. Compute the variance term,

$$\hat{\omega}_n^2 = \frac{1}{n} \sum_{i=1}^{n} (\ell_i^A - \ell_i^B)^2 - \left[\frac{1}{n} \sum_{i=1}^{n} (\ell_i^A - \ell_i^B) \right]^2.$$

Step 5. We have the following results ($\hat{\omega}_n$ is the square root of $\hat{\omega}_n^2$):

$$(i)\ T_n \equiv n^{-1/2} \frac{LR_n}{\hat{\omega}_n} \longrightarrow_d \mathrm{N}(0, 1) \qquad \text{if Specifications A and B are equivalent}$$

$$(ii)\ T_n \equiv n^{-1/2} \frac{LR_n}{\hat{\omega}_n} \longrightarrow_{a.s} \infty \qquad \text{if Specification A is better than B}$$

$$(iii)\ T_n \equiv n^{-1/2} \frac{LR_n}{\hat{\omega}_n} \longrightarrow_{a.s} -\infty \qquad \text{if Specification A is worse than B.}$$

In practice, one chooses the "significance" level of the test α (which really is equal to one minus the statistical confidence level), say $\alpha = 0.1$, and obtains the standard Normal upper quantile $c = z_{1-\alpha/2}$. If $T_n > c$, the null hypothesis is rejected in favor of Specification A. If $T_n < -c$, the null hypothesis is rejected in favor of Specification B. If $T_n \leq |c|$, one cannot discriminate between the two specifications given the data.

One may think that we could execute this test irrespective of whether we have applied maximum likelihood estimation or not, since the distributional assumption on the composite error term must be made regardless of the estimation method, and so we always know the assumed log-likelihood. But this is not valid: while the likelihood as a mathematical expression will indeed be known no matter whether we apply MLE or not, in order to compute the actual values, we will have to use the estimates of the unknown parameters; the estimates we will obtain if we use, say, COLS, will not necessarily be the same as those that would maximize the likelihood by an ML algorithm, due to finite sample variability. So the test requires that we estimate by using maximum likelihood. We allowed such "mixing" of estimation methods earlier, but only to obtain an informal diagnostic, not to execute a formal statistical test.

7.4 Application: Detecting and Testing 2TSF Models

In the empirical illustration of this chapter, we want to show how we can support the use of the 2TSF model; compare results between the two benchmark statistical specifications, Normal-Exponential and Normal-Half Normal; and apply the likelihood ratio model selection test.

We use part of an earnings panel-data set from Cornwell and Rupert (1988), specifically the data related to 1982, taken from the USA Panel Study of Income Dynamics. The data concerns heads of households between the ages of 24 and 71 in 1982, which reported a positive wage in non-farm employment. Applying the 2TSF model here has the purpose of measuring the effects that information deficiency of employers and workers has on the wage, compared to the full-information ("balanced") theoretical value (see the discussion in Chap. 3).

We estimated the model by OLS and computed statistical tests related to the skewness and excess kurtosis of the residuals. We applied the Godfrey and Orme (1991) test for zero skewness, which is robust against non-Normality. It gave a test statistic of 0.01687. This statistic is distributed asymptotically as a chi-square with one degree of freedom (χ_1^2), and the p-value is 0.9. So there is no evidence of skewness in the error term. We performed next a zero excess-kurtosis test. Under the null hypothesis of Normality, the sampling distribution of the excess kurtosis coefficient follows a zero-mean Normal with variance equal to $24/n$, n being the size of the sample. We obtained a χ_1^2 test statistic 2.847 with p-value of 0.092. Zero excess-kurtosis cannot be rejected at the 99% and 95% confidence levels, but it is rejected with 90% statistical confidence: there are indications of non-zero excess kurtosis in the error term, which is the distinguishing feature of a 2TSF model in the presence of symmetry. In all, we have the case of an essentially symmetric error term together with evidence that the effects investigated by the 2TSF specifications are nevertheless there.

We proceeded to estimate the model using both the NE and the NHN error specification, by maximum likelihood. For the NHN specification, while the reparameterization used earlier is convenient for theoretical results, for the application we estimated the three original parameters σ_v, σ_w, and σ_u, treating the other symbols as composite coefficients in the software script. This has the advantage of providing directly the standard errors on the parameters of interest. Table 7.1 contains the estimation results from the OLS estimation as well as from the two specifications of the 2TSF model.

Diagnostics and Statistical Tests for Model Selection We see that the near-symmetry of the error term distribution carries over in the 2TSF models, as evidenced by the closeness of the estimated values of the scale parameters of the one-sided error components under each specification. We also see that the three models produce virtually identical results related to the regressor coefficient estimates. In addition, the standard deviation of the estimated composite error term is identical in the two 2TSF specifications and fairly close to the corresponding OLS estimate. In terms of the maximized value of the log-likelihood (assumed Normal for the OLS case), the three models appear very close. Applying the non-nested LR test for the NE and NHN specifications, we obtained a value for the statistic T_n of 0.8038, to be compared against a quantile of the Standard Normal distribution. For a conventional confidence level, say 0.9 (with the 0.1 remaining uncertainty to be allocated equally at the two sides of the distribution), we have $z_{0.95} = 1.645$. The value of T_n falls within the interval $[-1.645,\ 1.645]$ so this test says that we cannot discriminate between the two models given the data. This accords with the obtained estimates.

Metrics Per Model We now look at our metrics (Table 7.2). Given that we examine information deficiencies, the core metrics here are M_5 and M_6: M_5 measures how the incomplete information of the "buyer" (employer) tends to push the wage above the full-information (balanced) outcome, while M_6 measures how the incomplete information of the "seller" (worker) tends to push the wage below the balanced level. We also compute M_7, the net effect.

We observe that under the Normal-Half Normal specification, these effects are estimated as being visibly stronger compared to their estimates under the Normal-Exponential specification, but that the net average effect on the transactions ends up virtually identical and near zero, something to be expected since the error density appears to be symmetric around zero.

Turning to individual metrics, we provide in Table 7.3 basic statistics of the net effect series $\{M_{7i}\}$.

Although the two densities have very similar location parameters, they differ in their scale and shape properties: under the NE specification, the standard deviation is 0.122, while under the NHN model, it is 0.174, 43% higher. Also, skewness and excess kurtosis differ markedly. To obtain a more wholesome picture, we show in Fig. 7.1 the kernel density estimates for the net effect series under the two models.

Here it is clear how the two specifications may differ: the distribution of the net effect for the NE model is much more concentrated and peaked than the

Table 7.1 Log wage specification for 1982, $n = 595$

		2TSF	
	OLS	Exponential	Half Normal
Constant	5.5901	5.5990	5.6103
	(0.1901)	(0.1844)	(0.19)
EXP	0.0294	0.0290	0.0293
	(0.0065)	(0.0065)	(0.0065)
EXP2	−0.0005	−0.0005	−0.0005
	(0.0001)	(0.0001)	(0.0001)
WKS	0.0034	0.0038	0.0036
	(0.0027)	(0.0025)	(0.0025)
OCC	−0.1615	−0.1649	−0.1639
	(0.0369)	(0.037)	(0.0373)
IND	0.0847	0.0839	0.0833
	(0.0292)	(0.0297)	(0.0306)
SOUTH	−0.0588	−0.0604	−0.0587
	(0.0309)	(0.0298)	(0.0301)
SMSA	0.1662	0.1659	0.1657
	(0.0296)	(0.0292)	(0.0296)
MS	0.0952	0.0943	0.0939
	(0.0489)	(0.0458)	(0.0469)
FEM	−0.3246	−0.3218	−0.3275
	(0.0607)	(0.0666)	(0.0679)
UNION	0.1063	0.1093	0.1061
	(0.0317)	(0.033)	(0.0336)
ED	0.0572	0.0561	0.0566
	(0.0066)	(0.0062)	(0.0062)
BLK	−0.1904	−0.1894	−0.1893
	(0.0544)	(0.0537)	(0.0551)
σ_v	–	0.2590	0.2198
		(0.0327)	(0.0951)
σ_w	–	0.1316	0.2637
		(0.0357)	(0.1116)
σ_u	–	0.1388	0.2876
		(0.0399)	(0.1040)
σ_ε	0.3256	0.3219	0.3219
$\ln \mathcal{L}$	−169.98	−168.77	−169.44
Residual skewness	−0.017	−0.015	−0.021
Residual ex. kurtosis	0.339	0.352	0.342

Asymptotic standard errors in parentheses. Description of variables: *LWAGE* logarithm of wage, *EXP* years of full-time work experience, *EXP2* square of EXP, *WKS* weeks worked, *OCC* 1 if occupation is blue-collar, *IND* 1 if in manufacturing; SOUTH, 1 if in the South, *SMSA* 1 if in a standard metropolitan statistical area, *MS* 1 if married, *FEM* 1 if female, *UNION* 1 if wage is set by a union contract, *ED* years of education, *BLK* 1 if black

Table 7.2 Information inefficiency metrics. Sample level

Metric	Formula	NE	NHN
M_5	$E(\exp\{w\}) - 1$	0.151	0.251
M_6	$1 - E(\exp\{-u\})$	0.122	0.194
M_7	$E(\exp\{w\}) \cdot E(\exp\{-u\}) - 1$	0.011	0.008

Table 7.3 Net effect M_7 individual series. Basic statistics

Statistic	NE	NHN
Mean	0.0112	0.0084
Median	0.0074	0.0029
St.Dev	0.1224	0.1748
Skewness	1.6229	0.7151
Excess Kurtosis	10.881	2.2464
Interquantile Range	0.1273	0.2270

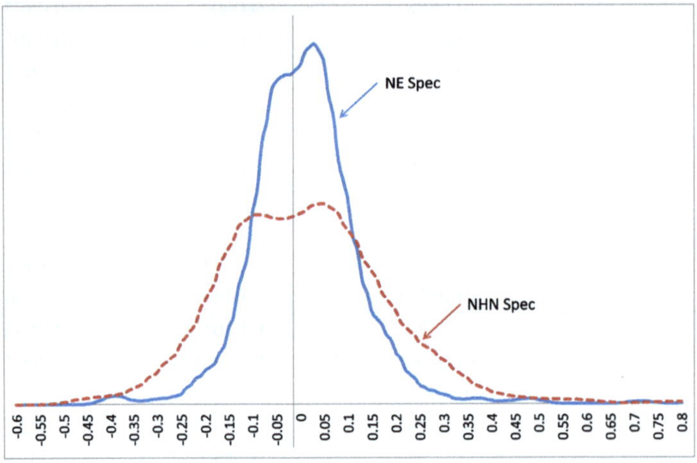

Fig. 7.1 Kernel density estimation of net effect of informational deficiencies on full-information wage

corresponding distribution for the NHN one. Nevertheless, this difference does not affect the *ranking* of the individuals involved in these transactions (or the ranking of each transaction). Two thirds of the observations change rank, but most move up or down just one position, and only one observation changed eight places from one specification to the other, which still is negligible given the size of the sample: to wit, Spearman's rank correlation coefficient between the rankings from the two specifications was calculated as 99.99%.

With this we conclude this empirical illustration and indeed this central chapter of our book. In the next chapters, we will abandon some of the assumptions under which the material here was derived, introducing heterogeneity, panel data estimation, and intra-error dependence.

Chapter 8
Incorporating Heterogeneity in the 2TSF Model

8.1 The Importance of Allowing Heterogeneity

There are a variety of reasons to account for heterogeneity in social science settings. Consider the canonical 2TSF model that we have discussed so far. This model assumes that the one-sided latent forces are iid ("identically and independently distributed"). This is a well-known piece of statistical parlance, but what are its consequences? Well, from an interpretational point of view, it implies that the latent forces are random and cannot be further explained inside the model; they can only be measured as to their effect, not as to why this effect is what it is. But naturally we would want to examine questions like "does having greater access to information for the buyer lead to lower information deficiency?" And "greater access" could be approximated by a number of different variables that are available as data. For example, the existence of digitized and publicly available data on local utility networks could help a prospective house buyer in assessing whether there may be issues with electricity supply, sewage, or Internet access in the area. Including such kind of covariates in the model could help us determine whether it actually does affect information deficiency (in the sense that, availability of information is one thing; actually exploiting this information is another).

The ability to answer such questions greatly increases the value of such models. Without accounting for heterogeneity in latent forces, we can only measure their effects on the outcome and the distribution of these effects over the sample. While this may be enough as initial impetus for policy, it cannot provide more focused guidance as to what kind of policies should be designed and implemented, to, say, mitigate information deficiency in the market, for example. Note that this is a different kind of "heterogeneity" than the one traditionally entertained when we have panel data available and which we take on in Chap. 9. Here we want to find ways to introduce heterogeneity in cross-sectional models.

A. Papadopoulos, C. F. Parmeter, *Two-Tier Stochastic Frontier Analysis for the Social Sciences*, https://doi.org/10.1007/978-3-031-81513-3_8

The setup that we have presented in Chap. 7 does not allow this on its own. But suppose we do have these ideas about certain factors available as data and possibly influencing our latent forces (usually called "determinants"), and we would like to explore the matter. It might seem natural to still use the models of Chap. 7 and then simply follow a sequential approach: estimate a basic 2TSF model, recover the metric of interest at the observation level (see Chap. 2), and then run some regression of these estimated metrics (as dependent variables) on a variety of observable data. However, this approach, known colloquially in the single-tier stochastic frontier literature as the "two-step" approach (Parmeter & Kumbhakar, 2014; Wang & Schmidt, 2002), suffers from a variety of statistical and intuitive shortcomings, the most prominent of which is simply that the researcher starts the analysis by assuming that the one-sided terms are iid and then proceeds to decompose/explain them through the use of available covariates. This is logically inconsistent, while practically it begs the question "if you had data available in the first place, why did you not use them in the first stage?" A better approach is to design a model from the start that allows for such heterogeneity.

A traditional method to acknowledge heterogeneity is to condition the sample itself on observables ("stratification"). This has been undertaken in the very first 2TSF article by Polachek and Yoon (1987), who broke their full dataset into smaller subsamples (based on observable characteristics) and then re-estimated the 2TSF model. One issue with such approach is that we potentially lose statistical efficiency because the subsamples can be quite small relative to the overall sample depending on the characteristic(s) of interest. Consider that, while splitting on gender may lead to roughly equal size subsamples, further subdividing by gender and marital status may produce vastly different sized subsamples, which then distorts statistical efficiency. This method essentially conditions the data set on characteristics to create smaller universes given these attributes.

Another approach is to condition the distribution moments on observables. Here we model directly the impact that these characteristics have on the one-sided terms in a likelihood setting, by conditioning the parameters of their distributions on these observables. The first known version of this in the 2TSF framework is Groot and Oosterbeek (1994). To allow incomplete information to depend on observable characteristics, Groot and Oosterbeek (1994) proposed modeling the parameters of the Exponential distributions of u and w as parametric functions. We caution readers to the specification proposed by Groot and Oosterbeek (1994) (as well as Groot & van den Brink 2007 and Tomini et al. 2012), given that the means of both u and w are modeled linearly, which cannot mimic the true mean of the distribution since both error terms are one-sided, and so their means should be guaranteed to be positive, something that their specification as linear functions does not. The more common approach to modeling the mean of the distribution, initially proposed by Kumbhakar and Parmeter (2009, 2010), is as an Exponential function of the observable characteristics, which ensures positivity. There are other mathematical formulations of course, one of which we illustrate through statistical software in Sect. 11.2.

8.2 Moment Dependence and Maximum Likelihood

To use the determinants of latent forces, represented by vectors of covariates z_w and z_u for which we have data series over the sample, we express the unknown distribution parameters of the latent forces no more as constants but as functions of these covariates, setting

$$\sigma_{wi} = \sigma_w(z_{wi}; \boldsymbol{\delta}_w), \qquad \sigma_{ui} = \sigma_u(z_{ui}; \boldsymbol{\delta}_u) \qquad i = 1, \ldots, n.$$

Then we insert these directly into the likelihood function. Compare the baseline 2TSF Normal-Exponential sample log-likelihood function, Eq. (7.13):

$$\ln L\left(\boldsymbol{x}; \boldsymbol{\theta}\right) = -n \ln\left(\sigma_w + \sigma_u\right) + \sum_{i=1}^{n} \ln\left[e^{a_{1i}} \Phi\left(b_{1i}\right) + e^{a_{2i}} \Phi\left(b_{2i}\right)\right],$$

with $\boldsymbol{\theta} = \{\boldsymbol{\beta}, \sigma_v, \sigma_w, \sigma_u\}$, to the likelihood where we allow determinants to influence each of the separate scale parameters:

$$\begin{aligned} \ln L\left(\boldsymbol{x}; \boldsymbol{\theta}\right) = &-\sum_{i=1}^{n} \ln\left(\sigma_w(z_{wi} + \sigma_u(z_{ui}; \boldsymbol{\delta}_u); \boldsymbol{\delta}_w)\right) \\ &+ \sum_{i=1}^{n} \ln\left[e^{a_{1i}(z_{ui};\boldsymbol{\delta}_u)} \Phi\left(b_{1i}(z_{ui}; \boldsymbol{\delta}_u)\right) + e^{a_{2i}(z_{wi};\boldsymbol{\delta}_w)} \Phi\left(b_{2i}(z_{wi}; \boldsymbol{\delta}_w)\right)\right]. \end{aligned} \tag{8.1}$$

with $\boldsymbol{\theta} = \{\boldsymbol{\beta}, \sigma_v, \boldsymbol{\delta}_w, \boldsymbol{\delta}_u\}$. We start with the same functional expression at the observation level, but components that previously were constant now vary per observation, and also, previous constants are now functions of data. This likelihood requires perhaps a more cumbersome optimization because there are additional parameters (one for each element of z_w and z_u instead of the two sigmas that we had previously). Note that the vectors z_w and z_u could be identical or differ fully or partially; this will depend on the application and the available data.

A natural question is the functional forms that we could use for $\sigma_w(z_w; \boldsymbol{\delta}_w)$ and $\sigma_u(z_u; \boldsymbol{\delta}_u)$. As already mentioned, these in all cases represent scale parameters of statistical distributions, and we must ensure that they remain positive throughout. Standard devices to ensure positivity are the absolute value, the square, and the Exponential function. We suggest using the Exponential function as in, say for two determinant-covariates for each latent force,

$$\begin{aligned} \sigma_w(z_w; \boldsymbol{\delta}_w) &= \exp\{\delta_{w0} + \delta_{w1} z_{w1} + \delta_{w2} z_{w2}\} \\ \sigma_u(z_u; \boldsymbol{\delta}_u) &= \exp\{\delta_{u0} + \delta_{u1} z_{u1} + \delta_{u2} z_{u2}\}. \end{aligned}$$

Note that we do *not* constrain the $\boldsymbol{\delta}$ coefficient vectors to be non-negative or nonpositive. We may very well include in the z vectors determinants that may affect positively or negatively the scale factor σ, with the interpretation depending on the conceptual framework. For example, if with the latent force w we measure the information deficiency of a buyer in the housing market, then, if z_{w1} represents an index of information accessibility, the higher the better, we logically expect that $\delta_{w1} < 0$: the greater accessibility to information, the lower the information deficiency. But we let the data decide on this one; we may find, for example, that $\delta_{w1} \approx 0$, which would indicate that buyers are not really hunting persistently for information to exploit.

Another thing to note is that, in the z vectors, we could include also regressors from the vector $\boldsymbol{x}$; nothing forbids this. On the contrary, this allows for some form of dependence between the error term and the regressors and may mitigate issues of endogeneity (see Sect. 14.1.4). Also, some z-variables could appear in both $\boldsymbol{z}_w$ and in $\boldsymbol{z}_u$. In all of these cases, this modeling approach, from a statistical point of view, renders the latent forces and so the whole composite error $\varepsilon = v + w - u$ heteroskedastic.[1] In such a situation, the variance-covariance matrix of the MLE *should* be the robust, "sandwich" estimator (Eq. 7.5).

Finally, note the existence of constant terms δ_{w0} and δ_{u0}, one in each functional specification. Given that we use the Exponential function, we can write, say for $\sigma_u(\boldsymbol{z}_u; \boldsymbol{\delta}_u)$,

$$\sigma_u(\boldsymbol{z}_u; \boldsymbol{\delta}_u) = \exp\{\delta_{u0} + \delta_{u1}z_{u1} + \delta_{u2}z_{u2}\} = \exp\{\delta_{u0}\} \cdot \exp\{\delta_{u1}z_{u1} + \delta_{u2}z_{u2}\}.$$

The term $\exp\{\delta_{u0}\}$ is modeled and will be estimated as a stand-alone constant throughout the sample, a common factor for all observations. Its theoretical justification is that it reflects a base value of σ around which the z-determinants exert their influence, and it could be the sole non-zero coefficient of the delta vector if it so happens that σ is homoskedastic after all (or at least, the specific determinants we assigned to it proved not to be relevant).

But there is a practical issue to consider: because another constant term already exists (the constant term β_0 of the regression function), and even though the delta constants enter the likelihood embedded in highly nonlinear expressions, this could create identification issues, practical if not theoretical, which in turn could contaminate the overall quality of estimation. Applied researchers should be aware of this danger of estimation failure or difficulty, and if it materializes, try different specifications of heterogeneity.

[1] Nothing stops us to model also the standard heteroskedasticity relation and make the σ_v parameter of the noise component a function of covariates too.

8.2.1 Moments and Metrics

From a theoretical point of view, the expression of the distribution parameters of w and u as functions of the z covariates makes the specification of the composite error term conditional on z. This implies that the *marginal* distributions of w and u are not Exponential any more (or Half-Normal, depending on the base specification that we have chosen). To be clear, here, assuming that we pivot off the NE model, the specification of the composite error term is

$$v \sim \mathrm{N}(0, \sigma_v), \qquad w|z_w \sim \mathrm{Exp}(\sigma_w(z_w)), \qquad u|z_u \sim \mathrm{Exp}(\sigma_u(z_u)).$$

In words, we specify that the *conditional* distributions of w and u are Exponential, a characteristic that does not transfer to their marginal laws (and the same holds for a NHN base specification). While the convolution of the three and the resulting log-likelihood will have the exact same mathematical structure of the standard NE model (a great convenience for empirical estimation), it will be a conditional composite error density.

How does this specification change affects the computation of the sample-level and individual metrics of Chap. 2? There, we used the unconditional (i.e., the marginal) expected value, since we knew the marginal distributions, as well as the conditional expectation given the composite error term ε. In Chap. 7, we provided every single mathematical expression needed for these computations.

Here, the situation is different: we do not know the marginal distributions of w and u, but we have specified conditional distributions for them given the z covariates. In turn, the conditional moments are functions of the parameters of the conditional distribution. It follows that they are themselves functions of the z covariates. But this has an important consequence: we do not need to use the conditional expectation given ε, because we have an *alternative* conditional expectation available—the one conditional on z. This is beneficial in that we do not introduce into the estimated conditional moments the distortion due to ε containing also signals related to its other two components.

Then, in this heterogeneity setting, all individual metrics of Chap. 2 are to be computed using $E(t|z)$, $t = w, u$ and not $E(t|\boldsymbol{\varepsilon})$. The mathematical expressions that we need to compute these conditional expectations are the *sample-level* unconditional expressions of Chap. 7, adjusted for the conditioning. In other words, the expressions for conditional expectations given ε are not needed here. For example, if we have assumed that $w|z$ follows the Half-Normal distribution, and we want to compute $E\left(\exp\{-w_i\}|z_{wi}\right)$, we will not use Eq. (7.43) but the simpler unconditional expression, individualized here as

$$E\left(e^{-w_i}|z_{wi}\right) = \frac{\Phi(-\sigma_w(z_{wi}))}{\phi(\sigma_w(z_{wi}))}\sqrt{\frac{2}{\pi}}, \qquad i = 1, \ldots, n.$$

This holds for all the conditional expectations and moments we need to compute for the metrics of Chap. 2.

The use of the conditional expectations given z instead of those given ε has a core argument to support it: we *resorted* in using, say, $E(w|\varepsilon)$, because that was the only source of information we had on w. We exploit ε to extract information on w just because the former includes the latter as a component. But $E(w|z)$ is a structural aspect of the model; the variables z are assumed to explain w through their effect on its moments. So its value should be considered more definite and relevant as a predictor of w.

But there is a risk, specifically for the Normal-Exponential 2TSF model: as discussed in detail in Sect. 7.2.1 if, say, the scale parameter $\sigma_w > 1$, then the unconditional moment of the exponentiated variable $E(e^w)$ (and the related $E(e^{w-u})$) does not exist. In the heterogeneity setup, this translates into the following: if for a specific observation j it so happens that $\sigma_w(z_{wj}) > 1$, then for this observation, the conditional expectation $E(e^{w_j}|z_{wj})$ does not exist, neither does $E(e^{w_j-u_j}|z_{wj})$. The good news is that this only affects the specific observation. So, and except if the phenomenon is widespread, we may fail to get information for a few observations, but we will still be able to obtain and examine the series of conditional expectations for the rest.

Sample-Level Metrics It is mathematically intractable to derive the expressions of the sample-level (unconditional) moments of w and u (and it would require knowing the distributions of the z variables also), but we can compute them as sample averages. To do that, we first invoke the "Law of Iterated Expectations" (or the "Tower Property") that links conditional expectations with unconditional expected values, say for w,

$$E(w) = E\left[E(w|z)\right].$$

If we use the NE specification, we have that $E(w|z) = \sigma_w(z_w; \boldsymbol{\delta}_w)$. Moreover, because the z vectors are random variables and also iid (and so ergodic), the outside expected values can be consistently estimated by sample averages. So

$$\widehat{E}(w) = \frac{1}{n}\sum_{i=1}^{n} \sigma_w(z_{wi}; \boldsymbol{\delta}_w).$$

If we use the NHN specification, we will have $E(w|z) = \sigma_w(z_w; \boldsymbol{\delta}_w)\sqrt{2/\pi}$ and so

$$\widehat{E}(w) = \sqrt{\frac{2}{\pi}}\frac{1}{n}\sum_{i=1}^{n} \sigma_w(z_{wi}; \boldsymbol{\delta}_w).$$

The same principle applies for the exponentiated measures, say

$$E\left(e^{-w}\right) = \frac{\Phi(-\sigma_w)}{\phi(\sigma_w)}\sqrt{\frac{2}{\pi}} \implies \widehat{E}\left(e^{-w}\right) = \sqrt{\frac{2}{\pi}}\frac{1}{n}\sum_{i=1}^{n} \frac{\Phi(-\sigma_w(z_{wi}))}{\phi(\sigma_w(z_{wi}))}.$$

Of course nothing forbids us to compute the conditional expectations given ε, but as already said, these introduce more inaccuracy in the computation.

8.3 The Scaling Property and NLS

Maximum likelihood estimation always relies on making distributional assumptions; it requires assuming specific distributions for the components of the composite error term. This then bears the risk of distributional misspecification, which in the context of two-tier stochastic frontier analysis, appears to increase since we have to make three such assumptions. However, if one is willing to make a less specific assumption regarding the nature of w and u, another way to model heterogeneity exists and can be viewed as "distribution-free."

8.3.1 The Scaling Property

If a distribution possesses the "scaling property," the conditional density has the same functional form as the unconditional (marginal) density, and all dependence is enclosed "inside" the distribution parameters. In the context of maximum likelihood estimation, the scaling property is convenient because if we want to introduce heterogeneity based on observables, we can achieve it and keep using familiar likelihoods and mathematical formulas. The scaling property relates mostly to single-parameter distributions, although it can be extended to distributions with, say, two parameters if we fix appropriately one of them.[2]

For our purposes here, we need only a weaker version of the scaling property. Specifically, we assume that our latent forces, say w, can be decomposed multiplicatively as

$$w_i = w_i^* \cdot \sigma_w(\boldsymbol{x}_i, z_{wi}), \qquad E(w_i^* | \{\boldsymbol{x}_i, z_i\}) = E(w_i^*).$$

Namely, we assume that w_i is the product of a non-negative random variable w_i^* with unitary variance, times some positive function of $(\boldsymbol{x}_i, z_{wi})$, and that w_i^* is mean-independent from (all) the observables in the sample, a purely idiosyncratic component. This setup is sufficient to allow consistent estimation using nonlinear least squares (NLS), without making any distributional assumption on v, w, or u.

As noted by Alvarez et al. (2006), the ease of interpretation and the ability to dispense with distributional assumptions are key benefits of adopting the scaling property. Imposition of correct distributional assumptions would produce efficient estimators through maximum likelihood, but it is rarely the case that information

[2] See Wang and Schmidt (2002) for a more detailed discussion.

on these distributions exists. Moreover, the simplicity of nonlinear least squares suggest that empirical researchers may be more comfortable arguing for imposition of the scaling property and avoiding issues of distributional specification altogether. And there is more. First, dependence between the two one-sided latent factors is allowed and it is automatically taken into account, as we will see, while to account for dependence through exact distributional assumptions makes things much more complicated (although workable, as we show in Chap. 10). The same holds for possible endogeneity of the regressors with respect to w and u (see Sect. 14.1.4). Second, the scaling property admits conditional heteroskedasticity, which is more plausible than homoskedasticity across most economic settings. Third, invoking the scaling property for w and u allows the distributions of each to differ. Lastly, the scaling property "upgrades" the latent forces from being components of a single composed error term to being additional explanatory variables, whose influence on the conditional mean of the model can be measured and explained. All of these features were pointed out by Parmeter (2018) in his proposal to use NLS to estimate the 2TSF model, a method that we present next.

8.3.2 NLS Estimation

Consider the standard 2TSF regression function:

$$y_i = \boldsymbol{x}_i'\boldsymbol{\beta} + v_i + w_i - u_i, \quad i = 1, \ldots, n.$$

Assume that along with the regressors $\boldsymbol{x}$, we also have available data (z_w, z_u) that we believe influence w and u, respectively, perhaps alongside (some or all) of the regressors $\boldsymbol{x}$. We assume that w and u follow distributions that have the scaling property, so

$$w_i = w_i^*\sigma_w(\boldsymbol{x}_i, z_{wi}), \qquad u_i = u_i^*\sigma_u(\boldsymbol{x}_i, z_{ui}).$$

Under the scaling property, w^* and u^* are independent from $(\boldsymbol{x}, z_w, z_u)$. Applying the Exponential modeling for $\sigma_w(\boldsymbol{x}_i, z_{wi})$ and for $\sigma_u(\boldsymbol{x}_i, z_{ui})$, we obtain

$$E(w_i|\boldsymbol{x}_i, z_{ui}, z_{wi}) = \mu_w^* \exp\{\boldsymbol{x}_i'\boldsymbol{\gamma}_w + z_{wi}'\boldsymbol{\delta}_w\}, \qquad \mu_w^* \equiv E(w_i^*), \tag{8.2}$$

$$E(u_i|\boldsymbol{x}_i, z_{ui}, z_{wi}) = \mu_u^* \exp\{\boldsymbol{x}_i'\boldsymbol{\gamma}_u + z_{ui}'\boldsymbol{\delta}_u\}, \qquad \mu_u^* \equiv E(u_i^*). \tag{8.3}$$

For the noise component v, assume

$$E(v_i|\boldsymbol{x}_i, z_{wi}, z_{ui}) = E(v_i) = 0,$$

i.e., that it is mean-independent from the data, and it has zero unconditional mean (the zero unconditional mean is guaranteed from the fact that the $\boldsymbol{\beta}$ vector includes a constant term).

Then taking the conditional expectation on the regression function, we have

$$
\begin{aligned}
E(y_i|\boldsymbol{x}_i, z_{ui}, z_{wi}) &= \boldsymbol{x}_i'\boldsymbol{\beta} + E[w_i|\boldsymbol{x}_i, z_{ui}, z_{wi}] - E[u_i|\boldsymbol{x}_i, z_{ui}, z_{wi}] \\
&= \boldsymbol{x}_i'\boldsymbol{\beta} + \mu_w^* \exp\{\boldsymbol{x}_i'\boldsymbol{\gamma}_w + z_{wi}'\boldsymbol{\delta}_w\} - \mu_u^* \exp\{\boldsymbol{x}_i'\boldsymbol{\gamma}_u + z_{ui}'\boldsymbol{\delta}_u\}.
\end{aligned} \tag{8.4}
$$

This is a most general formulation. Some of the elements in the gamma vectors could be set to zero, if some of the regressors are not deemed to affect the latent forces. We could also compact z_w and z_u into a comprehensive vector z, while some of the elements of the delta vectors would be set equal to zero. In fact this would better represent notation-wise the possibility of dependence between w and u. With these in mind, the conditional expectation function can be estimated by nonlinear least squares, by minimizing the objective function:

$$
n^{-1}\sum_{i=1}^{n}\left[y_i - \boldsymbol{x}_i'\boldsymbol{\beta} \;-\; \mu_w^* \exp\{\boldsymbol{x}_i'\boldsymbol{\gamma}_w + z_i'\boldsymbol{\delta}_w\} \;+\; \mu_u^* \exp\{\boldsymbol{x}_i'\boldsymbol{\gamma}_u + z_i'\boldsymbol{\delta}_u\}\right]^2, \tag{8.5}
$$

over $(\boldsymbol{\beta}, \boldsymbol{\gamma}_w, \boldsymbol{\gamma}_u, \boldsymbol{\delta}_w, \boldsymbol{\delta}_u, \mu_w^*, \mu_u^*)$. The parameters μ_w^* and μ_u^* are to set the scale of each of the individual means and can be subsumed inside the Exponential functions by including an intercept in both. They are the additional constant terms discussed earlier (if this is adopted, then we will be estimating their logarithms).

Note that the only thing left out of the estimating equation is the noise component v. As long as it is mean-independent from the data as assumed, then the NLS estimator will be consistent and asymptotically Normal. Note also that if there is dependence between the regressors and the latent forces w and u, this is captured by the estimating equation. Also, if there is dependence *between* w and u, this is also captured by the estimating equation. So this approach accommodates many stochastic relations that could complicate model-building and estimation if we were to handle them through distributional assumptions (although it can be done with established theoretical and applied tools as we discuss in Chap. 10 and Sect. 14.1.4).

Also, we clarify that this is *not* the "corrected" NLS estimation method that we presented in Chap. 7. There is nothing that needs to be corrected here, because the latent forces w and u are determined (around and alongside their mean level) by what is already included in the estimating equation (while in the COLS approach, where there are no determinants of latent forces, they are left outside of the estimating equation and there we still need distributional assumptions to obtain the parameters of interest).

Given the presence of heteroskedasticity in the composite error term, one has to obtain heteroskedasticity-robust standard errors for proper inference. Heteroskedasticity robust errors can be calculated by noting that the nonlinear least squares estimator is what is called an M-estimator. Consider an objective function $\Psi(y, \boldsymbol{x}, z, \boldsymbol{\theta})$, the derivative of which with respect to $\boldsymbol{\theta}$ is $\psi(y, \boldsymbol{x}, z, \boldsymbol{\theta})$. An M-estimator of $\boldsymbol{\theta}$, $\widehat{\boldsymbol{\theta}}$, is defined as the vector that satisfies

$$\sum_{i=1}^{n} \psi(y_i, \boldsymbol{x}_i, z_i, \widehat{\boldsymbol{\theta}}) = 0. \tag{8.6}$$

Nonlinear least squares and maximum likelihood are both types of M-estimators. As shown in White (1994, Thm. 6.10),

$$\sqrt{n}\left(\widehat{\boldsymbol{\theta}} - \boldsymbol{\theta}\right) \xrightarrow{d} N(0, S(\theta)), \tag{8.7}$$

where $S(\boldsymbol{\theta}) = H(\boldsymbol{\theta})D(\boldsymbol{\theta})H(\boldsymbol{\theta})$, with

$$H(\boldsymbol{\theta}) = \left(E[-\psi'(y, \boldsymbol{x}, z, \boldsymbol{\theta})]\right)^{-1} \text{ and } D(\theta) = \text{Var}[\psi(y, \boldsymbol{x}, z, \boldsymbol{\theta})].$$

$H(\boldsymbol{\theta})$ is the inverse of the expected value of the second derivative of the objective function, $\Psi(y, \boldsymbol{x}, z, \boldsymbol{\theta})$, and can be thought of as the Hessian matrix. Also, a natural estimator for $D(\boldsymbol{\theta})$ is the outer product of the empirical first derivatives of the objective function (or the outer product of gradient estimator):

$$\widehat{D} = n^{-1} \sum_{i=1}^{n} \psi(y_i, \boldsymbol{x}_i, z_i, \widehat{\boldsymbol{\theta}})\psi(y_i, \boldsymbol{x}_i, z_i, \widehat{\boldsymbol{\theta}})'. \tag{8.8}$$

These matrices can be calculated by available routines in various software packages. For example, in R, the `sandwich` package (Zeileis, 2004, 2006) can construct robust standard errors for a model estimated via nonlinear least squares through the `nls()` command, and so also for the 2TSF model with observable heterogeneity in the latent forces. In the `gretl` software, the command `nls` has a "robust" option for the same purpose.

8.3.3 *Metrics*

With the use of the z variables to specify the latent forces without assuming a distribution, we no longer possess theoretical moment expressions to compute the expected values of interest at the sample level. Instead, we have the individual, per observation values, without the need for conditioning on the error term. When the dependent variable is in its original units of measurement, the metrics per observation have already been given; they are Eqs. (8.2) and (8.3) for w and u, respectively. The sample-level metrics would use their sample mean.

When the dependent variable is in logarithms, the distribution sirens start to sing again. Here we would want to compute conditional expectations like

$$E(\exp\{w_i\}|\boldsymbol{x}_i, z_{ui}, z_{wi}) = E\left(\exp\{\sigma_w(\boldsymbol{x}_i, z_{wi}) \cdot w_i^*\}|\{\boldsymbol{x}_i, z_{ui}, z_{wi}\}\right).$$

Since we condition on $(\boldsymbol{x}_i, z_{ui}, z_{wi})$, the term $\sigma_w(\boldsymbol{x}_i, z_{wi})$ is fixed. Then this expected value is the moment-generating function of w_i^* evaluated at $\sigma_w(\boldsymbol{x}_i, z_{wi})$. But in order to derive this theoretically, we would need to make a distributional assumption for w_i^*, and we want to avoid that. We can escape the sirens by applying a first-order Taylor expansion on this expected value around the mean value μ_w^* which will give

$$E\left(\exp\{\sigma_w(\boldsymbol{x}_i, z_{wi}) \cdot w_i^*\}|\{\boldsymbol{x}_i, z_{ui}, z_{wi}\}\right) \approx \left[\exp\{e^{\boldsymbol{x}_i'\boldsymbol{\gamma}_w + z_{wi}'\boldsymbol{\delta}_w}\}\right]^{\mu_w^*}.$$

Likewise for all related conditional expectations, we have

$$E(e^{w_i}|\boldsymbol{x}_i, z_{ui}, z_{wi}) \approx \left[\exp\{e^{\boldsymbol{x}_i'\boldsymbol{\gamma}_w + z_{wi}'\boldsymbol{\delta}_w}\}\right]^{\mu_w^*}, \tag{8.9}$$

$$E(e^{-w_i}|\boldsymbol{x}_i, z_{ui}, z_{wi}) \approx \frac{1}{\left[\exp\{e^{\boldsymbol{x}_i'\boldsymbol{\gamma}_w + z_{wi}'\boldsymbol{\delta}_w}\}\right]^{\mu_w^*}}, \tag{8.10}$$

$$E(e^{u_i}|\boldsymbol{x}_i, z_{ui}, z_{wi}) \approx \left[\exp\{e^{\boldsymbol{x}_i'\boldsymbol{\gamma}_u + z_{ui}'\boldsymbol{\delta}_u}\}\right]^{\mu_u^*}, \tag{8.11}$$

$$E(e^{-u_i}|\boldsymbol{x}_i, z_{ui}, z_{wi}) \approx \frac{1}{\left[\exp\{e^{\boldsymbol{x}_i'\boldsymbol{\gamma}_u + z_{ui}'\boldsymbol{\delta}_u}\}\right]^{\mu_u^*}}, \tag{8.12}$$

and for their differences,

$$E(e^{w_i - u_i}|\boldsymbol{x}_i, z_{ui}, z_{wi}) \approx \frac{\left[\exp\{e^{\boldsymbol{x}_i'\boldsymbol{\gamma}_w + z_{wi}'\boldsymbol{\delta}_w}\}\right]^{\mu_w^*}}{\left[\exp\{e^{\boldsymbol{x}_i'\boldsymbol{\gamma}_u + z_{ui}'\boldsymbol{\delta}_u}\}\right]^{\mu_u^*}}, \tag{8.13}$$

$$E(e^{u_i - w_i}|\boldsymbol{x}_i, z_{ui}, z_{wi}) \approx \frac{\left[\exp\{e^{\boldsymbol{x}_i'\boldsymbol{\gamma}_u + z_{ui}'\boldsymbol{\delta}_u}\}\right]^{\mu_u^*}}{\left[\exp\{e^{\boldsymbol{x}_i'\boldsymbol{\gamma}_w + z_{wi}'\boldsymbol{\delta}_w}\}\right]^{\mu_w^*}}. \tag{8.14}$$

Notice the double exponentiation operation. Notice also that here, the negative measures are the reciprocals of the positive ones and that the exponentiated difference measure is the ratio of the two measures, which is not the case when we condition on the error term. This is because in the present setup, the conditioning is complete and separates the two exponentiated expressions. The sample means of these series will provide estimates for the sample-level corresponding values.

8.3.4 Identification Perils and Traps

For compactness, in this subsection, we join any regressors $\boldsymbol{x}$ affecting w or u with z in the vector $\tilde{z}$, which includes all regressors and all determinants for both w and u, while the parameter vectors $\boldsymbol{\delta}_w$ and $\boldsymbol{\delta}_u$ include here also the gamma coefficients. These vectors can be difficult, or even infeasible, to identify in certain settings. Moreover, imposing a priori restrictions on the coefficients may lead to useless results. To see this, note that the conditional mean of y is

$$E\,(y|\boldsymbol{x}, z) = \boldsymbol{x}'\beta - \mu_u^* e^{\tilde{z}'\boldsymbol{\delta}_u} + \mu_u^* e^{\tilde{z}'\boldsymbol{\delta}_w}.$$

Consider the case where we set a priori the same elements of $\boldsymbol{\delta}_w$ and $\boldsymbol{\delta}_u$ equal to zero. This reflects an argument that the exact same covariates affect w and u. Then if we attempt to maximize the objective function of the NLS estimator separately for $\boldsymbol{\delta}_w$ and for $\boldsymbol{\delta}_u$, we will obtain a gradient with linearly dependent elements, and hence a singular Hessian, and the estimation attempt will fail. In such a case, the conditional expectation that can be estimated is

$$E\,(y|\boldsymbol{x}, z) = \boldsymbol{x}'\beta + (\mu_w^* - \mu_u^*)e^{\tilde{z}'\boldsymbol{\delta}}.$$

Namely, we can only estimate a common coefficient vector $\boldsymbol{\delta}$. There are two problems here, one interpretational and the other technical. The interpretational problem is that by assuming a priori that the exact same variables affect w and u, we end up *imposing that each variable will affect w and u in the exact same way*, both as regards the sign of its coefficient, i.e., the direction of influence, and also the magnitude of its marginal effect (the value of the associated element of $\boldsymbol{\delta}$). And this, for all variables involved. But this may not align well with the structural/behavioral setup of the model. There may be a variable for which logic and reason indicate that it will affect, say, w positively but u negatively . Therefore, such "homogenizing" a priori restrictions should be avoided.

Moreover, from a technical point of view, we also find that we cannot identify separately μ_w^* and μ_u^*, only the difference $(\mu_w^* - \mu_u^*)$. And the closer this is to zero, the harder it will be to estimate $\boldsymbol{\delta}$, even if we are ok with assuming a common delta vector for both w and u.

So, as a matter of design, we should avoid using the same set of covariates for representing w and u in the objective function.

Another identification trap may form by the use of binary z-covariates. One can formally show that if the specification of the Exponential function $\exp\{\tilde{z}'\boldsymbol{\delta}\}$ includes just a constant and a binary z-variable, there is linear dependence in the associated

Jacobian and the estimation attempt will completely fail. This issue may arise also if we have more determinants in number, but all binary ones.

Finally, we return to the initial alert about the three constants in the NLS specification ($\beta_0, \mu_w^*, \mu_u^*$): although formally there exists non-linearity that should allow joint estimation of all three, there may be cases and samples where the estimator will fail to converge or produce very high standard errors, due to near-colinearity. This, for example, could happen if one of the Exponential functions behaves through the sample as "nearly" a vector of ones, i.e., having little variation, and around unity. Then it will become nearly colinear with the regressor associated with the constant term of the regression β_0, and estimation will suffer. There are no guaranteed ways to remedy such a situation; it will depend on the specifics of the data sample and the study at hand. Applied researchers could always revert back to the maximum likelihood framework developed in the first part of this chapter, if they encounter such a problem that they cannot solve or sidestep.

8.4 Application: The Housing Market in the USA

For the empirical application of this chapter, we revisit the housing market, but in a different country. Both Parmeter (2006) and Kumbhakar and Parmeter (2010) use data from the American Housing Survey (AHS) to ascertain the amount of incomplete information on both sides of the housing market. A key feature of the AHS is that not only is standard housing information collected (as in our earlier example focusing on Canada) but characteristics of the primary homeowner are also available. This allows linking incomplete information to various observable characteristics of the buyers and sellers, which is then used to estimate a two-tier stochastic frontier model accounting for incomplete information, applying maximum likelihood estimation to the Normal-Exponential distributional specification.

Kumbhakar and Parmeter's (2010) analysis proceeds in two stages. First, buyer and seller characteristics are ignored, which allows for a baseline of incomplete information to be constructed. In this case, any differences in incomplete information that are observed are interpreted as idiosyncratic. Second, buyer and seller characteristics are incorporated into the model to understand how these characteristics impact incomplete information.

Based on the same sample, we estimate a less general hedonic specification. The model of Kumbhakar and Parmeter (2010) included the determinants of information deficiency in *both* the hedonic part and the variance components for the Normal-Exponential specification they deploy. Here, we only include these variables into the error components. This also allows us to then compare hedonic estimates between OLS, NLS, and MLE (under both the Normal-Exponential and Normal-Half-Normal specifications). To estimate the two models via MLE, we first maximize the log-likelihood function using the OLS coefficient estimates as starting values (with 0s in place for all of the error components) using Nelder-Mead simplex optimization. We then use these estimates as a new set of starting values and randomly perturb them to create new starting values (called multistarting). For each multistart, we optimize using the BFGS algorithm. The set of estimates that produces the maximum (minimum) likelihood (sum of squared error) value is selected as our final set of estimates. We conduct five multistarts for each method. For NLS, we follow the same multistart procedure but optimization is performed using the Levenberg-Marquardt algorithm.[3]

We first discuss the main hedonic estimates before moving the discussion into information deficiency. The hedonic estimates across these four estimation approaches appear in Table 8.1. We follow both Harding et al. (2003) and Kumbhakar and Parmeter (2010) and include (in order) the logarithm of square footage, a dummy to indicate if square footage was top-coded, the number of bathrooms, the number of rooms, dummies to signify if the house was classified as single family attached or detached, dummies for a range of age categories (<5, 5–10, 10–15, and ≥30), dummies for location of the house (city center, urban/suburban, other urban and rural), a dummy if the house was deemed inadequate to live in upon sale of the house, a degree days code to capture weather in the area (integer coded from 1 to 6), and year of sale dummies and dummies for the size of the "metropolitan statistical area" (MSA) the house is located in.[4]

Nearly all of the coefficient estimates have the anticipated sign across all three specifications: larger houses, those with more bathrooms, and those that are newer sell for more, on average, for example. It appears from these estimates that the choice of model specification (Normal-Exponential, Normal-Half-Normal, or the scaling property/NLS) does not dramatically impact the shape of the hedonic price function, outside of OLS. Omission of the determinants of deficiency may lead to inconsistent estimation of the hedonic parameters via OLS if buyer/seller characteristics are correlated with observable structural housing characteristics. For example, the estimated impact of the size of the home is nearly 25% larger for the OLS model than the other three models that incorporate information deficiency. The smaller estimated intercept, while not interesting on its own from an interpretational standpoint, could signify that it is missing information deficiency that would lead to a larger value (as in the NE, NHN, and NLS models). A similar feature emerges for many of the other coefficient estimates: OLS produces consistently larger (in magnitude) numbers, which may hint at differences being driven by a failure to account for determinants of information deficiency.

To that end, Table 8.2 presents the coefficient estimates of the z covariates for the variance specifications for the three different models that incorporate them. These estimates differ from what is presented in Kumbhakar and Parmeter (2010) given that we use a different approach to optimization, and a different hedonic model

[3] The Levenberg-Marquardt algorithm is an iterative procedure that interpolates between the gradient descent method and the Gauss-Newton algorithm.

[4] An MSA is defined as "a geographic entity based on a county or a group of counties with at least one urbanized area with a population of at least 50,000 and adjacent counties with economic ties to the central area."

Table 8.1 Two-tier stochastic frontier estimates. Dependent variable is the house price in logarithms. Sample size is $n = 4{,}962$. Year of sale and metro size dummies were included in the model but are not reported here

	OLS	NE	NHN	NLS
Constant	6.203	7.745	8.121	7.733
	(0.187)	(0.861)	(0.140)	(0.033)
ln SqFt	0.453	0.303	0.305	0.327
	(0.028)	(0.100)	(0.018)	(0.034)
$D_{SqFt-TopCode}$	−0.215	−0.114	−0.130	−0.123
	(0.049)	(0.084)	(0.039)	(0.057)
# Baths	0.309	0.241	0.228	0.229
	(0.018)	(0.022)	(0.016)	(0.024)
# Rooms	0.025	0.023	0.010	0.016
	(0.007)	(0.006)	(0.006)	(0.009)
$D_{SF-attach}$	0.832	0.574	0.510	0.646
	(0.052)	(0.093)	(0.048)	(0.061)
$D_{SF-detach}$	0.896	0.647	0.593	0.726
	(0.033)	(0.067)	(0.035)	(0.039)
D_{City}	−0.103	−0.104	−0.119	−0.080
	(0.034)	(0.032)	(0.030)	(0.041)
$D_{UrbanSurban}$	0.0004	−0.011	−0.028	−0.014
	(0.032)	(0.027)	(0.028)	(0.037)
$D_{OtherUrban}$	−0.289	−0.305	−0.324	−0.266
	(0.038)	(0.042)	(0.034)	(0.045)
D_{Rural}	−0.345	−0.304	−0.316	−0.288
	(0.035)	(0.030)	(0.031)	(0.041)
$D_{Age<5}$	0.290	0.228	0.194	0.292
	(0.036)	(0.032)	(0.030)	(0.042)
$D_{Age5-10}$	0.161	0.115	0.092	0.140
	(0.029)	(0.023)	(0.024)	(0.034)
$D_{Age10-15}$	0.096	0.052	0.042	0.062
	(0.028)	(0.022)	(0.024)	(0.033)
$D_{Age>30}$	−0.032	−0.030	0.007	−0.013
	(0.023)	(0.027)	(0.020)	(0.027)
D_{Inadeq}	−0.031	−0.010	0.034	0.049
	(0.075)	(0.066)	(0.067)	(0.088)
Degree Days	−0.013	−0.012	−0.008	−0.010
	(0.007)	(0.006)	(0.006)	(0.008)

specification, and deploy a variety of distributional assumptions to identify the model. Generally, all three approaches yield similar insights regarding the direction of impact of determinants of deficiency, but the size of the effects differs. As we have stated in earlier chapters, we should not directly compare magnitudes across the three specifications since the coefficients do not have identical interpretations.

Table 8.2 Determinants of deficiency on housing prices across specifications

	Seller $\sigma_u(z_u)$			Buyer $\sigma_w(z_w)$		
	NE	NHN	NLS	NE	NHN	NLS
μ_u and μ_w	0.672 (0.111)	1.042 (0.063)	0.855	0.197 (0.067)	0.111	0.106 (0.081)
Out of town	–	–	–	0.043 (0.057)	0.066 (0.079)	0.112 (0.097)
First time	–	–	–	0.008 (0.052)	0.003 (0.033)	0.139 (0.163)
Income	−0.266 (0.037)	−0.293 (0.027)	−0.225	0.239 (0.038)	0.288 (0.017)	0.268 (0.066)
Own business	−0.264 (0.068)	−0.207 (0.042)	−0.249	0.067 (0.056)	0.121 (0.072)	0.157 (0.091)
Age	−0.273 (0.121)	−0.256 (0.048)	−0.28	0.461 (0.112)	0.718	0.695 (0.308)
Black	0.349 (0.104)	0.241 (0.078)	0.288 (0.036)	−0.56 (0.151)	−3.996 (0.565)	−0.655 (0.339)
Married	−0.16 (0.054)	−0.139 (0.041)	−0.109 (0.014)	−0.045 (0.059)	−0.008 (0.087)	−0.058 (0.184)
Single female	−0.029 (0.064)	−0.06 (0.047)	−0.05 (0.051)	0.002 (0.073)	0.127 (0.105)	0.077 (0.132)
Education	−0.37 (0.067)	−0.307 (0.035)	−0.279	0.301 (0.064)	0.458 (0.065)	0.409 (0.137)
Kids	−0.028 (0.055)	−0.022 (0.034)	0.01 (0.043)	−0.092 (0.043)	−0.128 (0.059)	−0.094 (0.086)

However, we can compare signs and we also note that for the Normal-Exponential and NLS settings, the coefficients are viewed as equivalent in interpretation. We will present the 2TSF metrics stemming from these three approaches shortly.

We first mention that, as is wont to happen in empirical work, several of our estimates (mainly for NLS but also for the Normal-Half-Normal setting) did not produce standard errors as the Hessian matrix possessed negative elements along the diagonal. An alternative would be to use the outer product of gradients discussed in Chap. 7. This can also be suggestive that the estimator had a difficult time identifying a maximum for this parameter. As such, care is warranted in interpreting these estimates.

All three models produce estimates with the same sign for each variable for both buyers and sellers (except for the constant and the "having kids" dummy for sellers). The coefficients that are measured at conventionally accepted statistical confidence also agree across the models: income, sellers owning a business, the buyers'/sellers' age or race, sellers' marital status, and buyer's/seller's years of education.

The variable "income," originally measured in USD yearly, is standardized—centered on its mean and divided by its standard deviation. Hence, when we talk about "income," we essentially refer to deviations from its sample mean measured in standard deviations. Qualitatively, it is "higher than average" income that has the direction of influence indicated by the sign of the coefficient (while "lower than average" will have the opposite effect since it will be a negative value that will reverse the effect indicated by the coefficient sign). Also, the variable "age" is *divided* by its mean (and not centered). Here, either above or below the mean, age will have the direction of influence indicated by the coefficient sign, but "higher than average" will increase quantitatively this effect, while "lower than average" will weaken it.

Table 8.2 shows that all three specifications suggest that both out-of-town and first-time buyers are not impacted by information deficiency due to these characteristics, which may seem surprising, but this is a role that real estate agents can serve to overcome. We also see opposing effects for nearly every variable (aside from the dummy for kids and the dummy for being married): what tends to decrease the strength and effects of the information deficiency of sellers (increasing the price), tends to increase the strength of information deficiency of buyers, *increasing the price also*: this, for example, happens for covariates income, own business, age, and education. Fundamentally, one thing we learn is that transactions where buyers and sellers have higher income, own a business, are older, and have higher education are characterized by a higher selling price due to a greater information gap between sellers and buyers in this subgroup, in favor of the sellers.

Contrast these findings with the estimates of Kumbhakar and Parmeter (2010, Table 6). There, they document (at the 99% confidence level) that none of the observable characteristics of buyers impact overall information deficiency aside from the buyer being black (we too find this as well as income, age, and education), while on the sellers' side, income, owning a business, age, and college education all lower the effects of information deficiency, which agrees with our results (having a negative coefficient, they lower the value of u, which is the carrier of seller's information deficiency). So, despite the difference in the specification, the two broadly agree.[5] These impacts on the sellers' side make economic sense: having more income, owning a business, being married, and having a college education all suggest an increased ability to obtain more information (either through patience, hiring a better agent, or sharing time with the sale of the house). It is somewhat surprising that almost nothing on the buyers' side showed up in their study as statistically significant, but one could easily suggest that information deficiency is exactly what real estate agents are purportedly supposed to mitigate. Moreover, "statistical insignificance" is a misnomer for high uncertainty as regards the direction of influence, rather than an assertion about whether an effect exists or not. On a technical note, including a variable both in the regressor matrix and as a determinant of deficiency may unavoidably lead to a "split-and-share" of its effect on the dependent variable, and it is debatable whether this is, or is not, an artificial result.

[5] As mentioned, Kumbhakar and Parmeter (2010) include buyer and seller characteristics in the main specification also.

2TSF Metrics We turn now to the 2TSF metrics for the latent forces.[6] The metrics were computed per model as described earlier in this chapter in all three models, namely, we used the conditional expectations given z for the Normal-Exponential and Normal-Half-Normal models (i.e., the unconditional formulas of the corresponding marginal distributions with the sigmas being now functions of the z covariates), while for the distribution-free specification estimated by NLS, we use the fully determined formulas (due to the presence of the z vectors). Since the dependent variable (house price) was in logarithms, we used the exponentiated formulas to compute them.

Then, reality hit. To begin with, some obtained values of the underlying conditional expectations series in the NE specification were unacceptable; specifically, some values for the M_2 and M_5 metrics were negative, indicating that the conditional expectations for the particular observations do not exist (see the discussion on the sample-level measures in Sect. 7.2.1), or that, in any case, we failed to compute a meaningful value for them. There were 6 such observations for the NE$/M_2$ metric and another 27 for the NE$/M_7$ metric. Not much, in a sample with size $n = 4{,}962$. These were excluded from the sample, and we mention that with them, some admissible but extreme values for the metrics in the NHN specification disappeared as well. But mathematically legitimate outliers, with totally unrealistic high values, were still with us for the M_2 and M_5 metrics in the NE model. To give an example, if the metric $M_2 = E(e^w) - 1$ is higher than 2, it means that the positive latent force w tends to *more than triple* the value of the dependent variable. And this is highly unlikely or, to put it another way, indicates a poor regression specification where we have failed to include core explanatory variables in it. Our canaries in the coal mine are these two metrics. So we excluded also from the sample all observations for which M_2 and/or M_5 were computed as being greater than 2. And to have an equal footing, these observations were not taken into consideration for the metrics of the NHN and the NLS models either. These were in total 280 observations, so in all we shed 313 observations or 6.3% of the total. Not small, but not catastrophic. Moreover, readers should not take this approach as a general advice on how to deal with outliers, which will ultimately depend also on the goals of a particular study. But here we present results in order to showcase the 2TSF models and their tools, so we are permitted such smoothing operations without much agony. After these interventions, Table 8.3 provides the summary statistics (mean, quartiles, and interquartile range) from the incomplete information measures for the Normal-Exponential, Normal-Half-Normal, and NLS estimation of our model.

We will examine the metrics in three pairs, (M_1, M_2), (M_5, M_6), and (M_7, M_{10}), based on the discussion in Chap. 2, where we have laid out the reasoning behind the comparability of the metrics in each pair. Note that the mean and the median are generally close. This is certainly due also to the exclusion of the extreme observations. Otherwise, there could be much larger differences.

---

[6] We note here that the appropriate metrics differ from that studied in Kumbhakar and Parmeter (2010).

Table 8.3 Incomplete information effects on housing prices across specifications

	Mean	Q_1	Median	Q_3	Q_3-Q_1
	$M_1 = 1 - E(e^{-w}\|z_w)$				
NE	0.243	0.199	0.237	0.282	0.084
NHN	0.177	0.126	0.170	0.224	0.098
NLS	0.222	0.160	0.209	0.271	0.110
	$M_2 = E(e^{u}\|z_u) - 1$				
NE	0.765	0.421	0.689	1.063	0.642
NHN	0.925	0.565	0.878	1.271	0.706
NLS	0.738	0.546	0.737	0.930	0.384
	$M_5 = E(e^{w}\|z_w) - 1$				
NE	0.540	0.330	0.449	0.649	0.319
NHN	0.270	0.157	0.233	0.344	0.188
NLS	0.298	0.191	0.265	0.371	0.180
	$M_6 = 1 - E(e^{-u}\|z_u)$				
NE	0.281	0.228	0.290	0.340	0.112
NHN	0.359	0.300	0.371	0.429	0.129
NLS	0.413	0.353	0.424	0.482	0.129
	$M_7 = E(e^{w-u}\|z_w, z_u) - 1$				
NE	0.114	−0.085	0.038	0.221	0.306
NHN	−0.182	−0.313	−0.215	−0.091	0.221
NLS	−0.235	−0.358	−0.263	−0.140	0.217
	$M_{10} = 1 - E(e^{u-w}\|z_w, z_u)$				
NE	−0.343	−0.586	−0.290	−0.057	0.529
NHN	−0.592	−0.901	−0.550	−0.259	0.642
NLS	−0.359	−0.557	−0.356	−0.163	0.394

If we go into the details of Table 8.3, we observe that the NE results are in disagreement everywhere with the NHN results, while in half of the cases, they appear broadly close to the NLS results. Moreover, it so happens that the Vuong test (see Sect. 7.3.2) rejects the Normal-Half-Normal specification in favor of the Normal-Exponential setting at all conventional levels of significance ($T = 6.22$ with a p-value of less than 0.00001).

So, we abandon the NHN model and stick with the other two. Their first two "points of agreement" are the pair of metrics (M_1, M_2), the "seller/buyer" surplus measures (in the standard microeconomic sense of the word). In a setting where the transaction price is determined under incomplete information, the w latent force is conceptually associated with the buyers (higher buyer information deficiency leads to a higher price), and the u factor with the sellers (higher seller information deficiency leads to a lower price). But as discussed in Chap. 2, the M_1 metric that involves w measures the surplus of the seller, the metric M_2 that involves u measures the surplus of the buyer, while both metrics are computed relative to the actual price. So, for the *median* seller (M_1), 20%–24% of the price received was due to the information deficiency of buyers, while the median buyer (M_2) enjoyed a shaving

Table 8.4 Statistics for $\sigma_w(z_w)$ and $\sigma_u(z_u)$

	Mean	Median	Min	Max	SD
	$\sigma_w(z_w)$				
NE	0.329	0.309	0.108	0.664	0.107
NLS	0.255	0.234	0.063	0.714	0.102
	$\sigma_u(z_u)$				
NE	0.403	0.408	0.048	0.666	0.131
NLS	0.542	0.551	0.090	0.864	0.142

off the price they would have to pay if sellers were fully informed equal to 69%–74% of the price they actually paid. At the same time, these two metrics could also be viewed as the markdown (M_1) and markup (M_2) factors to reach the lower and the upper price frontier, respectively: buyers were at most 24% away from reaching the lower price frontier, while sellers would need a price higher by at least 69% to reach the upper price frontier. In these relative terms, buyers are in a better position than sellers, in the sample.

As regards the metrics M_5 and M_6, the two models provide opposing pictures: the NE model suggests that buyers' deficiency tends to increase the full-information price P^{FI} by 45% (while the NLS model estimates M_5 to be just 29%), while sellers' deficiency tends to decrease P^{FI} by 29% per the NE model, while NLS puts that to 42% (M_6). It is remarkable that the sum of two measures per model are close—and $M_5 + M_6 = M_4$ is the volatility of P^{FI} due to the aggregate of information deficiencies. So, the two models broadly agree on the total volatility, but the NE specification assigns this volatility mainly to buyers' deficiency w, while the NLS models assigns it to sellers' deficiency u.

Combining the definition of the metrics, both models agree that $M_1 < M_2 \implies P < (P^{LF} + P^{UF})/2$, i.e., that the actual price is below the midpoint of the interval delineated by the lower and upper price frontiers (this interval is the stochastic price space in average terms). But the NE model suggests $M_5 > M_6 \implies P^{FI} < (P^{LF} + P^{UF})/2$, while NLS gives $M_5 < M_6 \implies P^{FI} > (P^{LF} + P^{UF})/2$.[7] So, from a qualitative angle, their disagreement lies in where the two models place the full-information price in relation to the midpoint of the price space: the NE model puts it below this midpoint, while NLS puts it above. Intuitively, this indicates that for the NE model, the latent force w (buyer deficiency) is stronger than sellers' deficiency u, while NLS suggests the reverse. Maybe looking at the sample-level statistics for $\sigma_w(z_w)$ and $\sigma_u(z_u)$ will provide some answers? Table 8.4 has the relevant information.

The statistics for $\sigma_w(z_w)$ and $\sigma_u(z_u)$ reveal that both models predict the w force to be *weaker* on average than u, with the statistics of the related series in general

[7] We encountered the relations between metrics $M_1 < M_2$ and $M_5 > M_6$ in the empirical application of Chap. 3, where—coincidence?—there too we were looking at the housing market, although of another country.

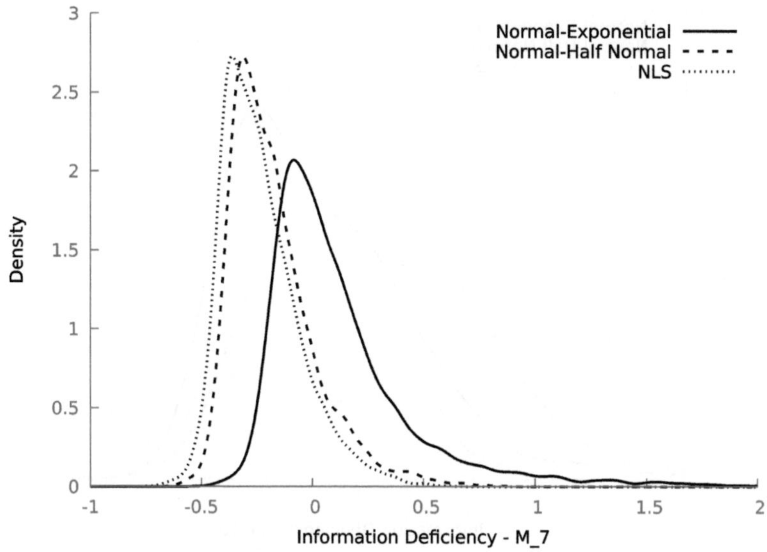

Fig. 8.1 Net effect on full-information price (metric M_7)

alignment between the models.[8] One thing that differs is that the distance between $E(w)$ and $E(u)$ is much smaller in the NE model (0.329–0.403) compared to the NLS estimates from the scaling model (0.255–0.542). Apparently, the NE seems in conflict with itself since based on the M_5 and M_6 metrics, one would expect that it would pick w as the stronger force. Moreover, the general conclusion that information deficiency hurts buyers less than sellers is in agreement with the results (in both models) related to the metrics M_1 and M_2. So, we decide at this point to abandon the NE model (which would also trouble us with conflicting results for the M_7 and M_{10} metrics) and continue this analysis based on the NLS specification.

For the net effect on the full-information price (metric M_7), the NLS results indicate that it is negative, tending to drag P^{FI} down by 26%, while for the net gain (metric M_{10}), it gives its median value as -35.6%: the influence of information deficiency results in favoring the buyers to the level of 35% of the actual price.

One should also pay attention to the other statistics of the metrics, for a more wholesome picture of this market: the volatility of P^{FI} due to information deficiencies (metric $M_4 = M_5 + M_6$ as mentioned) rises to 69% of P^{FI}. Given that both sides typically retain real estate agents, such strong informational effects are a bit surprising—or revealing.

We close this empirical application with two visual cues, where we return to a comparative approach. Figures 8.1 and 8.2 plot out the estimated kernel densities

[8] This result carries over even if we re-include all excluded observations, it does not depend on the smoothing of the sample that we have performed.

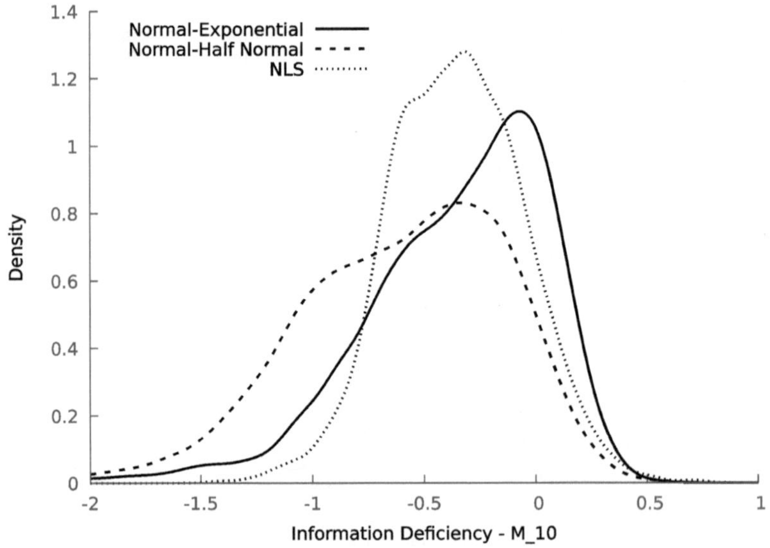

Fig. 8.2 Net gain as a proportion of actual price (metric M_{10})

corresponding to the M_7 and M_{10} metrics for the three specifications. For the M_7 metric, we see similar shapes for all three densities, but the two parametric distributions have thicker right tails and display more variation than the equivalent metric for NLS. On the other hand, we cannot but point out the strong similarity (both in location and in shape) between the NLS and the NHN densities.

For the M_{10} metric, all three densities have the majority of their mass in the negative domain, but they differ more in general. All three have negative skewness but NLS is very close to symmetric, and its estimates are much more concentrated, having shorter tails than the two distributions arising from the parametric specifications. But since this was an exhibition application, we have the luxury not to take sides.

Chapter 9
2TSF Panel Data Models

In single-tier stochastic frontier analysis, there was, and continues to be, an influx of research focused on panel data models. On the other hand, methodological research related to panel data modeling and estimation in a two-tier context is almost nonexistent, and the same holds for empirical applications.[1] We must clarify that many recent studies that apply the 2TSF model use a data sample that is a panel. But they invariably pool the data to estimate, and so they treat it as a cross-sectional sample, failing to exploit the panel data aspect and additional dimension.

In light of this, we take it upon ourselves in this chapter to develop fully the generalized panel data 2TSF model. Compared to the cross-sectional models we have presented up to now, the addition of unobservable heterogeneity (which is a standard feature of panel data models) *and* the additional dimension of time due to the panel data structure increase the number of unobservables of the model from three to six: individual heterogeneity, time-invariant (persistent) and time-variant latent forces, positive and negative, and of course the noise term, which varies across time and cross-sections.[2] We will see that the *estimation* method that we will choose will dictate in large part what we can learn for these unobservables and for which ones. The chapter includes also some additional avenues to combine panel data with two-tier stochastic frontier analysis and a comparative application where different models compete on the same data set.

[1] To be exact, we know of only two published papers that apply 2TSF panel data models.

[2] While the "panel structure" can go beyond two dimensions, and also, can be mapped to other kinds of dimensions that do not necessarily include "time," we stick with the benchmark "time/cross-section" combination.

A. Papadopoulos, C. F. Parmeter, *Two-Tier Stochastic Frontier Analysis for the Social Sciences*, https://doi.org/10.1007/978-3-031-81513-3_9

9.1 Why Is Access to a Panel Important?

Panel data tracks individual phenomena and measurements over time. These repeated observations offer a host of benefits relative to a single snapshot (a cross-section). Perhaps the most important benefit of panel data is the ability to account for various time constant effects, commonly referred to as latent heterogeneity. If we are studying a panel of workers with a 2TSF structure of latent forces in mind, and our concern is that the workers' incomplete information is impacting their wages, we would want to account for fixed idiosyncratic aspects (such as innate ability or work ethic) over time to ensure that they are not being conflated with incomplete information. A panel offers a variety of approaches to separate such heterogeneity from the phenomena that interest us.

Another, somewhat obvious, benefit of access to panel data is that it increases the sample size. This can lead to more informed and robust inference than one would have with just a cross-section. As we will discuss later in this chapter, if we want to allow unobserved heterogeneity into our model and estimate it while also accounting for the two-tier structure, we will have to make some modifications to the traditional estimation approaches commonly deployed with panel data.

A third benefit of panel data is that it can allow analysis of dynamic relationships. For example, there could be learning over time by both workers and firms, and this may lead to interesting interactions between both sides of the market that would be nearly impossible to detect in a cross-sectional setting. Further, we may have dynamics associated with our latent forces: a strong downward push in a previous time period may influence the magnitude of today's downward push. The potential of panel data models to capture such 2TSF dynamics is left for future research.

9.2 The Generalized 2TSF Panel Data Model

The benchmark regression equation that is used with panel data is[3]

$$y_{it} = \alpha_i + \beta_0 + \boldsymbol{x}'_{it}\boldsymbol{\beta} + \varepsilon_{it}, \qquad E(\varepsilon_{it}) = 0, \qquad i = 1 \dots n, \quad t = 1 \dots T. \tag{9.1}$$

The component α_i is a time-invariant individual effect, something that differentiates each cross-section from the rest, but stays constant through time. It is treated

[3] We acknowledge that in the panel data literature, the cross-sectional size of the sample is predominately denoted by "N" rather than "n," but we use the latter to keep the notation across the book consistent. The total sample size is therefore nT.

as a random variable and we do not have data on it.[4] The term ε_{it} is the regression error. This model is sometimes called the "one-way error component model."

An extension of it is the "two-way error component model," where we have the regression specification:

$$y_{it} = \alpha_i + v_t + \beta_0 + \boldsymbol{x}'_{it}\boldsymbol{\beta} + \varepsilon_{it}.$$

Here the term v_t has been added, which is common to all cross-sections but varies through time. It, too, is treated as a random variable and is unobservable. It represents forces that affect all individuals in the sample. If, for example, we look at the behavior of firms in a specific market, v_t could represent exogenous macroeconomic shocks that affect all firms.

We see that the regression formulation for panel data recognizes unobservable components that look somewhat like what stochastic frontier analysis incorporates in its models. But there is a big difference: the main concern of the traditional panel data framework is how to estimate the regressor effects, the betas in the above formulation, treating α_i and v_t as hurdles to be overcome. There are two main approaches on this matter, the "random-effects" (RE) framework and the "fixed-effects" (FE) framework.[5] In RE estimation, we handle the unobservables by assuming that α_i and v_t have zero mean and are mean-independent from the regressors. In the FE approach, we acknowledge the possibility of correlation of these components with the regressors (endogeneity), and then we apply various transformations on the regression and the data, such as centering the variables ("de-meaning"), or first-differencing, that bypass the endogeneity problem. Here we do not need to make assumptions on α_i and v_t, but at the same time, we do not learn anything about them.

In two-tier stochastic frontier analysis, estimating and characterizing the effects of such unobservables is of core importance. Not only do we recognize their existence, but we are specifically interested in them. And even more than that, we do not identify them fully with our, familiar by now, w and u latent forces, but we keep α_i conceptually separate. Why not also v_t? Well, from a practical point of view, this could fatally increase the number of unobservables included in the model. From a historical point of view, the two-way error component model has been much less used; hence, v_t did not loom large in the minds of scholars transforming panel data models for stochastic frontier analysis (which, as we already said, was almost exclusively done for the single-tier variety anyway). Also, common forces, such as

[4] Sometimes, α_i is treated as an "individual intercept," hence as a constant and not as a random variable.

[5] The terms random effects and fixed effects have their origins in a time where the modeling issue was whether to treat the (individual heterogeneity) "effects" α_i as constants (hence "fixed") or as random variables (hence "random"). This issue has become moot, once it was realized that in the FE estimation methodology, we can remain agnostic as regards the nature of α_i, since it is decisively swept out of the model. But the names have stuck and are universally used, so we use them as labels, sometimes preferring the acronyms "RE" and "FE" rather than the full words.

those that v_t wants to represent, can in many cases be proxied by available data, (e.g., a measure of the regulatory environment of a market or the growth rate of DGP).[6] On the other hand, unobservables related to individuals (persons, firms, etc.) are in many cases beyond reach in the sense that we do not have available reliable proxies for them.

For these reasons, the comprehensive 2TSF panel data model that we will develop focuses on modeling and estimating unobservable forces related to cross-sections, persistent and time-varying, and it does not include a time-varying common component, although it does keep α_i distinct. Specifically, the 2TSF model arises by taking Eq. (9.1) and decomposing the error term ε_{it} as follows:

$$y_{it} = \alpha_i + \beta_0 + \boldsymbol{x}_{it}'\boldsymbol{\beta} + [v_{it} + (\gamma_i + w_{it}) - (\eta_i + u_{it})]. \tag{9.2}$$

The bracketed terms represent the error term ε of the regression, which is now a composite. In this formulation, γ_i is the time-invariant component of w, and η_i is the time-invariant component of u. Note carefully that *all four are non-negative random variables with non-zero means.* These expected values are fixed numbers and are, in general, distinct. For α_i and v_{it}, we maintain the standard assumption that they both have a zero mean. In all, the location assumptions on the unobservables are

$$0 \neq E(\gamma_i) \neq E(w_{it}) \neq 0, \qquad 0 \neq E(\eta_i) \neq E(u_{it}) \neq 0, \quad E(\alpha_i) = E(v_{it}) = 0.$$

We will use μ, appropriately indexed, to denote these expected values.

This model is the natural extension of the four-component model proposed by Colombi et al. (2014) and Kumbhakar et al. (2014) ("model 6" in the latter paper). These studies assumed that the inefficiency components (time-varying and persistent) are Half-Normal random variables, in effect making the four-component error term (including the heterogeneity element) a sum of two Skew Normal random variables. The Skew Normal belongs to the greater Closed Skew Normal family of distributions, and so sums of Skew Normals remain a Skew Normal. This allowed Colombi et al. (2014) to obtain a closed-form log-likelihood and implement direct maximum likelihood estimation. Kumbhakar et al. (2014) noted that this was "quite involved" and proposed a two-stage procedure. In our case, we cannot take the road of Colombi et al. (2014), for the following simple reason: in the four-component setup, there are two zero-mean variables and two non-negative. In our six-component setup, we still have two zero-mean variables but *four* non-negative ones. So, we cannot create three pairs similar to the two pairs of variables that Colombi et al. (2014) had, allowing them to obtain a closed-form likelihood based on the multivariate Closed Skew Normal distribution. And apart

[6] A simple noncausal way to incorporate aggregate time effects is by the inclusion in the model of "time-period" dummies that replace the universal constant and operate as time-period-specific intercepts. This is what the World Happiness Report people are doing; see the application in Chap. 5.

from that, we should constrain ourselves to assume Half-Normal variables and not Exponential, since the Normal-Exponential pairs do not belong to a known closed distribution family, i.e., one that is preserved under addition/subtraction of variables. But the Exponential is the benchmark 2TSF distributional assumption for the latent forces. So, we will take the Kumbhakar et al. (2014) route and examine two-stage estimation procedures.

Before elaborating on estimation methods, we need to validate the model from an interpretational point of view. Suppose that w and u represent some form of deficiency, say our benchmark concept of informational deficiency. What is the purpose, and the usefulness, of decomposing deficiency into time-varying and persistent components? Why does this matter? They are both inefficiencies, are they not? It is their sum that determines the full inefficiency, say related to w, and is it not their sum that we are ultimately interested in? The same questioning applies to any other kind of latent forces that w and u may represent.

There are two responses to these questions. First, depending on the situation, it may be the case that, after all, the persistent effects are *not of the same nature* as the time-varying effects. Here certainly we would want to measure them separately, for obvious reasons. Second, even if they are of the same nature (e.g., information deficiencies), by construction the time-invariant/persistent inefficiency becomes part of the core profile of the entities under study, because, at least in the short- and mid-term, it is not expected to change. On the other hand, the time-varying components, by the very fact that they are fluctuating, are subject to influence and change. Hence, this decomposition *provides a realistic boundary on our ability to intervene*. It may be the case that both persistent and time-varying components represent informational deficiency, but the first component we can only measure, the second, we can attempt to reduce it. And this alone is reason enough to want to be able to measure them separately.

The model is completed by the traditional heterogeneity component α_i, and the noise term v_{it}, whose roles remain the same as in the conventional panel data framework. So, the full model in principle is impressively rich, having persistent and time-varying latent forces with positive and negative effects, other unobserved heterogeneity, and unmodeled forces in v_{it}. This leads us naturally to the topic of estimation methods, because they prove to have different consequences on what we can actually learn about all these unobservables.

9.2.1 Fixed-Effects Two-Stage Iterated Estimation

The FE estimation method in a standard panel data framework operates by transforming the original model (9.1) in a way that the resulting estimating equation does not include the heterogeneity term α_i. The main two such transformations are first-differencing the data, or using them in deviations from their time mean form.

For the 2TSF panel data model (9.2), the approach needs to be rather more elaborate than that, and it is implemented in two stages. This is the approach taken by Polachek and Yoon (1996), the first 2TSF study to apply panel data estimation, and we present it next.[7]

The first estimation stage is a standard fixed-effects estimation routine. To use it validly, we have to eliminate from the estimating equation the time-invariant components, and we also necessarily accept that the error term of the estimating equation has a zero mean. Inspecting the regression model Eq. (9.2), these requirements necessitate the definition of the composite term:

$$\alpha_i^* = \alpha_i + \gamma_i - \eta_i + \beta_0 + \mu_w - \mu_u.$$

We have brought out of the error term the time-invariant components, we have thrown in the mix the individual heterogeneity component and the universal regression constant, and we have added and subtracted the means of the time-varying components. This term α_i^* remains time-invariant, although it is a sum of "apples and oranges," it has no meaningful interpretation. But if we insert it into Eq. (9.2) while compensating for the μ's, we get

$$\begin{aligned} y_{it} &= \alpha_i^* + \boldsymbol{x}'_{it}\boldsymbol{\beta} + [v_{it} + (w_{it} - \mu_w) - (u_{it} - \mu_u)] \\ &= \alpha_i^* + \boldsymbol{x}'_{it}\boldsymbol{\beta} + \varepsilon_{it}^*, \qquad E(\varepsilon_{it}^*) = 0. \end{aligned} \tag{9.3}$$

With these theoretical transformations, we obtain an estimating equation with a single time-invariant component, the time-varying part of the relation $\boldsymbol{x}'_{it}\boldsymbol{\beta}$, and a zero-mean error term. Namely, we have transformed the generalized 2TSF panel data model into a standard panel data one. If one compares Eqs. (9.3)–(9.1), the only difference is that α_i^* of Eq. (9.3) does not have a zero mean as α_i of Eq. (9.1) does. But this is inconsequential, because α_i^* is still time-invariant, and it will be eliminated by the FE estimator before estimation, as we discuss below.

First Stage We can now execute the first-stage estimation step using the standard FE estimator (the "Within" estimator). To do this we de-mean the variables by their *time* mean:

$$y_{it}^* = y_{it} - \bar{y}_i, \qquad \bar{y}_i = \frac{1}{T}\sum_{t=1}^{T} y_{it}, \qquad i = 1 \dots n,$$

and the same for the regressors. This is equivalent to de-mean the whole equation across the time dimension, which results in eliminating the composite time-invariant α_i^*.[8] We can now run pooled OLS on this transformed data sample to obtain $\widehat{\boldsymbol{\beta}}^{FE}$ of

[7] Polachek and Yoon (1996) did not assume a separate "individual heterogeneity" term α_i, but they identified it with the difference $\gamma_i - \eta_i$. So, strictly speaking, theirs was a "*five*-component" 2TSF panel data model. But this makes no practical difference, since their fixed-effects estimation approach does not provide any useful result on the time-invariant components of the model, eliminating all irrespective of their number.

the estimating equation:

$$y_{it}^* = \mathbf{x}_{it}^{*\prime}\boldsymbol{\beta} + \varepsilon_{it}^*. \tag{9.4}$$

In applied practice, since the previous transformations were theoretical in order for us to know what each transformed mathematical symbol represents, to implement this stage, we just upload our data to the software and we ask it to run a fixed-effects panel data estimation using the Within estimator. The software will automatically de-mean the variables and execute and produce $\widehat{\boldsymbol{\beta}}^{FE}$.

But our goal in the 2TSF context is not to retain the FE estimates for the betas. The purpose here is to obtain estimates for the elements of the series $\{\alpha_i^*\}$. We do that using the time averages:

$$\widehat{\alpha}_i^* = \bar{y}_i - \bar{\mathbf{x}}_i'\widehat{\boldsymbol{\beta}}^{FE}, \qquad i = 1, \dots, n.$$

Second Stage In the second stage, we return (partly) back to Eq. (9.3), and we transform it yet again by taking out of the error term ε_{it}^* the unknown mean values μ_w and μ_u while using $\widehat{\alpha}_i^*$ in place of α_i^*:

$$y_{it} = \widehat{\alpha}_i^* - \mu_w + \mu_u + \mathbf{x}_{it}'\boldsymbol{\beta} + \varepsilon_{it}, \quad \varepsilon_{it} = v_{it} + w_{it} - u_{it}. \tag{9.5}$$

This equation has an estimate for all time-invariant components of the model, in $\widehat{\alpha}_i^*$. This includes *also* the footprint of any time-invariant regressors originally included in the regressor matrix, and this is why this equation is *not* Eq. (9.3) "fully" (with just the error term retransformed), but only "partly": as all other time-invariant components of the model, any time-invariant regressors have been lost to us under the cloak of $\widehat{\alpha}_i^*$. So, the regressor matrix in this second stage will include only those regressors that exhibit time variation.

Moreover, the composite error term ε_{it} has the familiar 2TSF structure, and we can make a 2TSF distributional assumption for it (see Chap. 7) that will give the density $f_\varepsilon(\varepsilon;\, \boldsymbol{\theta})$, where here $\boldsymbol{\theta}$ stands for all the composite distribution parameters. We then pool the data and apply maximum likelihood, where the log-likelihood contribution for observation it is

$$\ell_{it}(\boldsymbol{\beta}, \boldsymbol{\theta} | y_{it}, \mathbf{x}_{it}, \widehat{\alpha}_i^*) = \ln f_\varepsilon(y_{it} - \widehat{\alpha}_i^* + \mu_w(\boldsymbol{\theta}) - \mu_u(\boldsymbol{\theta}) - \mathbf{x}_{it}'\boldsymbol{\beta}\,;\, \boldsymbol{\theta}).$$

For the sample log-likelihood, we sum ℓ_{it} over i and over t.

Note that here we use the non-centered versions of the dependent variable and the time-varying regressors. We then re-estimate $\boldsymbol{\beta}$ jointly with $\boldsymbol{\theta}$; we do not profile the likelihood using $\widehat{\boldsymbol{\beta}}^{FE}$. We have written $\mu_u(\boldsymbol{\theta})$ and $\mu_w(\boldsymbol{\theta})$ deliberately to indicate

[8] In some cases, the regressor matrix may also include regressors that vary per cross-section but are time-invariant. They will be eliminated too.

generically that these mean values will be functions of the assumed composed error distribution parameters. In practice, not all elements of $\boldsymbol{\theta}$ will be present in each μ. If we have assumed the benchmark Normal-Exponential 2TSF distribution, each mean value equals the single scale parameter of the one-sided component that the mean represents:

$$\ell_{it}(\boldsymbol{\beta}, \sigma_v, \sigma_w, \sigma_u | y_{it}, \boldsymbol{x}_{it}, \widehat{\alpha}_i^*) = \ln f_\varepsilon^{\rm NE}\left(y_{it} - \widehat{\alpha}_i^* + \sigma_w - \sigma_u - \boldsymbol{x}_{it}'\boldsymbol{\beta}\,;\, \sigma_v, \sigma_w, \sigma_u\right).$$

But if we assume the Normal-Half-Normal 2TSF specification, the mean values equal the corresponding scale parameter times the factor $\sqrt{2/\pi}$.

$$\begin{aligned}\ell_{it}(\boldsymbol{\beta}, \sigma_v, \sigma_w, \sigma_u | y_{it}, \boldsymbol{x}_{it}, \widehat{\alpha}_i^*) \\ = \ln f_\varepsilon^{\rm NHN}\left(y_{it} - \widehat{\alpha}_i^* + \sqrt{\frac{2}{\pi}}\sigma_w - \sqrt{\frac{2}{\pi}}\sigma_u - \boldsymbol{x}_{it}'\boldsymbol{\beta}\,;\, \sigma_v, \sigma_w, \sigma_u\right).\end{aligned}$$

Other distributional assumptions would give more involved expressions for $\mu_w(\boldsymbol{\theta})$ and $\mu_u(\boldsymbol{\theta})$. The sigma parameters are identifiable, because they exist also in other places inside the density expression, and in a nonsymmetrical way.

Polachek and Yoon (1996) further iterate the second stage, by obtaining anew estimates for α_i^* through

$$\widehat{\alpha}_i^* = \bar{y}_i - \bar{\boldsymbol{x}}_i'\widehat{\boldsymbol{\beta}}^{MLE}, \qquad i = 1, \ldots, n,$$

and rerunning maximum likelihood to estimate $\boldsymbol{\beta}$ again, repeating the process until it converges to a fixed point, in the sense of obtaining the same estimates, or nearly the same according to some distance criterion.

Once these estimation steps have been completed, we have the following maximum likelihood estimates and predictors at our disposal:

1. Estimates for the distribution parameters $\boldsymbol{\theta}$ of the time-varying error components, usually $(\sigma_v, \sigma_w, \sigma_u)$;
2. Estimates for the regression coefficients of the time-varying regressors;
3. Predictions for the series $\{\alpha_i^*\}$, using time-means of the data and the final maximum likelihood estimates of the betas;
4. And so, a panel of residuals $\{\widehat{\varepsilon}_{it}\}$ that *should not* include the series $\{\alpha_i^*\}$[9].

Given all these, we can compute individual measures for the time-varying latent forces:

$$E(w_{it}|\widehat{\varepsilon}_{it}), \quad E(u_{it}|\widehat{\varepsilon}_{it}), \quad E\left(e^{\pm w_{it}}|\widehat{\varepsilon}_{it}\right), \quad E\left(e^{\pm u_{it}}|\widehat{\varepsilon}_{it}\right), \quad E\left(e^{\pm(w_{it}-u_{it})}|\widehat{\varepsilon}_{it}\right),$$

[9] Anticipate an alert in a while, as regards what are the residuals that we want and what a software package may store automatically as "residuals" in panel data models.

using the conditional expectation expressions that relate to the chosen distribution of the error term (available in Chap. 7 or the benchmark 2TSF specifications), without any modification.

But *we have learned nothing about the three original time-invariant components* $\alpha_i, \gamma_i, \eta_i$ (or the universal regression constant, or the time-invariant regressors). All these live only inside α_i^*, but we have no individual information on them or a way to decompose the latter into its constituent components. Moreover, α_i^* does not conform as-is to any meaningful interpretation, since it sums components that do not belong together.

This is why these elements are treated as "incidental," which brings to the surface an additional, technical argument in favor of FE estimation: the primary argument for the FE approach is that the random variables included in α_i^* may be correlated with the regressors inducing endogeneity, so we need to take them out of the picture in order to preserve the desirable properties of the estimator. But there is also an argument as regards the number of unknowns we attempt to estimate: if we wanted to directly predict the series of $\{\alpha_i^*\}$, we would be estimating n parameters, and so their number increases with the cross-sectional sample size. Plus $\boldsymbol{\beta}$, plus the number of the parameters of the composite error distribution. And our sample size is nT, so the magnitude of T becomes important.

With this link to the "incidental parameters" problem, we see that Polachek and Yoon (1996) in essence estimated what was later termed, in the single-tier stochastic frontier setting, the "true fixed-effects" (TFE) panel stochastic frontier model (Greene, 2005a). The theory behind the incidental parameters problem essentially requires a decoupling from those parameters whose dimension does not depend on n and those that do. For a linear model, using the "within" transformation allows consistent estimation of $\boldsymbol{\beta}$. However, in other settings, particularly nonlinear ones, it is typically much harder to eliminate the incidental parameters problem and an asymptotic bias exists that needs to be handled.

Greene's (2005a) approach was to ignore the incidental parameter problem and estimate $\boldsymbol{\beta}$ and α_i^* directly via maximum likelihood. In our setting, this would essentially amount to taking Eq. (9.5), and instead of using $\widehat{\alpha}_i^*$, we would include n dummy variables as additional parameters to be estimated. Greene (2005a) reports simulations with panels of dimensions $n = 500$ and $T = 5$, where the variance estimates of the composite error are poor.[10]

Because we want to learn things about the time-persistent components, we will consider next RE estimation. In that section, we will have to face the issue of incidental parameters, and we will provide some rules of thumb from the statistical community as regards a desired minimum ratio of sample size to the number of estimated parameters, and hence a minimum absolute magnitude for the T dimension of the sample, in order to avoid these issues (which are similar to what is known as "overfitting" in other contexts).

---

[10] Belotti and Ilardi (2018) proposed a marginal simulated maximum likelihood estimator to rectify this issue in the single-tier panel data frontier setting.

9.3 Random-Effects Two-Stage Estimation

In this section, we extend the RE two-stage methodology of Kumbhakar et al. (2014) to be applicable for the 2TSF panel data model.[11] We repeat for convenience Eq. (9.2):

$$y_{it} = \alpha_i + \beta_0 + \boldsymbol{x}_{it}'\boldsymbol{\beta} + [v_{it} + (\gamma_i + w_{it}) - (\eta_i + u_{it})].$$

Let a superscript "c" denote that the variable is centered on its mean, $x^c = x - E(x)$. We transform the specification by adding and subtracting mean values as follows:

$$\beta_0^* \equiv \beta_0 + \mu_\gamma - \mu_\eta + \mu_w - \mu_u,$$
$$\alpha_i^* \equiv \alpha_i + \gamma_i^c - \eta_i^c, \quad E(\alpha_i^*) = 0,$$
$$\varepsilon_{it}^c \equiv v_{it} + w_{it}^c - u_{it}^c, \quad E(\varepsilon_{it}^c) = 0.$$

Then the regression equation can be written as

$$y_{it} = \alpha_i^* + \beta_0^* + \boldsymbol{x}_{it}'\boldsymbol{\beta} + \varepsilon_{it}^c. \tag{9.6}$$

Importantly, note that here α_i^* has a different definition than the one it had in the FE framework. Specifically, it has been kept separate from the regression constant, while the included means are those of the two non-zero-mean time-invariant ones, not of the two time-varying latent forces, as was the case in the FE setup. This results in it having mean zero, and this will eventually prove critical.

First-Stage Estimation Equation (9.6) is a standard random-effects panel data model, and can be estimated as usual: we upload our data to the software and tell it to run panel data RE estimation (again, all the previous transformations are theoretical and do not affect the data). From this model, we keep as final the estimates $\widehat{\boldsymbol{\beta}}^{\text{RE}}$, but the estimate for β_0^* will be adjusted later on to arrive at the estimate for the universal constant β_0, if one is interested in its value.

Also, we will obtain a predicted series for $\{\widehat{\alpha}_i^*\}$ over cross-sections. Software packages usually compute these automatically in RE estimation and are available to retrieve.

Moreover, we will have a panel of residuals $\{\widehat{\varepsilon}_{it}^c\}$. ***Alert***: readers are strongly advised to read carefully the documentation of the software they are using, because

[11] As mentioned in Papadopoulos (2021d), the first to propose a random-effects six-component 2TSF model was a master thesis, "Six-Component Panel Stochastic Frontier Model" by Pang-Yu Wang (2017, National Taiwan University). But they provided wrong formulas for a 2TSF model using the Truncated Normal distribution, and they did not go into the required level of detail as regards the nuts and bolts of estimation.

what is automatically stored as "residuals" after the estimation of an RE panel data model may differ from what is needed here. For example, in the `gretl` software, the stored "residuals" equal our $\widehat{\alpha}_i^* + \widehat{\varepsilon}_{it}^c$ (the logic being that both are unobservables and also not part of the systematic part of the regression). Thankfully, the series $\widehat{\alpha}_i^*$ is also stored individually, so in order to obtain $\widehat{\varepsilon}_{it}^c$ (which is what we want), we subtract from the panel stored as "residuals" the former. Just remember that $\widehat{\varepsilon}_{it}^c$ is a panel, while $\widehat{\alpha}_i^*$ is a cross-section, so we will have

$$\text{for } i \neq j \in \{1, \ldots, n\}$$

$$\widehat{\varepsilon}_{it}^c = \widehat{\varepsilon}_{it}^{\text{gretl}} - \widehat{\alpha}_i^*, \qquad t = 1, \ldots, T$$

$$\widehat{\varepsilon}_{jt}^c = \widehat{\varepsilon}_{jt}^{\text{gretl}} - \widehat{\alpha}_j^*, \qquad t = 1, \ldots, T.$$

The same warning applies to FE estimation: read carefully the software documentation.

Second-Stage Estimation In the second stage, we perform *two separate* and unrelated estimation procedures. Consider again

$$\varepsilon_{it}^c \equiv v_{it} + w_{it}^c - u_{it}^c \implies \varepsilon_{it}^c + \mu_w - \mu_u = v_{it} + w_{it} - u_{it} \equiv \varepsilon_{it},$$

$$\alpha_i^* \equiv \alpha_i + \gamma_i^c - \eta_i^c \implies \alpha_i^* + \mu_\gamma - \mu_\eta \equiv \alpha_i + \gamma_i - \eta_i \equiv \psi_i^\alpha.$$

The rightmost expressions in both cases have the 2TSF structure. So, the densities for ε_{it} and for ψ_i^α are 2TSF densities (assuming that α_i also follows a Normal distribution). But the *argument* of these 2TSF densities will be $\varepsilon_{it}^c + \mu_w - \mu_u$ and $\alpha_i^* + \mu_\gamma - \mu_\eta$, respectively. Moreover, as regards the persistent effects, we can see here why it is critical for α_i^* to have zero mean: if this was not the case, to obtain the 2TSF structure, we would need to adjust it by something additional than just μ_γ and μ_η, something that most likely would be non-estimable, while μ_γ and μ_η are estimable, since they are functions of the parameters of the 2TSF density that we will assume.

Time-Varying Latent Forces To estimate the distribution parameters of the time-varying latent forces, we assume a 2TSF density and we estimate these parameters by maximum likelihood *pooling* the (predicted) data sample $\{\widehat{\varepsilon}_{it}^c\}$. There is no regression equation here. Also, because the argument of the density is not just the variable but it is augmented by the means, the expressions used must be adjusted accordingly.

To stick with the benchmark 2TSF Normal-Exponential model, recall from Chap. 7, Eq. (7.13), that the cross-sectional likelihood function for $\varepsilon_i = y_i - \beta_0 - \boldsymbol{x}_i' \boldsymbol{\beta}$ is

$$\mathcal{L} = -n \ln(\sigma_w + \sigma_u) + \sum_{i=1}^{n} \ln\left(e^{a_{1i}} \Phi(b_{1i}) + e^{a_{2i}} \Phi(b_{2i})\right),$$

where

$$a_{1i} = \frac{\sigma_v^2}{2\sigma_u^2} + \frac{\varepsilon_i}{\sigma_u}, \quad b_{1i} = -\left(\frac{\varepsilon_i}{\sigma_v} + \frac{\sigma_v}{\sigma_u}\right),$$

$$a_{2i} = \frac{\sigma_v^2}{2\sigma_w^2} - \frac{\varepsilon_i}{\sigma_w}, \quad b_{2i} = \frac{\varepsilon_i}{\sigma_v} - \frac{\sigma_v}{\sigma_w}.$$

Here, because we "observe" $\widehat{\varepsilon}_{it}^c$ while the above density is the density of ε_{it} and not of $\widehat{\varepsilon}_{it}^c$, the pooled log-likelihood is

$$\mathcal{L} = -nT\ln(\sigma_w + \sigma_u) + \sum_{t=1}^{T}\sum_{i=1}^{n} \ln\left(e^{a_{1it}}\Phi(b_{1it}) + e^{a_{2it}}\Phi(b_{2it})\right), \tag{9.7}$$

but one should use

$$a_{1it} = \frac{\sigma_v^2}{2\sigma_u^2} + \frac{\widehat{\varepsilon}_{it}^c + \mu_w - \mu_u}{\sigma_u}, \quad b_{1it} = -\left(\frac{\widehat{\varepsilon}_{it}^c + \mu_w - \mu_u}{\sigma_v} + \frac{\sigma_v}{\sigma_u}\right)$$

$$a_{2it} = \frac{\sigma_v^2}{2\sigma_w^2} - \frac{\widehat{\varepsilon}_{it}^c + \mu_w - \mu_u}{\sigma_w}, \quad b_{2it} = \frac{\widehat{\varepsilon}_{it}^c + \mu_w - \mu_u}{\sigma_v} - \frac{\sigma_v}{\sigma_w}.$$

In this particular 2TSF specification, it so happens that $\mu_w = \sigma_w$ and $\mu_u = \sigma_u$. But with other specifications, we should remember to include the expressions for the means, not the scale parameters.

Individual Predictions for Time-Varying Effects The above procedure will give us estimates for σ_v, σ_w, and σ_u. The conditional expectation formulas of Chap. 7 are directly applicable, as formulas. But in those formulas, we must insert the predicted ε_{it} and *not* $\widehat{\varepsilon}_{it}^c$ that we currently have available. We can obtain the former by

$$\widehat{\varepsilon}_{it} = \widehat{\varepsilon}_{it}^c + \widehat{\mu}_w - \widehat{\mu}_u.$$

So, we can eventually predict the latent forces at the panel-observation level. Again we stress that μ_w and μ_u are equal to σ_w and σ_u only if we assume that w and u follow Exponential distributions.

Persistent Latent Forces This is the unique contribution of RE estimation. The methodology is exactly the same as for the time-varying effects. We assume a 2TSF density for $\psi_i^\alpha = \alpha_i + \gamma_i - \eta_i$, where α_i is the zero-mean component. The "sample" size here is equal to the n dimension of the panel. We implement the density we choose as before, inserting $\widehat{\alpha}_i^* + \mu_\gamma - \mu_\eta$ where the argument ψ_i^α of the density must be. So, for example, for the NE specification, we will have the log-likelihood:

$$\mathcal{L} = -n \cdot \ln(\sigma_\gamma + \sigma_\eta) + \sum_{i=1}^{n} \ln\left(e^{a_{1i}}\Phi(b_{1i}) + e^{a_{2i}}\Phi(b_{2i})\right), \tag{9.8}$$

with

$$a_{1i} = \frac{\sigma_\alpha^2}{2\sigma_\eta^2} + \frac{\widehat{\alpha}_i^* + \mu_\gamma - \mu_\eta}{\sigma_\eta}, \qquad b_{1i} = -\left(\frac{\widehat{\alpha}_i^* + \mu_\gamma - \mu_\eta}{\sigma_\alpha} + \frac{\sigma_\alpha}{\sigma_\eta}\right)$$

$$a_{2i} = \frac{\sigma_\alpha^2}{2\sigma_\gamma^2} - \frac{\widehat{\alpha}_i^* + \mu_\gamma - \mu_\eta}{\sigma_\gamma}, \qquad b_{2i} = \frac{\widehat{\alpha}_i^* + \mu_\gamma - \mu_\eta}{\sigma_\alpha} - \frac{\sigma_\alpha}{\sigma_\gamma}.$$

We run maximum likelihood without a regression function, using $\widehat{\alpha}_i^*$ as our "data," and we obtain estimates for the distribution parameters of the assumed specification, from which we can also compute μ_γ and μ_η, if they are not estimated directly. With them, we have characterized the marginal distribution of the persistent latent forces.

Note that in principle, the 2TSF distributional assumption here need not be identical to the previous one. This may have a structural justification. For example, the researcher may have reasons to believe that the real-world situation under study is such that the time-varying components are correlated, while the time-invariant components are independent but have their mode away from zero. In such a case, the 2TSF density for the time-varying components could be one of those presented in Chap. 10, while for the persistent components, we could select the Generalized Exponential (see Chap. 14). This is an additional advantage of the RE model: it allows structural differentiation between the time-varying and the time constant latent forces.

Individual Predictions for Persistent Effects We can obtain individual predictions at the cross-sectional level, by using the standard conditional expectation formulas that here have the form, say, $E(\gamma_i|\widehat{\psi}_i^\alpha)$ $i = 1, \dots, n$, while we retrieve ψ_i^α by

$$\widehat{\psi}_i^\alpha = \widehat{\alpha}_i^* + \widehat{\mu}_\gamma - \widehat{\mu}_\eta.$$

Note that since there is no regression equation, the conditioning variable is not a "residual" from estimation, but the predicted variable whose density we used for the log-likelihood.

Universal Constant Finally, we can retrieve the universal constant of the regression through

$$\beta_0^* \equiv \beta_0 + \mu_\gamma - \mu_\eta + \mu_w - \mu_u \implies \widehat{\beta}_0 = \widehat{\beta}_0^* - \widehat{\mu}_\gamma + \widehat{\mu}_\eta - \widehat{\mu}_w + \widehat{\mu}_u. \tag{9.9}$$

9.3.1 Incidental Parameters and Sample Size

We saw that the RE estimation framework estimates fully the six component 2TSF panel data model. How reliable are its estimates, given the large number of parameters that we are estimating? To remind the reader, we have a panel data sample of total size nT, and in the first stage, we are estimating n heterogeneity effects and K coefficients, while we expect that T will be smaller than n, usually much smaller. Harrell (2015, pp. 72–74), reviewing studies that examine the issue of "number of parameters and sample size" in general statistical setups, finds that 10–20 observations per estimated coefficient appear a safe general rule of thumb, depending also on the strength of some measure of "signal-to-noise" in the data. So, and as long as the number of regressors is much smaller than the cross-sectional dimension, it is not advisable to implement RE estimation unless we have $T \geq 10$. This is not so difficult to achieve anymore. For example, in the "Going Green in China" empirical application of Chap. 5, the panel had dimension $T = 11$, so one could apply proper random-effects panel data estimation, without strong objections as to the possible detrimental effects due to the estimation of the incidental (time-invariant) parameters.

9.3.2 Method-of-Moments Estimation

The two estimation procedures in the second stage of the RE model can be carried out using a method-of moments (MoM) approach, instead of maximum likelihood. Recall that from the RE estimation in the first stage (which essentially implements some variant of least-squares methodology), we obtain residual series for the regression error and for the time-invariant effects. Going to the second stage, instead of running maximum likelihood, we could make our 2TSF distributional assumptions, obtain the related cumulant equations, and either solve the resulting system of equations (the "COLS" methodology of Sect. 7.1.2) or, even better, execute generalized method of moments to obtain also standard errors for the estimated distributional parameters (see Sect. 7.1.4).

But we have to pause for a moment and examine whether we can apply the methodology as-is or whether it needs modification.

Time-Varying Latent Forces For the time-varying forces, what we have available from the first stage is the panel of $\widehat{\varepsilon}^c_{it}$, whose theoretical definition is

$$\varepsilon^c_{it} \equiv v_{it} + w^c_{it} - u^c_{it}.$$

The theoretical cumulant equations are (under assumed independence)

$$\kappa_2(\varepsilon^c) = \kappa_2(v) + \kappa_2(w^c) + \kappa_2(u^c) \quad = \kappa_2(v) + \kappa_2(w - \mu_w) + \kappa_2(u - \mu_u)$$
$$\kappa_3(\varepsilon^c) = \kappa_3(w^c) - \kappa_3(u^c) \quad = \kappa_3(w - \mu_w) + \kappa_3(u - \mu_u)$$
$$\kappa_4(\varepsilon^c) = \kappa_4(w^c) + \kappa_4(u^c) \quad = \kappa_4(w - \mu_w) + \kappa_4(u - \mu_u).$$

Now, μ_w and μ_u are constants, and a fundamental property of cumulants of order greater than one, is, for some constant c, that

$$r > 1, \qquad \kappa_r(x + c) = \kappa_r(x).$$

So, the theoretical cumulants on the right-hand side are identical to the cumulants of the uncentered variables. If we want to assume a 2TSF Normal-Exponential distribution, this would imply, for example, that

$$\text{2TSF NE:} \qquad \kappa_2(w^c) = \kappa_2(w) = \sigma_w^2,$$

while if we decide to assume a 2TSF Normal-Half-Normal specification,

$$\text{2TSF NHN:} \qquad \kappa_2(w^c) = \kappa_2(w) = \left(1 - \frac{2}{\pi}\right)\sigma_w^2.$$

Likewise for the other cumulants. So, the equations are the same as in the simple cross-sectional setup. The left-hand sides in the cumulant equations will be the empirical cumulants of the available pooled panel $\{\widehat{\varepsilon}_{it}^c\}$.

Individual Predictions After obtaining the estimates for the 2TSF distribution parameters, we can correct $\{\widehat{\varepsilon}_{it}^c\}$ to obtain the uncentered variable as shown previously and then again use the mathematical formulas of Chap. 7 as-is for the conditional expectations.

Persistent Latent Forces For the persistent latent forces, what we have available as (predicted) data from the first stage is the cross-section $\{\widehat{\alpha}_i^*\}$, whose theoretical definition is

$$\alpha_i^* \equiv \alpha_i + \gamma_i^c - \eta_i^c.$$

The situation is exactly the same as with the time-varying framework above.

So in all cases, we conclude that a method-of-moments estimation of the distribution parameters in the second stage, as an alternative to maximum likelihood, does not require any specially adjusted formulas, as regards the latent forces, compared to those available for the simple cross-sectional setup.

9.3.3 *Modeling Correlation Between Regressors and Heterogeneity*

In the RE framework, our estimates will be asymptotically consistent under assumptions of mean independence between the regressors and the individual heterogeneity term, and there is always concern that these assumptions may be violated in the data, leading to endogeneity and inconsistent estimates. The seminal work of Mundlak (1978) provides a way to handle this problem head-on: if there is correlation between the individual heterogeneity term and the regressors, we might as well model it.

Let $j = 1, \ldots, K$ be the index for the regressors we have. So, x_{jit} denotes the realization of regressor j related to cross-section i at time period t. Let the time average for this regressor and this cross-section be written:

$$\bar{x}_{ji} = T^{-1} \sum_{t=1}^{T} x_{jit} \,.$$

Include all such time averages for cross-section i in a vector, $\bar{\boldsymbol{x}}_i = (\bar{x}_{1i}, \bar{x}_{2i}, \ldots, \bar{x}_{Ki})'$. Define a vector of constants, common for all cross-sections, $\boldsymbol{\phi} = (\phi_1, \ldots, \phi_K)'$. Then we model the unobserved heterogeneity of cross-section i as

$$\alpha_i = \bar{\boldsymbol{x}}_i' \boldsymbol{\phi} + \zeta_i, \qquad E(\zeta_i) = 0. \tag{9.10}$$

We can now reasonably assume that the component ζ_i is uncorrelated with $\boldsymbol{x}$, being essentially the "pure" unobservable heterogeneity component, totally idiosyncratic, while the heterogeneity that relates to the regressors is assumed to have a common structure for all cross-sections, but changing in value as the time averages of the regressors change per cross-section. In a sense, Mundlak (1978) exploited the time dimension of the panel to perform *reverse instrumentation*: instead of instrumenting the endogenous regressors, he instrumented the unobservable that is correlated with them.[12]

Inserting this expression into the six-component 2TSF panel data model Eq. (9.2), we have, substituting for α_i and re-arranging,

$$y_{it} = \zeta_i + \bar{\boldsymbol{x}}_i' \boldsymbol{\phi} + \beta_0 + \boldsymbol{x}_{it}' \boldsymbol{\beta} + [v_{it} + (\gamma_i + w_{it}) - (\eta_i + u_{it})]. \tag{9.11}$$

At the price of having to estimate K more parameters (the $\boldsymbol{\phi}$ vector), we have a six-component 2TSF panel data model without detrimental correlations, and we can implement the RE two-stage estimation procedure, doing with ζ_i all the transformations that we did before with α_i.

[12] He was, after all, Yair Mundlak, the pioneer of panel data econometrics.

9.4 Computing and Interpreting Metrics in the 2TSF Panel Data Model

The M-metrics we compute based on the definitions of Chap. 2 must be put in context, being metrics developed *relative to some base*. In the cross-sectional setting, we have a single latent force on each side, and we counterfactually include and exclude it to obtain certain meaningful measurements of the situation. Moreover, we measure in relation either to the "actual" outcome or in relation to the "balanced" outcome, where the latent forces have zero effect. In the panel data setting, we have time-invariant individual effects and also time-varying and persistent latent forces on both sides. How are we going to compute and interpret the metrics we have defined? It proves that the situation is very different depending on whether we estimate a logarithmic equation (dependent variable in logarithms) or one in the original measurement scale.

Logarithmic Estimating Equation Suppose that we want to compute the M_1 metric for the time-varying component w_{it}. How is this computed and how is it interpreted?

The abstract definition of this metric is

$$M_1 \equiv E\left(\frac{Y - Y^{LF}}{Y}\right).$$

What constitutes the "lower frontier" in the panel data setting when examining the time-varying w_{it}? In order to use the same computational formulas, we must have

$$\begin{aligned} &M_1(w_{it}) \\ &= E\left(\frac{G(\boldsymbol{x})\cdot\exp\{v_{it}+\alpha_i+\gamma_i-\eta_i+w_{it}-u_{it}\}-G(\boldsymbol{x})\cdot\exp\{v_{it}+\alpha_i+\gamma_i-\eta_i-u_{it}\}}{G(\boldsymbol{x})\cdot\exp\{v_{it}+\alpha_i+\gamma_i-\eta_i+w_{it}-u_{it}\}}\right) \\ &= 1 - E(e^{-w_{it}}). \end{aligned}$$

But we would get *exactly the same result* if we wrote, say

$$\begin{aligned} M_1(w_{it}) &= E\left(\frac{G(\boldsymbol{x})\cdot\exp\{v_{it}+w_{it}-u_{it}\}-G(\boldsymbol{x})\cdot\exp\{v_{it}-u_{it}\}}{G(\boldsymbol{x})\cdot\exp\{v_{it}+w_{it}-u_{it}\}}\right) \\ &= 1 - E(e^{-w_{it}}). \end{aligned}$$

The reader can verify that this perhaps surprising property holds for all metrics.[13] This essentially comes from the multiplicative way that the unobservables interact with the systematic part. It implies that each metric computed with its usual formulas for each unobservable provides a measure for its effects that is not colored by which other unobservables have been included in the computational formula, as long as of course they are included, or excluded, in an internally consistent way. Put another way, it is misleading, actually wrong, to say "$M_1(w_{it})$ gives us the contribution of w_{it} to the outcome, but conditional on the presence and effects of the time-invariant unobservables,"—because we just saw that this is not true. The time-invariant unobservables could be set equal to zero, and still the $M_1(w_{it})$ metric would give the same result.

This should be considered as comforting, because otherwise the interpretation of the metrics could become seriously entangled and confusing—or more challenging intellectually.

Estimating Equation Is in Levels ... And indeed, this "invariance" result related to the M-metrics does not hold when the estimating equation is in levels. Here the additive relation between $G(\boldsymbol{x})$ and the unobservables destroys the property. We leave it as a useful exercise for the readers to contemplate what happens and how the M-metrics can and should be used in the case of a panel data model where we estimate a six-component 2TSF model with the dependent variable in its original measurement units.

9.5 Heterogeneity in the 2TSF Panel Data Model

9.5.1 Heterogeneity in the Time-Varying Latent Forces

Implementing the spirit of the models in Chap. 8, and under the assumption that the distributions assumed for the time-varying latent forces have the scaling property, we can express the time-varying latent forces as functions of covariates and an independent random variable:

$$w_{it} = w_{it}^* e^{z'_{w,it}\delta_w}, \qquad u_{it} = u_{it}^* e^{z'_{u,it}\delta_u}.$$

Consequently, by applying the mathematical manipulations of Chap. 8 on the original 2TSF panel data regression Eq. (9.2), we get

$$y_{it} = (\alpha_i + \gamma_i - \eta_i + \beta_0) + \boldsymbol{x}'_{it}\boldsymbol{\beta} + \mu_w^* e^{z'_{w,it}\delta_w} - \mu_u^* e^{z'_{u,it}\delta_u} + \varepsilon_{it}^*,$$

[13] Except for the metric M_3-"relative strength," which is a pairwise comparison and does not involve a frontier to begin with.

$$\varepsilon_{it}^{*} = v_{it} + (w_{it}^{*} - \mu_{w}^{*})e^{z_{w,it}'\delta_w} - (u_{it}^{*} - \mu_{u}^{*})e^{z_{u,it}'\delta_u}.$$

This is a regression equation with a nonlinear (in parameters) part and an error term that is zero-mean and mean-independent from the regressors as well as the z variables. To remain in the FE spirit, we need to eliminate the time constant terms. We can do that by first-differencing the equation leading to

$$\Delta y_{it} = \Delta \mathbf{x}_{it}'\boldsymbol{\beta} + \mu_{w}^{*}\left(e^{z_{w,it}'\delta_w} - e^{z_{w,it-1}'\delta_w}\right) - \mu_{u}^{*}\left(e^{z_{u,it}'\delta_u} - e^{z_{u,it-1}'\delta_u}\right) + \Delta\varepsilon_{it}^{*}, \tag{9.12}$$

where for any variable y, $\Delta y_{it} = y_{it} - y_{it-1}$. This model can be estimated using generalized nonlinear least squares to account for the covariance structure that exists in $\Delta\varepsilon_{it}^{*}$.

After estimation, we will have available estimates for $\boldsymbol{\beta}$ attached to time-varying regressors, and also estimates for μ_{w}^{*} and μ_{u}^{*}, and the corresponding $\boldsymbol{\delta}$s. With these, we can also compute predicted panels for the time-varying latent forces:

$$E(\widehat{w}_{it}|z_{it}) = \widehat{\mu}_{w}^{*}e^{z_{w,it}'\widehat{\delta}_w}, \qquad E(\widehat{u}_{it}|z_{it}) = \widehat{\mu}_{u}^{*}e^{z_{u,it}'\widehat{\delta}_u}.$$

We do not get estimates for $\boldsymbol{\beta}$ that relate to time-invariant regressors, and we still do not learn anything about the time-invariant unobservables. This is still the FE framework, although with more nuance in how the time-varying latent forces are determined. The z-conditional expectations of Chap. 8 are also applicable.

9.5.2 Nonlinear Heterogeneity

Exhibiting much needed and duly appreciated modesty, researchers have by now acknowledged that whatever the specification of a structural model, possibly important unmodeled factors are bound to exist influencing visibly the phenomenon under study (although not necessarily catastrophically for the empirical study under way). The code name for this in the panel data context is "individual heterogeneity," and it has become commonplace to invoke it and handle it at the technical level without exploring what actually may be the structural setup that leads to heterogeneity worthy of being separated from noise.

Das and Polachek (2017) is an exception to this tendency. Their goal was to estimate *gross* flows in and out of employment, which are rightfully considered to be as important for policy as net flows are. By building their model from scratch using identities, logic, and reasonable mathematical manipulations, they arrived at a panel data model (with cross-sections being further characterized by three group identifiers—US state, age, and gender), where the net change in employment (in percentage terms) is shown to be a function of population changes and of the proportion of each group population that is employed. But this proportion is also shown to be the difference of two unobservable terms, the first related to gross inflows to employment, minus the second which is related to the gross outflows from employment. Moreover, the first one depends organically on a group-specific parameter. With a conventional zero-mean noise component included, they arrived at the 2TSF error structure. Specifically, their panel data specification is

$$y_{it} = x_{it} + v_{it} + w_{it}(e^{\theta_{0i}} - 1) - u_{it}, \tag{9.13}$$

where y_{it} captures growth in the labor force, x_{it} are changes in the population, while u_{it} is the labor market exit and w_{it} is the labor market entry. Note that there is no "systematic part of the regression" as usual, since y_{it} and x_{it} are data and there is no regression coefficient attached to x_{it}. In fact, we have essentially a determination of the composite error term by data, and then the goal is to estimate its various components. This is clear if we write

$$y_{it} - x_{it} = \varepsilon_{it}, \qquad \varepsilon_{it} = v_{it} + w_{it}(e^{\theta_{0i}} - 1) - u_{it}, \tag{9.14}$$

with the left-hand side being pure data. Apart from that, the difference from the traditional panel data model is the absence of an additive heterogeneity term, which now is present nonlinearly in one of the components of the error term. This, in turn, leads to the positive component of ε_{it} no longer being identically distributed with the negative component. Nevertheless, $\theta_{0i} \geq 0$ represents the steady-state value of (a deterministic function of) the rate of labor force participation (per group); hence, it is a constant over time periods. In such a case, under the assumption that w_{it} is Exponentially distributed, so too is $w^*_{it} = w_{it}(e^{\theta_{0i}} - 1)$, but with a mean that varies over data groups: $E(w^*_{it}) = E(w_{it}) \cdot (e^{\theta_{0i}} - 1)$, and θ_{0i} will change as the group changes. So the distribution remains an Exponential, with a different scale parameter.

This is an example of heterogeneity arising organically, which also determines how it appears in the model. Here the time-varying and persistent components are linked in a nonlinear way, and the one influences the effects that the other has.

As regards the estimation method, the likelihood function is nearly identical to that in Polachek and Yoon (1996) except that instead of linear heterogeneous effects, the mean for w_{it} changes with i. For large i, this makes direct estimation infeasible as the incidental parameter problem again arises. To avoid this, Das and Polachek (2017) also propose a two-stage approach. First, they transform their benchmark model so that θ_{0i} is identified (and can be estimated). Second, these estimates of θ_{0i} are placed back into the likelihood function which is then maximized over the remaining parameters.

9.6 Application: The Original 2TSF Panel Data Study Extended

We apply here the six-component model to the data from Polachek and Yoon (1996), allowing both persistent and time-varying latent forces. Polachek and Yoon's (1996) data is taken from the Panel Study of Income Dynamics (PSID) for 838 workers for the period 1968–1984 (a balanced panel). The authors account for time effects by deflating wages by the Consumer Price Index. Their empirical model uses as regressors years of education, years of experience, and months of tenure, plus the squares of experience and tenure with the dependent variable the logarithm of deflated wages. We convert tenure into years as well, to reduce the number of decimal places in our estimates.

Polachek and Yoon (1996) estimated three different versions of their model. First, they ignored any structure inherent in either latent heterogeneity or the error term and estimated the model via ordinary least squares. Second, they ignored the latent heterogeneity completely and estimated a pooled cross-section two-tier stochastic frontier. Lastly, they estimate by the FE method the full model, in the way that we presented in detail earlier.

First, as a replication exercise, we estimate these three models, and then we extend our reach by applying RE two-stage estimation using both maximum likelihood and method-of-moments estimation of the six-component model. We note that the time dimension is $T = 17$, which exceeds sufficiently the minimum sample-size rule of thumb for RE estimation ($T \geq 10$). Our estimates appear in Table 9.1. There are several important features. First, our OLS estimates, unsurprisingly, are identical to those reported in Polachek and Yoon (1996) as are the standard errors (though it should be noted that they reported asymptotic t-statistics in their Table 9.1). All of the coefficients have the anticipated signs, and we see inverted U-shapes for both experience and tenure on earnings (i.e., a positive coefficient for the level variable and a negative one for the square).

If we move to the pooled cross-sectional results, again our estimates are virtually identical. The intercept is slightly different (0.006 vs. 0.01639), but otherwise we are exact to three decimal places for the key variables. We also see near equality of the three variance components. Also, Polachek and Yoon (1996) report a log-likelihood value of -7733.35, whereas our likelihood function is -7730.32. Quite remarkable similarity between codes written nearly 30 years apart. Turning to the fixed-effects two-stage estimation (where we note that there are no estimates for the time-invariant regressors, the universal constant and education), again our estimates are virtually identical. Our estimate for experience is 0.0349, whereas that reported in Polachek and Yoon (1996) is 0.0363, while our estimate for tenure is 0.000187, whereas that reported in Polachek and Yoon (1996) is 0.000194.[14]

[14] We have divided our FE coefficient estimate on tenure shown in Table 9.1 by 12 to account for the difference in scaling.

Table 9.1 Estimates of the 2TSF earnings function

	OLS	Pooled	Fixed effects	RE-MLE	RE-MoM
Constant	− 0.0217	0.0060	–	− 0.0139	
	(0.0265)	(0.0264)	–	(0.0540)	
Educ	0.0691	0.0682	–	0.0695	
	(0.0013)	(0.0013)	–	(0.0038)	
Exper	0.0369	0.0358	0.0349	0.0383	
	(0.0015)	(0.0014)	(0.0013)	(0.0015)	
Exper^2	− 0.0006	− 0.0006	− 0.0005	− 0.0006	
	(0.0000)	(0.0000)	(0.0000)	(0.0000)	
Tenure	0.0102	0.0094	0.0022	0.0032	
	(0.0014)	(0.0013)	(0.0007)	(0.0010)	
Tenure^2	− 0.0003	− 0.0002	− 0.0001	− 0.0001	
	(0.0000)	(0.0000)	(0.0000)	(0.0000)	
σ_v	–	0.2535	0.0339	0.0603	[Type II error]
	–	(0.0083)	(0.0058)	(0.0052)	
σ_w	–	0.2504	0.1855	0.1918	0.2320
	–	(0.0060)	(0.0022)	(0.0022)	
σ_u	–	0.2540	0.1856	0.1891	0.2200
	–	(0.0079)	(0.0022)	(0.0021)	
σ_α	–	–	–	0.2294	0.2317
	–	–	–	(0.0212)	
σ_γ	–	–	–	0.1649	0.1923
	–	–	–	(0.0174)	
σ_η	–	–	–	0.1650	0.1293
	–	–	–	(0.0177)	

Moving to the variance component estimates for the fixed-effects framework, we again see near-universal agreement. Polachek and Yoon (1996) report estimates of 0.0322, 0.1939, and 0.1778, whereas we have 0.0339, 0.1856, and 0.1855, respectively. Moreover, their reported likelihood value of − 311.46 is quite similar to what we obtained, − 311.763.

Next, we extend the work on this data set, estimating now using the random-effects methodology of the proposed six-component model. First, our main parameter estimates are in line with the estimates from the other three columns in terms of magnitude and economic and statistical significance. Second, the RE error components for the time-varying effects are quite similar to those found in the FE setting, though σ_v is double its size compared to that estimated under the fixed-effects framework. Here too, the individual time-varying terms have roughly the same variation, a consistent finding with those stemming from the pooled and fixed-effects approaches as well. However, under RE estimation, we can decompose the variation in latent heterogeneity into that which is idiosyncratic and that which resides on either side of the market. For the time-persistent components, we obtain an estimate of σ_α that is larger than either σ_γ or σ_η by nearly 40%.

We also estimated the sigma coefficients using method-of-moments, based on the random-effects model. The estimates do differ but they are not very far from the RE-MLE ones. But the main issue here is that we encountered a Type II error (Olson et al., 1980) when estimating the variance of our noise component. Namely, the system of cumulant equations gave a negative value for $\widehat{\sigma}_v$, which is inadmissible. This makes this bundle of estimates questionable.

The analysis of Polachek and Yoon (1996) stopped at average measures, since back then the formulas for individual metrics were not available. But we can go even further and assess the impact of information deficiency on both sides of the labor market; that which is time constant (persistent) and that which is time varying, by computing the related conditional expectations.

As mentioned, the exact same mathematical formulas are used to compute conditional expectations in all three models since in all cases, eventually we have assumed a proper, not modified in any way, 2TSF Normal-Exponential specification. Assuming that the true model is the six-component one, from an interpretational point of view, the conditioning variable in all three models may be the "error term," but this error term (and its predictor) incorporates different variables. To facilitate comparative thinking, we list below these theoretical relations:

$$\begin{aligned}
\text{Pooled ignoring fixed effects}: \varepsilon_{it}^{\mathrm{P}} &= (v_{it} + w_{it} - u_{it}) + (\alpha_i + \gamma_i - \eta_i)\\
\text{Fixed Effects} - \text{Time varying}: \varepsilon_{it}^{\mathrm{FE}} &= v_{it} + w_{it} - u_{it}\\
\text{Random Effects} - \text{Time varying}: \varepsilon_{it}^{\mathrm{RE}} &= v_{it} + w_{it} - u_{it}\\
\text{Random Effects} - \text{Time constant}: \psi_i^{\alpha} &= \alpha_i + \gamma_i - \eta_i .
\end{aligned}$$

These relations explain why the estimated sigmas in the pooled model (see again Table 9.1) are higher than the corresponding estimates in the FE and RE model. In fact, from the above expressions, we should expect that the estimated variance/standard deviation of the error term in the pooled model should be close to the estimated variance *of the sum* of the two RE triplets:

$$\left(\sigma_v^P\right)^2+\left(\sigma_w^P\right)^2+\left(\sigma_u^P\right)^2 \approx \left(\sigma_v^{RE}\right)^2+\left(\sigma_w^{RE}\right)^2+\left(\sigma_u^{RE}\right)^2+(\sigma_\alpha)^2+(\sigma_\gamma)^2+(\sigma_\eta)^2 .$$

And indeed, the left-hand side evaluates to 0.1832, while the right-hand side equals 0.1914, which is pretty close.

In the figures, we use everywhere the symbols "w" and "u" in the definitions of the various metrics (upper right boxes in each subfigure). They are to be understood as representing different variables per model. For example, in the "pooled" subfigures, w stands for $w+\gamma$, while in the RE-time constant subfigures, w stands for γ. Further, for our RE kernel density plots, we use the estimates based on

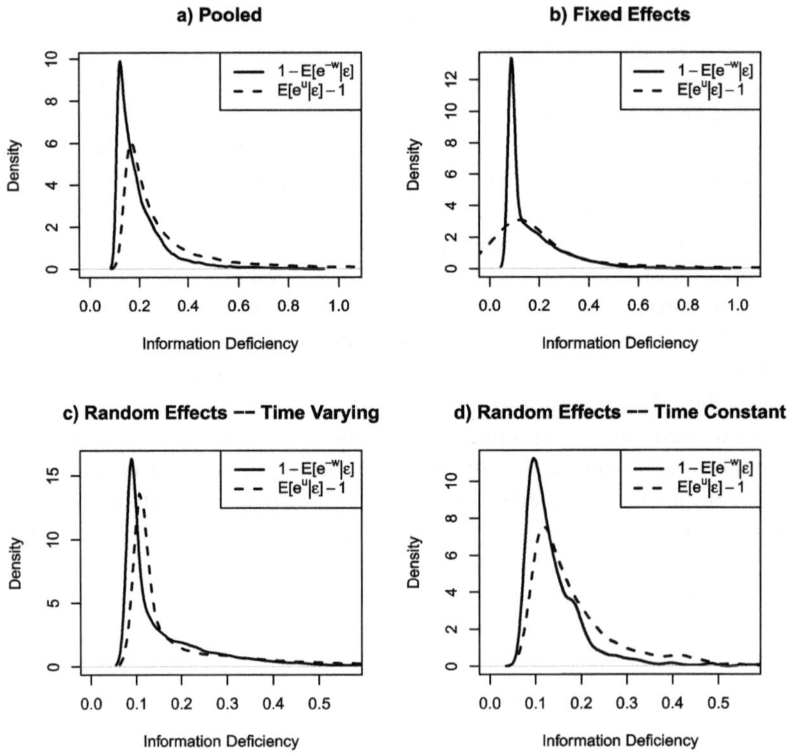

Fig. 9.1 Kernel densities for the M_1 and M_2 metrics

maximum likelihood for the residual series (since method-of-moments estimation produced a Type II error).

To start, Fig. 9.1 plots the estimated kernel densities for the M_1 and M_2 metrics, the relative gain/loss due to deficiency. We plot these estimated metrics for each of our models: pooled, fixed, and random effects, the latter being subdivided into persistent and time-varying effects. For both the pooled and random effects, the relative size of the M_1/M_2 metrics and the shape of the densities are quite similar. The majority of the effects are between 5% and 25%. Still, note that the scaling of the two subfigures, (a) and (c), is different: the metrics based on the "pooled" model (benchmark 2TSF NE model) have densities that are less concentrated and their right tail is more persistent. This aligns with their mean values (the estimates) being higher than the corresponding mean estimates for the RE model. For the FE estimates, we see a similar shape, but the distance between the modes of M_1 and M_2 here is much higher than in the Pooled and RE settings. This is most likely due to an overall compression of the heterogeneity since the time constant effects are removed as a whole rather than accounted for individually.

Figure 9.2 is identical to what is presented in Fig. 9.1 but using the M_5 and M_6 metrics, the markups/markdowns relative to the balanced outcome due to

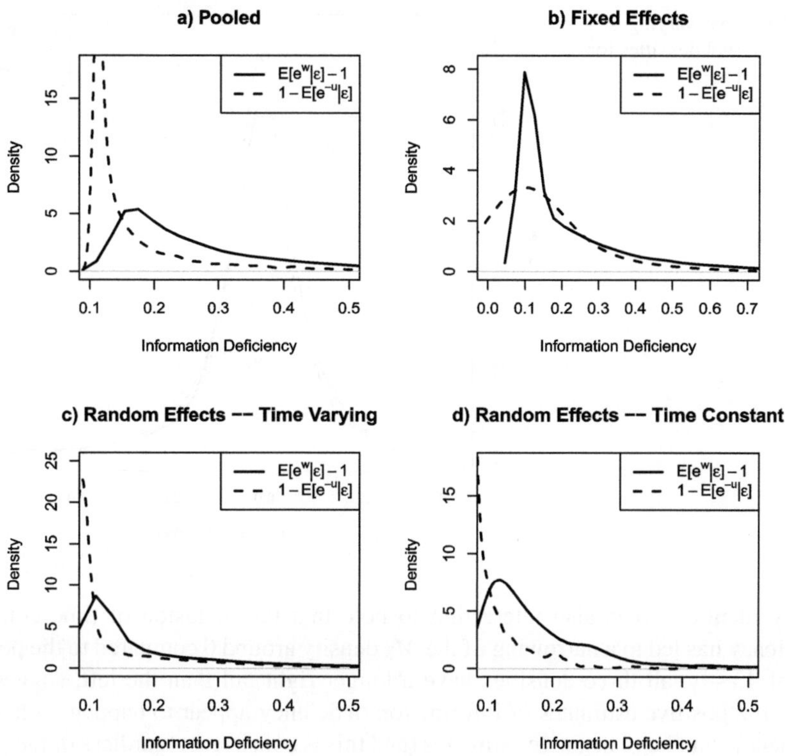

Fig. 9.2 Kernel densities for M_5 and M_6 metrics

information deficiency. Again, we plot these estimated metrics for each of our models: pooled, fixed, and random effects, with the random effects being further subdivided into persistent and time-varying effects.

Here we see some interesting patterns. First, due to the worker (seller) information deficiencies, we see markdowns from full-information wages across all of the models that are on the order of 10–20% and all have roughly the same shape (a large spike at/near zero with a thin right-hand tail). Where differences arise is in the markups due to the information deficiency of employers (buyers). The pooled model has a mode near 20%, while for both persistent and time-varying markups in the random-effects setting, the mode is smaller, near 10%. The estimated markups in the fixed effect setting are similar in shape, but the impact has switched. Whereas in both pooled and RE settings, the M_6 metric has the higher mode, M_5 has the higher mode in the FE setting.

Figure 9.3 plots the combined markup/markdown for the time-varying component of the random-effects model against the pooled cross-section and the fixed-effects specification. This is metric M_7 giving the proportional distance of actual wage from the full-information wage. We see again that all three models have similar shapes, modes, and overall effects, but the RE and FE models are

Fig. 9.3 Time-varying effects. Kernel densities for the M_7 metric

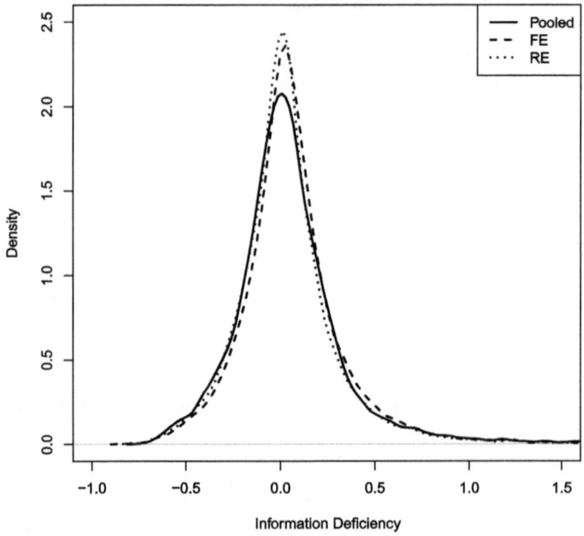

nearly identical. It is also interesting to note that the inclusion of time constant deficiency has led to a narrowing of the M_7 density around 0 compared to the pooled model. Lastly, all three densities have a longer right tail than the left, suggesting that large positive estimates of information deficiency appear to happen with more frequency than large negative estimates (and this is consistent regardless of the panel specification).

Figure 9.4 plots the net gain in wages (metric M_{10}), again for the three models, using only the time-varying component of the random-effects model. We see that all three models have similarly shaped kernel densities with the FE and RE being nearly identical. All three models have a mode near 0, the implication being that the impact of deficiency on wages in the labor market (across both sides) is likely 0 or close to it. However, there is also a much longer left tail in these three densities which does suggest that with more regularity, there are workers whose wages are in some sense harmed by deficiency in a way that employers are not.

It is insightful to contrast the long *right* tail of the M_7 metric with the long *left* tail of the M_{10} one: net information deficiency outliers related to the full-information wage tend relatively more to be positive (M_7), increasing the actual wage above the former; but since we do not have a long right also in the M_{10} metric, this implies that in cases with high M_7, the actual wage is already high pointing toward high earners. On the other hand, net information deficiency outliers as regards gains relative to actual wage (M_{10}) have the opposite tendency and are negative. Since we do not observe a long left tail in the M_7 metric, this implies that in cases with negative M_{10}, the actual wage is low, pointing to low earners: this information asymmetry as regards high and low earners, with high earners being better informed, accords well with casual observation, although it could also mask bargaining power effects

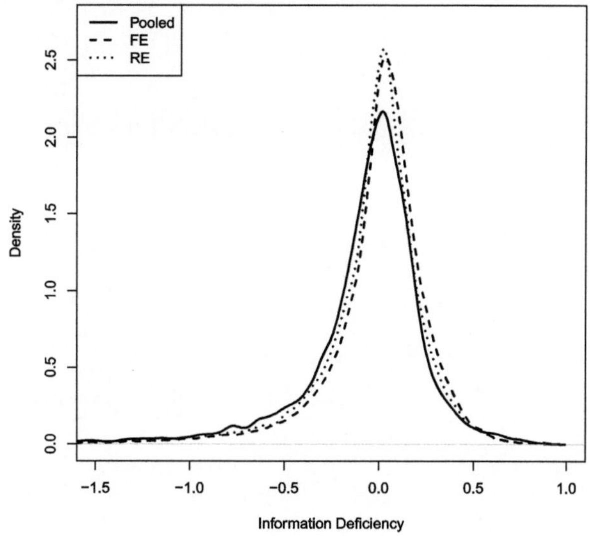

Fig. 9.4 Time-varying effects. Kernel densities for the M_{10} metric

(see Chap. 4), to which we will eventually return for the empirical application of Chap. 10.

With these observations, we conclude our empirical illustration of the six-component 2TSF panel data model.

Fig. 9.8 Time-varying ... kernel densities for the ...

Chapter 10
Allowing for Intra-error Dependence

Up to now, we have assumed that the components of the error term (v, w, u) are jointly independent. But there are cases where such independence is not likely to hold: for example, technologies that increase availability of information in the labor market (e.g., online platforms for job posting and seeking) will increase informational efficiency for both employers and job seekers—and this will make w and u dependent. Also, we have seen in the Nash bargaining 2TSF model of Chap. 4 that the relative bargaining power links the two one-sided components together, creating statistical dependence. Meanwhile, studies that incorporated intra-error dependence in single-tier stochastic frontier models (Amsler et al., 2014; El Mehdi & Hafner, 2014; Smith, 2008) have shown that ignoring dependence may not be safe. The same was shown for the 2TSF framework by Papadopoulos et al. (2021) in Monte Carlo simulations.

In this chapter, we maintain the assumption that the noise component v is independent and a Normal random variable, but we consider the case of dependence between the two competing latent forces w and u. The chapter draws mainly from Papadopoulos (2018, 2024a) and Papadopoulos et al. (2021).

We present four different models, all to be estimated via maximum likelihood. In the first, w and u jointly follow an Exponential-like bivariate distribution due to Freund (1961). In the second, they are assumed to follow a correlated Bivariate Truncated-Normal (BTN) distribution. These two models can be seen as the natural extensions of the benchmark Normal-Exponential and Normal-Half-Normal specifications that we presented in Chap. 7. We have applied the BTN model in Sect. 4.2. In the third model, we use a Copula to construct the likelihood function of the composed error term, which is a semiparametric way of modeling dependence and requires the use of simulated maximum likelihood. In the final model, we present a conditioning technique that is applicable when the dependence between the two latent forces is carried by the presence of a third variable. While what we

A. Papadopoulos, C. F. Parmeter, *Two-Tier Stochastic Frontier Analysis for the Social Sciences*, https://doi.org/10.1007/978-3-031-81513-3_10

develop here is tailored for the Nash bargaining model of Chap. 4, it can be used more generally.[1]

Before presenting the models, we note a limitation when attempting to model dependence between two non-negative random variables. Specifically, they can have any level of positive correlation, but if the correlation between them is negative, Pearson's correlation coefficient cannot reach the value -1. Intuitively, as one variable goes to ∞, the other cannot, by definition, go below 0. So, a "good" distributional specification in the sense of accommodating correlation should allow for full positive correlation and as much negative correlation as possible. Moran (1967) showed that the lower bound for negative correlation depends only on the marginal distributions, but the *actual* maximum allowed degree of negative correlation will also depend on the *joint* distribution of the two random variables.

10.1 The Correlated Exponential 2TSF Model

Freund (1961) proposed the following joint density for two non-negative random variables:[2]

$$f_{wu}(w,u) = \begin{cases} ab' \exp\left\{-b'u - \left(a+b-b'\right)w\right\} & 0 < w < u \\ a'b \exp\left\{-a'w - \left(a+b-a'\right)u\right\} & 0 < u < w \end{cases}, \quad a, a', b, b' > 0. \tag{10.1}$$

We consider the case where $a+b-a' \neq 0$ and $a+b-b' \neq 0$. The corresponding marginal distributions are

$$f_w(w) = \frac{(a-a')(a+b)}{a+b-a'} \exp\{-(a+b)w\} + \frac{a'b}{a+b-a'} \exp\left\{-a'w\right\}, \tag{10.2}$$

$$f_u(u) = \frac{(b-b')(a+b)}{a+b-b'} \exp\{-(a+b)u\} + \frac{ab'}{a+b-b'} \exp\left\{-b'u\right\}. \tag{10.3}$$

These densities are mixtures of Exponentials and can be monotonically declining or have their mode away from zero, depending on the parameter values. Thus, they are more flexible than the Exponential or Half-Normal distributions that both have zero modes.

---

[1] With so many, and somewhat complicated, models, reuse of some coefficient symbols and notation to mean different things per model was unavoidable. So in this chapter, each model is an island, notation-wise.

[2] The distribution is known as "Freund's Bivariate (Mixture) Exponential extension"; see Kotz et al. (2000, pp. 355–362) and Balakrishnan and Lai (2009, ch. 10.3), for a survey of the literature, properties, and applications of it.

Defining the parameter $m \equiv a/(a+b)$ for reasons that will be discussed in a while, the means and variances are

$$E(w) = \frac{a' + b}{a'(a+b)} = \frac{1}{a+b} + \frac{1-m}{a'},$$

$$\mathrm{Var}(w) = \frac{a'^2 + 2ab + b^2}{a'^2(a+b)^2} = \frac{1}{(a+b)^2} + \frac{1-m^2}{a'^2}, \tag{10.4}$$

and

$$E(u) = \frac{b' + a}{b'(a+b)} = \frac{1}{a+b} + \frac{m}{b'},$$

$$\mathrm{Var}(u) = \frac{b'^2 + 2ab + a^2}{b'^2(a+b)^2} = \frac{1}{(a+b)^2} + \frac{m(2-m)}{b'^2}. \tag{10.5}$$

The covariance and Pearson's correlation coefficient are

$$\mathrm{Cov}(w, u) = \frac{a'b' - ab}{a'b'(a+b)^2} = \frac{1}{(a+b)^2} - \frac{m(1-m)}{a'b'}, \tag{10.6}$$

$$\rho = \frac{a'b' - ab}{\sqrt{\left(a'^2 + 2ab + b^2\right)\left(b'^2 + 2ab + a^2\right)}}. \tag{10.7}$$

Looking at the expression for the correlation coefficient, we see that the model also allows for the case of only higher-order dependence with zero linear dependence, since the covariance will be zero if $a'b' = ab$ which is a weaker condition than $\left(a = a', b = b'\right)$ that is required for full independence.

The correlation coefficient lies in $(-1/3, 1)$. This is much wider than, for example, what holds for the Gumbel (1960) bivariate Exponential density that uses the Farlie-Morgenstern-Gumbel construction, where the correlation coefficient is restricted to lie in the interval $(-1/4, 1/4)$, or for another proposed joint distribution in the same paper in which the correlation can only be negative. So, apart from providing marginals that are more flexible than the NE and NHN specifications, this distribution dominates other "Exponential-based" joint distributions proposed in the literature as regards the extent of correlation that it can accommodate.

10.1.1 The Distribution of the Composite Error Term

Assume that (w, u) follow jointly Freund's Bivariate Exponential extension, and let $v \sim N(0, \sigma_v)$, independent of the other two. Then the density of the 2TSF composite error term $\varepsilon = v + w - u$ is

$$f_{\varepsilon}(\varepsilon) = \frac{a}{a+b} b' \exp\left\{\frac{1}{2}\sigma_v^2 b'^2\right\} \exp\left\{b'\varepsilon\right\} \Phi\left(-\frac{\varepsilon}{\sigma_v} - \sigma_v b'\right)$$
$$+ \frac{b}{a+b} a' \exp\left\{\frac{1}{2}\sigma_v^2 a'^2\right\} \exp\left\{-a'\varepsilon\right\} \Phi\left(\frac{\varepsilon}{\sigma_v} - \sigma_v a'\right).$$

If (a', b') are taken to be the reciprocal of the means of two Exponential random variables (i.e., as rate parameters), $f_{\varepsilon}(\varepsilon)$ looks very much like the benchmark 2TSF Exponential specification (see Chap. 7), apart for the terms $a/(a+b) = m$ and $b/(a+b) = 1 - m$ that multiply the two components of the sum. It is through these terms that statistical dependence is represented. If we set $(a' = a, b' = b)$, we obtain the density under independence. So, the 2TSF Correlated Exponential composite error density nests the independence case.

10.1.2 Identification

Due to the manner in which the parameters a and b enter the expression, they cannot be separately identified and estimated using maximum likelihood, because they create perfect colinearity in the score of the log-likelihood.

Using the mixture coefficient $m = a/(a+b)$ and defining the shorthands

$$\omega_2 \equiv \frac{\varepsilon}{\sigma_v} + b'\sigma_v, \quad \omega_3 \equiv \frac{\varepsilon}{\sigma_v} - a'\sigma_v,$$

we can rewrite the density as

$$f_{\varepsilon}(\varepsilon) = \sqrt{2\pi}\phi(\varepsilon/\sigma_v)\left[mb' \exp\left\{\frac{1}{2}\omega_2^2\right\}\Phi(-\omega_2) + (1-m)a' \exp\left\{\frac{1}{2}\omega_3^2\right\}\Phi(\omega_3)\right]. \tag{10.8}$$

The four parameters (σ_v, a', b', m) are now identifiable.

Non-identification of (a, b) is not a serious setback. Even though we will not be able to estimate the marginal moments of the one-sided error components, we will estimate a host of other characteristics of the sample that are more important as they pertain to the characterization of a real-world situation.

Specifically, from the marginal mean and variance expressions, we see that the parameters a and b affect them as location-shifters only. This carries over to the individual metrics for each observation i. Consequently, we will still be able to rank the observations of the sample based on the individual predictors of w_i and u_i. This is important in order to detect subsets of the sample where the effects of the one-sided error terms are more pronounced and/or asymmetric, by stratifying the sample and observing the rankings present in the various strata. In addition, we are still able to estimate the magnitude of the net effect that the one-sided terms combined exert on the outcome variable. Moreover, we will be able to test whether the two one-sided error components are actually dependent or not and to estimate whether they tend to move in the same direction or not (positive/negative correlation).

The cumulative distribution function is

$$F_{\varepsilon}(\varepsilon) = \Phi\left(\varepsilon/\sigma_v\right) + \sqrt{2\pi}\cdot\phi\left(\varepsilon/\sigma_v\right)\Big[m\exp\left\{\frac{1}{2}\omega_2^2\right\}\Phi(-\omega_2) - (1-m)\exp\left\{\frac{1}{2}\omega_3^2\right\}\Phi(\omega_3)\Big]. \tag{10.9}$$

This will be needed if one wants, in addition, to account for regression endogeneity by using a Copula approach rather than instrumental variables (see Chap. 14.1.4).

One can show (see Papadopoulos, 2018, ch. 4) that contrary to the case of independence, the sign of the skewness of the composed error term ε is not necessarily the same as the sign of $E(w-u)$, in the presence of dependence. Moreover, when the one-sided components follow the same distribution, under independence the property $E(w-u) = 0$ would imply a symmetric composite error distribution, but not when dependence is present. These results highlight the perils of ignoring dependence in an otherwise correct specification.

10.1.3 Prediction of w and u

Due to the identification issue, sample-level and individual metrics are computable only up to a location-shift or up to scale.

The conditional expectation formulas for predicting observation-specific levels of w and u are

$$E(w\,|\varepsilon) - \frac{1}{a+b} = \frac{\sigma_v(1-m)a'\exp\left\{\frac{1}{2}\omega_3^2\right\}\left[\omega_3\Phi(\omega_3)+\phi(\omega_3)\right]}{mb'\exp\left\{\frac{1}{2}\omega_2^2\right\}\Phi(-\omega_2)+(1-m)a'\exp\left\{\frac{1}{2}\omega_3^2\right\}\Phi(\omega_3)}, \tag{10.10}$$

and

$$E(u\,|\varepsilon) - \frac{1}{a+b} = \frac{\sigma_v mb'\exp\left\{\frac{1}{2}\omega_2^2\right\}\left[\phi(\omega_2)-\omega_2\Phi(-\omega_2)\right]}{mb'\exp\left\{\frac{1}{2}\omega_2^2\right\}\Phi(-\omega_2)+(1-m)a'\exp\left\{\frac{1}{2}\omega_3^2\right\}\Phi(\omega_3)}. \tag{10.11}$$

The right-hand sides are computable, and provide us with all the information about the distribution of the conditional expected values, except their exact location. Such a shift does not affect the rankings of individual observations, or the relative allocation of probability mass, or properties like the variance and the skewness of these distributions. Moreover, we can calculate the difference $E(w\,|\varepsilon) - E(u\,|\varepsilon)$ and obtain the estimated net effect on the outcome variable for each observation.

We turn now to prediction of $e^{\pm w}$ and $e^{\pm u}$, which will be needed for our metrics when the dependent variable is in logarithms. The corresponding conditional expected values to be used for measures at the observation level are

$$\left(1 \mp \frac{1}{a+b}\right) E\left(e^{\pm w}|\varepsilon\right) = \tag{10.12}$$

$$\frac{mb' \exp\left\{\frac{1}{2}\omega_2^2\right\} \Phi\left(-\omega_2\right) + (1-m)\,a' \exp\left\{\frac{1}{2}\left(\omega_3 \pm \sigma_v\right)^2\right\} \Phi\left(\omega_3 \pm \sigma_v\right)}{mb' \exp\left\{\frac{1}{2}\omega_2^2\right\} \Phi\left(-\omega_2\right) + (1-m)\,a' \exp\left\{\frac{1}{2}\omega_3^2\right\} \Phi\left(\omega_3\right)},$$

and

$$\left(1 \mp \frac{1}{a+b}\right) E\left(e^{\pm u}|\varepsilon\right) = \tag{10.13}$$

$$\frac{mb' \exp\left\{\frac{1}{2}\left(\omega_2 \mp \sigma_v\right)^2\right\} \Phi\left(\pm\sigma_v - \omega_2\right) + (1-m)\,a' \exp\left\{\frac{1}{2}\omega_3^2\right\} \Phi\left(\omega_3\right)}{mb' \exp\left\{\frac{1}{2}\omega_2^2\right\} \Phi\left(-\omega_2\right) + (1-m)\,a' \exp\left\{\frac{1}{2}\omega_3^2\right\} \Phi\left(\omega_3\right)}.$$

The existence of $E\left(e^{w}|\varepsilon\right)$ and $E\left(e^{u}|\varepsilon\right)$ requires $a + b > 1$, which is not unreasonable, since both parameters essentially function as *rate* parameters in an Exponential distribution. If the inequality holds in reverse, the distributions would become too fat-tailed. Here too, the right-hand sides are computable, and so we can rank the observations.

10.1.4 The Distribution of the Difference $z = w - u$

Due to the non-identification of the parameters a and b, the moments of the marginal distributions are not identifiable, as is clear from Eqs. (10.4) and (10.5). In light of this, the distribution of their difference becomes a crucial source of information. For $z = w - u$, we have the density

$$f_z(z) = \begin{cases} mb' \exp\left\{b'z\right\} & z < 0 \\ (1-m)\,a' \exp\left\{-a'z\right\} & z \geq 0. \end{cases} \tag{10.14}$$

We see that the distribution of the difference is fully characterized by the three identifiable parameters. As regards the measures of interest, we obtain

$$E\left(e^{w-u}\right)=\frac{mb'}{(b'+1)}+\frac{(1-m)\,a'}{(a'-1)}. \tag{10.15}$$

At the observation level, we have

$$E\left(e^{w-u}|\varepsilon\right)=$$
$$\frac{mb'\exp\left\{\frac{1}{2}(\omega_2+\sigma_v)^2\right\}\Phi\left(-\omega_2-\sigma_v\right)+(1-m)\,a'\exp\left\{\frac{1}{2}(\omega_3+\sigma_v)^2\right\}\Phi\left(\omega_3+\sigma_v\right)}{mb'\exp\left\{\frac{1}{2}\omega_2^2\right\}\Phi\left(-\omega_2\right)+(1-m)\,a'\exp\left\{\frac{1}{2}\omega_3^2\right\}\Phi\left(\omega_3\right)}. \tag{10.16}$$

For the negative $-z=u-w$, we have

$$E\left(e^{u-w}\right)=\frac{mb'}{(b'-1)}+\frac{(1-m)\,a'}{(a'+1)}, \tag{10.17}$$

and

$$E\left(e^{u-w}|\varepsilon\right)=$$
$$\frac{mb'\exp\left\{\frac{1}{2}(\omega_2-\sigma_v)^2\right\}\Phi\left(-\omega_2+\sigma_v\right)+(1-m)\,a'\exp\left\{\frac{1}{2}(\omega_3-\sigma_v)^2\right\}\Phi\left(\omega_3-\sigma_v\right)}{mb'\exp\left\{\frac{1}{2}\omega_2^2\right\}\Phi\left(-\omega_2\right)+(1-m)\,a'\exp\left\{\frac{1}{2}\omega_3^2\right\}\Phi\left(\omega_3\right)}. \tag{10.18}$$

10.1.5 Model Diagnostics and a Specification Test

Sign of Covariance Our statistical model comes from a theoretical model that incorporates structural/behavioral assumptions. In most cases, these assumptions will dictate whether the dependence between the two latent forces makes them move in the same direction or in the opposite. But even if the theoretical model does not impose such a sign restriction, learning the direction of co-movement is always important.

The model provides us with this information. From the equations of the conditional means, we can estimate $\text{Cov}\left(E\left(w\,|\varepsilon\right), E\left(u\,|\varepsilon\right)\right)$, since we are missing only a constant factor. One can show that

$$\text{sign}\left\{\text{Cov}\left(w,u\right)\right\}=\text{sign}\left\{\text{Cov}\left(E\left(w\,|\varepsilon\right), E\left(u\,|\varepsilon\right)\right)\right\}. \tag{10.19}$$

So, we can obtain the sign of the left-hand side by computing the right-hand side, and we can do that since $E(w|\varepsilon)$ and $E(u|\varepsilon)$ are estimable up to a constant.

A Formal Specification Test Freund's Bivariate Exponential extension nests independence when $a = a'$ and $b = b'$. This implies that if the two one-sided components are independent, we will have

$$m \equiv \frac{a}{a+b} = \frac{a'}{a'+b'} \Rightarrow (1-m)\,a' = mb'.$$

This equality is a necessary condition for independence, and it is formed by identifiable parameters. So, we can formulate an asymptotically valid statistical test with null hypothesis $\mathrm{H_0} : (1-m)\,a' = mb'$, which in practice will be based on the difference $T = \left(1-\hat{m}\right)\hat{a}' - \hat{m}\hat{b}'$. If the null hypothesis is rejected, we have statistical support for dependence. But if the null hypothesis is not rejected, there is still the possibility that dependence exists, and the test is inconclusive. Asymptotically, we obtain

$$\sqrt{n}T\,|_{\mathrm{H_0}} \rightarrow (1-m)\left[\sqrt{n}\,(\hat{a}'-a')\right] - m\left[\sqrt{n}\left(\hat{b}'-b'\right)\right] - (a'+b')\left[\sqrt{n}\,(\hat{m}-m)\right]. \tag{10.20}$$

Given the asymptotic properties of the maximum likelihood estimator, the limit of the right-hand side of Eq. (10.20) is a scaled sum of three dependent zero-mean Normal random variables that follows a multivariate Normal distribution. The estimated variance of T under $\mathrm{H_0}$ is

$$\begin{aligned}\widehat{\mathrm{Var}}\,(T) = &\left(1-\hat{m}\right)^2 \widehat{\mathrm{Var}}\,(\hat{a}') + \hat{m}^2\widehat{\mathrm{Var}}\left(\hat{b}'\right) + \left(\hat{a}'+\hat{b}'\right)^2 \widehat{\mathrm{Var}}\,(\hat{m}) \\ &- 2\left(1-\hat{m}\right)\hat{m}\widehat{\mathrm{Cov}}\left(\hat{a}', \hat{b}'\right) \\ &- 2\left(1-\hat{m}\right)\left(\hat{a}'+\hat{b}'\right)\widehat{\mathrm{Cov}}\,(\hat{a}', \hat{m}) + 2\hat{m}\left(\hat{a}'+\hat{b}'\right)\widehat{\mathrm{Cov}}\left(\hat{m}, \hat{b}'\right).\end{aligned} \tag{10.21}$$

Then we obtain a chi-square test:

$$\hat{q} \equiv \frac{\left(\sqrt{n}T\right)^2}{\widehat{\mathrm{Var}}\left(\sqrt{n}T\right)} = \frac{T^2}{\widehat{\mathrm{Var}}\,(T)} = \frac{\left(\left(1-\hat{m}\right)\hat{a}' - \hat{m}\hat{b}'\right)^2}{\widehat{\mathrm{Var}}\,(T)} \xrightarrow[\mathrm{H_0}]{d} \chi^2_1. \tag{10.22}$$

The $\hat{q}$ statistic in Eq. (10.22) can be computed using the maximum likelihood estimates and the elements in their variance-covariance matrix.

With this, we conclude the presentation of the 2TSF Correlated Exponential specification. We have shown that despite an identification issue, the specification provides important information related to the two latent forces: their net effect on the dependent variable, the ranking of individual observations on each one-sided component as well as on the net effect, the estimated direction of their co-movement (sign of covariance), and a statistical test of whether they are dependent or not.

10.2 The Correlated Truncated Normal 2TSF Model

In this section, we assume that $v \sim N(0, \sigma_v^2)$, while

$$(w, u) \sim N_+(\mathbf{0}, D), \qquad D = \begin{bmatrix} \sigma_u^2 & \sigma_{uw} \\ \sigma_{uw} & \sigma_w^2 \end{bmatrix}.$$

Namely, the latent forces follow jointly a bivariate Normal distribution, truncated from below at 0. We note that the elements in the covariance matrix represent variances and covariance of the full-support variates, not of the truncated versions that we are interested in.

Given independence between v and (w, u), we have $f(v, w, u) = f(v) \cdot f(w, u)$. Let $\chi = (w, u)$. Using also $\varepsilon = v + w - u \implies v = \varepsilon + u - w$, this results in the joint density

$$f(v, \chi) = \frac{[\Pr(u \geq 0, w \geq 0)]^{-1}}{2\pi |D|^{1/2} \sqrt{2\pi} \sigma_v} e^{-(1/2)\chi' D^{-1} \chi} \cdot e^{-(1/2)(\varepsilon + u - w)^2 / \sigma_v^2},$$

where $\sigma_{uw} = \rho \sigma_u \sigma_w$, ρ being the correlation coefficient of the full-support Normals. Integrating out w and u, the density of the composite error term $\varepsilon = v + w - u$ is

$$f_\varepsilon(\varepsilon) = \frac{2\pi}{\cos^{-1}(-\rho)} \frac{1}{\sigma_\varepsilon} \phi(\varepsilon / \sigma_\varepsilon) \cdot \Phi_2(a_w \varepsilon, -a_u \varepsilon; r), \tag{10.23}$$

where $\Phi_2(, ; r)$ is the bivariate standard Normal distribution function with correlation coefficient r. This looks pleasantly compact and clear, but underneath lies an intricate web of composite coefficients. Using the primal parameters $\sigma_v, \sigma_w, \sigma_u$, and ρ, the following shorthands live inside this composite error density:

$$\sigma_\varepsilon^2 = \sigma_v^2 + \sigma_w^2 + \sigma_u^2 - 2\rho \sigma_u \sigma_w, \quad r = \frac{\rho \sigma_v^2 + (1 - \rho^2) \sigma_u \sigma_w}{\psi_u \psi_w},$$

$$\psi_u^2 = \sigma_v^2 + (1 - \rho^2) \sigma_u^2, \quad \psi_w^2 = \sigma_v^2 + (1 - \rho^2) \sigma_w^2,$$

$$a_w = \frac{\rho \sigma_w - \sigma_u}{\sigma_\varepsilon \psi_w}, \quad a_u = \frac{\rho \sigma_u - \sigma_w}{\sigma_\varepsilon \psi_u}.$$

Note that the sigma parameters here are *not* the standard deviations of the variables involved, only their scale parameters. From the above density, one can obtain the log-likelihood of the sample and proceed with maximum likelihood estimation.

Specifying the model in terms of the primal parameters offers an important advantage of formally testing for the existence of dependence, since it hinges on whether ρ is statistically zero or not. So, the dependence aspect of the specification is tested together with estimation. Also, although not immediately obvious, one can verify numerically that for $\rho = 0$, the model nests the Half-Normal specification of the 2TSF model (see Chap. 7).

10.2.1 The Marginal Distributions of w and u

In the univariate case, the zero-mean Normal distribution truncated from below at zero is equivalent to a Half-Normal. But as Horrace (2005) has shown, the marginals of a multivariate Truncated Normal are not, in general, Truncated Normals. And indeed, what we find is that the marginal distributions of w and u are Skew Normals truncated from below at zero, each having the same "skew" parameter (which is a function of the correlation coefficient of the underlying Normals), and each having as scale parameter the standard deviation of each of the original Normals.[3] An interesting result is that these marginals become Half-Normals either when the correlation coefficient ρ is zero or when it tends to unity.[4]

The Truncated Skew Normal distribution has been studied by Jamalizadeh et al. (2009). The marginals obtained have the form, say for w,

$$f(w) = \frac{2\pi}{\cos^{-1}(-\rho)}\frac{1}{\sigma_w}\phi(w/\sigma_w)\Phi\left(\frac{\rho}{\sqrt{1-\rho^2}}(w/\sigma_w)\right). \tag{10.24}$$

Fo the u variable, we just have to change subscripts. Papadopoulos et al. (2021) have determined through numerical simulations that for $\rho > 0$, the marginal density has a strictly positive mode, while for $\rho < 0$, its mode is at zero. Note that the dependence parameter ρ is present in the marginal densities in contrast to cases like the multivariate Normal where the correlation coefficient does not appear in the marginal distributions. Of course, in a univariate setting, ρ would not be interpreted as a "dependence parameter" but only as a skewness parameter. This implies that to follow a Bivariate Truncated Normal, the two variables must have the same skewness parameter and may differ only in scale. The mean and variance are

[3] A closely related result, for a different truncation schema, was obtained by Arnold et al. (1993).

[4] The HN density result for the zero-correlation case was also mentioned in Horrace (2005).

$$E(w)=\frac{(\pi/2)(1+\rho)}{\cos^{-1}(-\rho)}\sqrt{\frac{2}{\pi}}\sigma_w,\quad \mathrm{Var}(w)=\left[1+\frac{\rho\sqrt{1-\rho^2}}{\cos^{-1}(-\rho)}-\frac{(\pi/2)(1+\rho)^2}{(\cos^{-1}(-\rho))^2}\right]\sigma_w^2. \tag{10.25}$$

To cover the case where we estimate a model with the dependent variable in logarithms, we need the expressions for $E(e^{sw})$ and $E(e^{su})$ where $s \in \{-1, 1\}$. This is essentially the moment-generating function (MGF) of the variables. Using the MGF of the Skew Normal truncated from below at 0 as given in Jamalizadeh et al. (2009), we have

$$E(e^{st}) = \frac{2\pi}{\cos^{-1}(-\rho)} \cdot e^{\sigma_t^2/2} \cdot \left(\Phi\left(s\rho\sigma_t\right) - \Phi_2\left(s\rho\sigma_t,\ -s\sigma_t;\ -\rho\right)\right) \tag{10.26}$$

$$s \in \{-1, 1\}, \quad t = w, u.$$

Turning to their correlation structure, the covariance and correlation coefficients are

$$\mathrm{Cov}(w,u) = \frac{\cos^{-1}(-\rho)\left[\rho\cos^{-1}(-\rho)+\sqrt{1-\rho^2}\right]-(\pi/2)(1+\rho)^2}{\left[\cos^{-1}(-\rho)\right]^2}\sigma_w\sigma_u. \tag{10.27}$$

$$\mathrm{corr}(w,u) = \frac{\cos^{-1}(-\rho)\left[\rho\cos^{-1}(-\rho)+\sqrt{1-\rho^2}\right]-(\pi/2)(1+\rho)^2}{\cos^{-1}(-\rho)\left[\cos^{-1}(-\rho)+\rho\sqrt{1-\rho^2}\right]-(\pi/2)(1+\rho)^2}. \tag{10.28}$$

The important information we learn from the correlation expression is the range of $\mathrm{Corr}(w, u)$: as $\rho \to 1$, it also tends to unity. But as $\rho \to -1$, we obtain $\mathrm{corr}(w, u) \approx -0.217$. This is the maximum negative correlation that the Bivariate Truncated Normal specification can accommodate.

10.2.2 Conditional Expectations for w and u

For the positive latent force w, we first set

$$\lambda_w = \frac{\sigma_v^2\rho + \sigma_w\sigma_u(1-\rho^2)}{\psi_u\sigma_v\sqrt{1-\rho^2}}, \quad \gamma_w = \frac{\sigma_u\sqrt{1-\rho^2}}{\psi_u\sigma_v}.$$

Then we obtain, using the composite coefficients defined earlier,

$$E(w|\varepsilon) = \frac{\psi_u \sigma_w \cdot \Omega_1}{\sigma_\varepsilon \Phi_2\left(a_w\varepsilon, -a_u\varepsilon; r\right)}, \tag{10.29}$$

with

$$\Omega_1 = \frac{\lambda_w}{\sqrt{b_w^2+\lambda_w^2}}\phi\left(\frac{b_w\gamma_w + a_u\lambda_w}{\sqrt{b_w^2+\lambda_w^2}}\varepsilon\right)\Phi\left(\frac{\gamma_w\lambda_w - a_u b_w}{\sqrt{b_w^2+\lambda_w^2}}\varepsilon\right) + \phi(a_u\varepsilon)\Phi(-\gamma_w\varepsilon)$$
$$-\left[\Phi\left(-a_u\varepsilon\right) - \Phi_2\left(\frac{b_w\gamma_w + a_u\lambda_w}{\sqrt{b_w^2+\lambda_w^2}}\varepsilon,\ -a_u\varepsilon;\ \delta = \frac{-\lambda_w}{\sqrt{b_w^2+\lambda_w^2}}\right)\right]\cdot a_u\varepsilon.$$

For the case when the dependent variable is in logarithms, we have

$$E(e^{\pm w}|\varepsilon) = \frac{\exp\left\{(\pm\sigma_w)(-a_u/b_w)\varepsilon + \sigma_w^2/2b_w^2\right\}\cdot\Omega_2}{\Phi_2\left(a_w\varepsilon, -a_u\varepsilon; r\right)}, \tag{10.30}$$

with

$$\Omega_2 = 1 - \Phi(-\gamma_w\varepsilon)\cdot\Phi\left(a_u\varepsilon - \frac{(\pm\sigma_w)}{b_w}\right)$$
$$-\Phi_2\left(\gamma_w\varepsilon,\ \frac{(a_u\lambda_w + b_w\gamma_w)\varepsilon - (\pm\sigma_w)(\lambda_w/b_w)}{\sqrt{\lambda_w^2+b_w^2}};\ \frac{b_w}{\sqrt{\lambda_w^2+b_w^2}}\right).$$

Turning to the negative latent force u, we set

$$\lambda_u = \frac{\sigma_v^2\rho + \sigma_w\sigma_u(1-\rho^2)}{\psi_w\sigma_v\sqrt{1-\rho^2}},\quad \gamma_u = \frac{\sigma_w\sqrt{1-\rho^2}}{\psi_w\sigma_v},$$

Then we have

$$E(u|\varepsilon) = \frac{\psi_w\sigma_u\cdot\Omega_3}{\sigma_\varepsilon\cdot\Phi_2\left(a_w\varepsilon, -a_u\varepsilon; r\right)}, \tag{10.31}$$

$$\Omega_3 = \frac{\lambda_u}{\sqrt{b_u^2+\lambda_u^2}}\phi\left(\frac{b_u\gamma_u + a_w\lambda_u}{\sqrt{b_u^2+\lambda_u^2}}\varepsilon\right)\Phi\left(\frac{a_w b_u - \gamma_u\lambda_u}{\sqrt{b_u^2+\lambda_u^2}}\varepsilon\right) + \phi(a_w\varepsilon)\Phi\left(\gamma_u\varepsilon\right)$$
$$+\Phi_2\left(\frac{b_u\gamma_u + a_w\lambda_u}{\sqrt{b_u^2+\lambda_u^2}}\varepsilon,\ a_w\varepsilon;\ \frac{\lambda_u}{\sqrt{b_u^2+\lambda_u^2}}\right)\cdot a_w\varepsilon,$$

and

$$E(e^{\pm u}|\varepsilon) = \frac{\exp\left\{\pm\sigma_u(a_w/b_u)\varepsilon + \sigma_u^2/2b_u^2\right\}\cdot\Omega_4}{\Phi_2\left(a_w\varepsilon, -a_u\varepsilon; r\right)}, \tag{10.32}$$

$$\Omega_4 = \Phi(\gamma_u\varepsilon)\Phi\left(a_w\varepsilon + \frac{(\pm\sigma_u)}{b_u}\right) + \Phi\left(\frac{(\pm\sigma_u)\left(\lambda_u/b_u\right) + (b_u\gamma_u + a_w\lambda_u)\varepsilon}{\sqrt{b_u^2+\lambda_u^2}}\right)$$
$$- \Phi_2\left(\gamma_u\varepsilon, \frac{(\pm\sigma_u)(\lambda_u/b_u) + (b_u\gamma_u + a_w\lambda_u)\varepsilon}{\sqrt{b_u^2+\lambda_u^2}}; \frac{b_u}{\sqrt{\lambda_u^2+b_u^2}}\right).$$

10.2.3 *The Distribution of the Difference $z = w - u$*

In order to calculate the net effect of the two components at the observation level, we can use $E(w|\varepsilon) - E(u|\varepsilon)$ when the dependent variable is estimated in levels since the linearity of the expected value is not affected by dependence. However, when the dependent variable enters the model in logarithmic form, we need $E\left(e^{w-u}|\varepsilon\right)$. This requires access to the distribution of $z = w - u$, since in the presence of dependence, even for the unconditional expectation, we have $E\left(e^z\right) = E\left(e^{w-u}\right) \neq E\left(e^w\right)E\left(e^{-u}\right)$.

The density of z is

$$f_Z(z) = \frac{2\pi}{\cos^{-1}(-\rho)}\frac{1}{s_{wu}}\phi(z/s_{wu})\cdot\begin{cases}\Phi\left(\frac{(\sigma_w-\sigma_u\rho)}{\sigma_u\sqrt{1-\rho^2}}(z/s_{wu})\right) & z\leq 0,\\[2ex] \Phi\left(-\frac{(\sigma_u-\sigma_w\rho)}{\sigma_w\sqrt{1-\rho^2}}(z/s_{wu})\right) & z>0,\end{cases} \tag{10.33}$$

with $s_{wu}^2 \equiv \sigma_w^2 + \sigma_u^2 - 2\rho\sigma_u\sigma_w$. Each branch of the density is a Truncated Skew Normal. Note also that the density is continuous at zero, although with a kink. The unconditional mean of e^z is

$$E(e^{w-u}) = \frac{2\pi e^{s_{wu}^2/2}}{\cos^{-1}(-\rho)}\cdot\left[\Phi_2\left(-(\rho\sigma_u-\sigma_w), -s_{wu}; \frac{\rho\sigma_u-\sigma_w}{s_{wu}}\right)\right.$$
$$\left. + \Phi(\rho\sigma_w-\sigma_u) - \Phi_2\left(\rho\sigma_w-\sigma_u, -s_{wu}; \frac{-(\rho\sigma_w-\sigma_u)}{s_{wu}}\right)\right]. \tag{10.34}$$

The observation-specific formula is

$$E(e^{w-u}|\varepsilon) = \exp\left\{\left(\frac{s_{wu}^2}{\sigma_\varepsilon^2}\right)\left(\frac{\sigma_v^2}{2}+\varepsilon\right)\right\}\cdot\frac{\Omega_5}{\Phi_2\left(a_w\varepsilon, -a_u\varepsilon; r\right)}, \tag{10.35}$$

where

$$\begin{aligned}\Omega_5 =&\Phi_2\left(-a_u(\varepsilon+\sigma_v^2), -\frac{(\varepsilon+\sigma_v^2)s_{wu}}{\sigma_v\sigma_\varepsilon}; \frac{\sigma_v\sigma_\varepsilon}{s_{wu}}a_u\right)+\Phi\left(a_w(\varepsilon+\sigma_v^2)\right)\\ &-\Phi_2\left(a_w(\varepsilon+\sigma_v^2), -\frac{(\varepsilon+\sigma_v^2)s_{wu}}{\sigma_v\sigma_\varepsilon}; -\frac{\sigma_v\sigma_\varepsilon}{s_{wu}}a_w\right).\end{aligned}$$

Turning to $-z = u - w$, we have

$$\begin{aligned}E(e^{u-w}) =& \frac{2\pi e^{s_{wu}^2/2}}{\cos^{-1}(-\rho)}\cdot\left[\Phi_2\left(\rho\sigma_u-\sigma_w, s_{wu}; \frac{\rho\sigma_u-\sigma_w}{s_{wu}}\right)\right.\\ &\left.+\Phi(-(\rho\sigma_w-\sigma_u))-\Phi_2\left(-(\rho\sigma_w-\sigma_u), s_{wu}; \frac{-(\rho\sigma_w-\sigma_u)}{s_{wu}}\right)\right],\end{aligned} \tag{10.36}$$

and

$$E(e^{u-w}|\varepsilon) = \exp\left\{\left(\frac{s_{wu}^2}{\sigma_\varepsilon^2}\right)\left(\frac{\sigma_v^2}{2}-\varepsilon\right)\right\}\cdot\frac{\Omega_6}{\Phi_2\left(a_w\varepsilon, -a_u\varepsilon; r\right)}, \tag{10.37}$$

where

$$\begin{aligned}\Omega_6 =&\Phi_2\left(-a_u(\varepsilon-\sigma_v^2), -\frac{(\varepsilon-\sigma_v^2)s_{wu}}{\sigma_v\sigma_\varepsilon}; \frac{\sigma_v\sigma_\varepsilon}{s_{wu}}a_u\right)+\Phi\left(a_w(\varepsilon-\sigma_v^2)\right)\\ &-\Phi_2\left(a_w(\varepsilon-\sigma_v^2), -\frac{(\varepsilon-\sigma_v^2)s_{wu}}{\sigma_v\sigma_\varepsilon}; -\frac{\sigma_v\sigma_\varepsilon}{s_{wu}}a_w\right).\end{aligned}$$

While certainly complicated, these formulas are straightforward to compute, once they are coded. The main advantage of the BTN model is that it does not suffer from identification issues as the Correlated Exponential one.

10.3 Using Copulas to Model Intra-error Dependence

Copulas are being deployed more and more in various scientific fields (e.g., in the traditional stochastic frontier literature, see Prokhorov, 2024), as a flexible way to model dependence, and much easier to work with compared to conventional multivariate distributions with dependence. In our case, we make use of them as follows: for the composed error term $\varepsilon = v+w-u$, we assume that v is independent from (w, u), and that these two are dependent, having marginal densities $f_w(w)$ and $f_u(u)$ and marginal distribution functions $F_w(w)$ and $F_u(u)$. Then their joint density can be written:

$$f_{wu}(w, u) = f_w(w) \cdot f_u(u) \cdot c\big(F_w(w),\ F_u(u)\big).$$

Here, $c\big(F_w(w),\ F_u(u)\big)$ is the copula *density*, having as arguments the marginal distribution functions of (w, u). The Copula itself is a distribution function. Its density is the cross partial derivative of its arguments. The flexibility of using the copula density is that it captures dependence in full, as a multiplicative factor to the product of the marginal densities. So, we can specify well-known marginal distributions, and then select a Copula to represent the dependence. Note that in the two previous models, the specified marginal distributions were *not* of the "widely known-widely studied" variety. This is what we reclaim by using Copulas.

The marginal density of the composite error ε is by definition

$$f_\varepsilon(\varepsilon) = \int_0^\infty \int_0^\infty f_{\varepsilon,w,u}(\varepsilon, w, u) du dw.$$

Applying a change-of-variables and using the additivity of the relationship, we can write the joint density in the integral as

$$f_{\varepsilon,w,u}(\varepsilon, w, u) = f_{v,w,u}(\varepsilon - w + u, w, u) = f_v(\varepsilon - w + u) f_{wu}(w, u).$$

We have also used the assumption that v is independent from (w, u). Combining all of the results, we obtain

$$f_\varepsilon(\varepsilon) = \int_0^\infty \int_0^\infty f_v(\varepsilon - w + u) f_w(w) f_u(u) c\big(F_w(w),\ F_u(u)\big) du dw. \quad (10.38)$$

Making assumptions about the Copula and the marginal distributions, we obtain the log-likelihood that can be estimated using simulated maximum likelihood, since we do not expect that the integral above possesses a closed-form solution.

10.3.1 Simulated Maximum Likelihood Estimation

At present, a closed-form density for the 2TSF model with intra-error dependence using Copulas does not exist. Thus, to estimate the parameters of the model, we must rely on simulated methods. We can manipulate the density of the composite error as follows:

$$\begin{aligned}
f_\varepsilon(\varepsilon) &= \int_0^\infty \int_0^\infty f_v(\varepsilon - w + u) f_w(w) f_u(u) \cdot c\big(F_w(w),\, F_u(u)\big) du dw \\
&= \int_0^\infty \int_0^\infty \Big(f_v(\varepsilon - w + u) \cdot c\big(F_w(w),\, F_u(u)\big)\Big) \cdot f_w(w) f_u(u) du dw \\
&\equiv E_{W \perp U}\big[f_v(\varepsilon - w + u) \cdot c\big(F_w(w),\, F_u(u)\big)\big]. \qquad (10.39)
\end{aligned}$$

This expectation can be evaluated, for any given value ε_i, by drawing a large number S of random realizations from the assumed marginal distributions of w and u, respectively (Exponential, Half-Normal, Gamma, etc.) coupled with the chosen copula density. Note that the expectation is with respect to $f_w(w) f_u(u)$ which is the joint density of (w, u) as if they were independent. So, our (w_s, u_s) draws should be independent for the same s.[5] This results in

$$f_\varepsilon(\varepsilon_i) \approx S^{-1} \sum_{s=1}^{S} f_v(\varepsilon_i - w_s + u_s) c\big(F_w(w_s),\, F_u(u_s)\big)\big), \ \forall i. \qquad (10.40)$$

For our purposes, we replace ε_i with $y_i - \boldsymbol{x}_i' \boldsymbol{\beta}$ (from the regression model that we are estimating). Thus, our simulated log-likelihood function is

$$\begin{aligned}
\mathcal{L} &= \sum_{i=1}^{n} \log(f_\varepsilon(\varepsilon_i)) \\
&= \sum_{i=1}^{n} \log\left(S^{-1} \sum_{s=1}^{S} f_v(y_i - \boldsymbol{x}_i' \boldsymbol{\beta} - w_s + u_s) c\big(F_w(w_s),\, F_u(u_s)\big) \right). \qquad (10.41)
\end{aligned}$$

The reader may have noticed that in the two expressions above, the random draws are *not* indexed with the observation index i. This means that we need only draw a *single* vector of length S, and use it for each observation. The reason we do not need to draw n such vectors is that they are used here to compute an expected value, they participate in the likelihood as weights to the expression whose mean we want to compute, and we do not rely in any deeper or critical way on their stochastic nature. Moreover, being "enclosed" in a sample mean, they do not affect the assumption of the sample being iid, certainly not asymptotically (and indicative simulations have verified this). So, we can save on computational burden and time by using the same vector for each observation and each iteration of the estimation procedure.

For those that want nevertheless to create n such vectors, we stress that, in such a scenario, each observation must have its own vector, *fixed* across all iterations—

[5] Of course, this procedure can be used with the "independence copula." In other words, one can apply simulated maximum likelihood in a model where there is no dependence between w and u.

otherwise the log-likelihood will stop being a continuous function of the unknown parameters, and most likely the estimation algorithm will not converge.

But there is another reason why creating just one vector to simulate all observations is beneficial. There are two common approaches to engaging in simulated methods: take either random draws or Halton numbers. Halton numbers are not random draws but are known to cover the support more evenly, resulting in better finite sample performance. But Halton draws become delicate to work with if we want to have available a large number of them. So, if we wanted to create n vectors of length S, we may need a very large count of Halton numbers and then various dependencies arise and require due care.

All statistical software that we know of are able to provide either random draws or Halton numbers. Note that Halton numbers mimic a Uniform $U(0, 1)$ random variable. So, in order to obtain w_s and u_s, we use the inverse distribution function method, $w_s = F_w^{-1}(h_s)$, for h_s a Halton number, $s = 1, \ldots, S$. Experience has shown that the length S when using Halton numbers can be as short as $1/10$ of the required length when using random draws. Greene (2003, pp. 185–186) contains a concise and useful discussion on the simulated MLE methodology (although the article is concerned with a specific single-tier SF model). As a practical tip, Greene (2003) uses Halton numbers setting $S = 200$ for estimating simulations, which would be considered too low if the estimation were to be done using a random number generator.

In practice, we feed the estimation algorithm with starting values for the unknown parameters (the $\boldsymbol{\beta}$'s, the parameters of the marginal distributions, and the parameters of the copula density), as well as our series of length S holding the bivariate vectors (w_s, u_s) from their marginal distributions, and of course with the data. This allows the algorithm to compute an initial value of the log-likelihood and iterate from there to convergence.

10.3.2 Choosing the Copula

For the bivariate case, there is a constantly growing menu of available Copulas. Three of them appear particularly flexible for empirical applications: the Gaussian, Frank, and Clayton Copulas.

The Gaussian Copula
For the Gaussian Copula (Song, 2000), the copula density is

$$c\big(F_w(w), F_u(u)\big) = \frac{1}{\sqrt{1-\rho^2}} \exp\left\{\frac{-\rho^2(w_*^2 + u_*^2) + 2\rho w_* u_*}{2(1-\rho^2)}\right\}, \tag{10.42}$$

where $w_* = \Phi^{-1}\left(F_w(w)\right)$ and $u_* = \Phi^{-1}\left(F_u(u)\right)$.

The correlation coefficient ρ is the dependence parameter, and it relates to the doubly transformed random variables. This correlation coefficient represents the maximum possible value of the linear correlation in absolute terms between the original variables (Klaassen & Wellner, 1997). Correlation ranges in $(-1, 1)$, a desirable property and one not often possessed by other Copulas. If ρ is statistically estimated as zero, it signals that, after all, there is no statistical dependence between the random variables involved.

There is theoretical and simulation evidence that the Gaussian Copula is robust to misspecification, in particular against other symmetric Copulas (Papadopoulos, 2022; Park & Gupta, 2012; Prokhorov & Schmidt, 2009). Nevertheless, it is always advisable to have alternatives, so we present next two additional Copulas, the (symmetric) Frank Copula and the (asymmetric) Clayton Copula. They both belong to the Archimedean group of Copulas (see Nelsen, 2006, ch. 4) but they have special properties: in particular, they are the only ones in the group that are "comprehensive" (namely, they reach both the upper and lower Frechet bound). In practice, this means that unlike many Archimedean Copulas, they can represent negative dependence, a needed property in the 2TSF framework as already discussed. They also nest the Independence (or "product") Copula, which means that in their case, the existence of dependence is tested jointly with estimation.

The Frank Copula

The Frank Copula was introduced by Frank (1979) and has been studied by Nelsen (1986) and Genest (1987). Compared to its initial expression, the current incarnation incorporates a transformation related to the dependence parameter α of the initial formulation, $\alpha = \exp\{-\theta\}$, and a change of logarithm base so that the convenient natural logarithm is used. Given these, the Frank copula density $c(F_w, F_u)$ can be written:

$$c(F_w, F_u) = \frac{\theta\left(1 - e^{-\theta}\right)\exp\{-\theta(F_w + F_u)\}}{\left[\left(e^{-\theta} - 1\right) + \left(\exp\{-\theta F_w\} - 1\right)\left(\exp\{-\theta F_u\} - 1\right)\right]^2}, \tag{10.43}$$

with $\theta \in \mathbb{R} \setminus \{0\}$. For $\theta \in (-\infty, 0)$, we have negative dependence, while for $\theta \in (0, \infty)$, we have positive dependence. We obtain independence as $\theta \to 0$.

The parameter θ is related to the general association measures Kendall's tau and Spearman's rho by the expressions

$$\tau(\theta) = 1 + \frac{4}{\theta}\left(D_1(\theta) - 1\right), \quad \rho_s(\theta) = 1 + \frac{12}{\theta}\left(D_2(\theta) - D_1(\theta)\right),$$

where for an integer k,

$$D_k(\theta) = \frac{k}{\theta^k}\int_0^\theta \frac{t^k}{e^t - 1}dt,$$

are Debye functions. For the Frank Copula, both measures range in $(-1, 1)$. The Frank Copula is symmetric as is the Gaussian Copula.

The Clayton Copula

This Copula was introduced by Clayton (1978), although Stigler's Law of Eponymy applies. The standard expression for its density in the bivariate case is

$$c(F_w, F_u) = (1+\theta)\frac{\left(F_w^{-\theta} + F_u^{-\theta} - 1\right)^{-2-1/\theta}}{\left(F_w F_u\right)^{1+\theta}} \cdot I\{F_w^{-\theta} + F_u^{-\theta} > 1\}, \tag{10.44}$$

with $\theta \in (-1, \infty) \setminus \{0\}$. Kendall's tau equals $\tau(\theta) = \theta/(\theta+2)$. The Clayton Copula is an example of an asymmetric Copula, with "tail dependence." For $\theta > 0$, we have lower tail dependence. This property should not be confused with the direction of co-movement: under lower tail dependence, and since in our case both w and u are non-negative, values close to zero correlate, and we expect to see a small u realization if w is realized close to zero. Evidently, tail dependence can be linked to the structural characteristics of each case, providing additional guidance in choosing the Copula. We obtain independence as $\theta \to 0$.

10.3.3 Conditional Expectations for w and u

Apart from estimating the parameters of the model, we are interested in obtaining predictors for the one-sided components per observation, as before. We want to obtain, say for w,

$$E(g(w)|\varepsilon) = \int_0^\infty g(w) f(w|\varepsilon) dw, \quad g(w) = w,\ e^{\pm w}.$$

We have $f(w|\varepsilon) = f_{\varepsilon,w}(\varepsilon, w)/f_\varepsilon(\varepsilon)$ and using the same technique as before,

$$\begin{aligned} f_{\varepsilon,w}(\varepsilon, w) &= \int_0^\infty f_{\varepsilon,w,u}(\varepsilon, w, u) du \\ &= \int_0^\infty f_v(\varepsilon - w + u) f_u(u) f_w(w) c\big(F_w(w), F_u(u)\big) du \end{aligned}$$

So,

$$\begin{aligned} E(g(w)|\varepsilon) &= \frac{1}{f_\varepsilon(\varepsilon)} \int_0^\infty \int_0^\infty \left[g(w) f_v(\varepsilon - w + u) c\big(F_w(w), F_u(u)\big)\right] \\ &\quad \times f_u(u) f_w(w) du dw \\ &= \frac{1}{f_\varepsilon(\varepsilon)} E_{W \perp U}\left[g(w) f_v(\varepsilon - w + u) c\big(F_w(w), F_u(u)\big)\right]. \end{aligned}$$

This expression will be computed after estimation. So, we have available estimates for the unknown coefficients, as well as a series of residuals, $\{\widehat{\varepsilon}_i\}$, and we can apply again simulation methods to compute the expression. We have (for each observation i)

$$E(g(w_i)|\varepsilon_i) \approx \frac{1}{f_\varepsilon(\widehat{\varepsilon}_i)} \cdot S^{-1} \sum_{s=1}^{S} g(w_s) f_v(\widehat{\varepsilon}_i - w_s + u_{si}) c\big(F_w(w_s),\ F_u(u_s)\big). \tag{10.45}$$

We do not have analytical results as to whether we can use the same vector of draws we used in the estimation stage, or whether we should generate a new set of draws. Since here no iterations are involved, and since these draws attempt to mimic the general behavior of the random variables and not a sample-specific one, perhaps the prudent thing to do would be to draw a new set, since the vector used in estimation has influenced the residuals.

Analogously for u, we obtain, for $g(u) = u,\ e^{\pm u}$,

$$E(g(u)|\varepsilon) =$$

$$\frac{1}{f_\varepsilon(\varepsilon)} \int_0^\infty \int_0^\infty \big[g(u) f_v(\varepsilon - w + u) c\big(F_w(w), F_u(u)\big)\big] f_u(u) f_w(w) du dw,$$

and so we can compute it by

$$E(g(u_i)|\varepsilon_i) \approx \frac{1}{f_\varepsilon(\widehat{\varepsilon}_i)} \cdot S^{-1} \sum_{s=1}^{S} g(u_s) f_v(\widehat{\varepsilon}_i - w_s + u_s) c\big(F_w(w_s),\ F_u(u_s)\big). \tag{10.46}$$

10.3.4 The Distribution of the Difference $z = w - u$

We want to obtain the density of $z = w - u$ in order to compute $E(e^{w-u})$ and $E(e^{w-u}|\varepsilon)$ (and also those with the negative exponents). The hurdle we face is that this density has two branches, for $z \leq 0$ and for $z > 0$, and the integral expression for the case $z \leq 0$ has $-z$ in the integration limits. This situation is not friendly to the use of simulated estimation of expected values as we relied on previously and would like to continue using. But we can bypass the need for this density and compute the expected values, if we can obtain draws from the distribution of z.

The general expressions of the magnitudes that interest us are

$$E(e^{w-u}) = \int_{-\infty}^{\infty} e^z f_z(z) dz = E(e^z). \tag{10.47}$$

and, since here $\varepsilon = v + z$,

$$E(e^{w-u}|\varepsilon) = \int_{-\infty}^{\infty} e^z f_{z|\varepsilon}(z|\varepsilon)dz = \frac{1}{f_\varepsilon(\varepsilon)} \int_{-\infty}^{\infty} e^z f_{\varepsilon,z}(\varepsilon, z)dz.$$

We have $f_{\varepsilon,z}(\varepsilon, z) = f_v(\varepsilon - z) f_z(z)$ so,

$$E(e^{w-u}|\varepsilon) = \frac{1}{f_\varepsilon(\varepsilon)} \int_{-\infty}^{\infty} f_z(z)e^z f_v(\varepsilon-z)dz = \frac{1}{f_\varepsilon(\varepsilon)} E_Z\big[e^z f_v(\varepsilon-z)\big]. \tag{10.48}$$

So, both can be computed as expected values over z, and we could achieve that if we could have draws from $z = w - u$, while taking into account the fact that (w, u) are dependent, and that they are assumed to follow a specific Copula. If we had such a series of draws $\{z_s\}$ of length S, we would compute

$$E(e^{w-u}) \approx S^{-1} \sum_{s=1}^{S} e^{z_s}, \tag{10.49}$$

and, for each observation i,

$$E(e^{w_i-u_i}|\varepsilon_i) \approx \frac{1}{f_\varepsilon(\widehat{\varepsilon}_i)} S^{-1} \sum_{s=1}^{S} \big[e^{z_s} f_v(\widehat{\varepsilon}_i - z_s)\big]. \tag{10.50}$$

Evidently, for computing $E(e^{-z})$, we can use

$$E(e^{u-w}) \approx S^{-1} \sum_{s=1}^{S} e^{-z_s}, \tag{10.51}$$

and

$$E(e^{u_i-w_i}|\varepsilon_i) \approx \frac{1}{f_\varepsilon(\widehat{\varepsilon}_i)} S^{-1} \sum_{s=1}^{S} \big[e^{-z_s} f_v(\widehat{\varepsilon}_i - z_s)\big]. \tag{10.52}$$

And in fact, we can obtain such z-draws, using the conditional Copula method to generate bivariate vectors with dependence (Nelsen, 2006, p. 41), for the Clayton and Frank Copulas.[6] This method will give us pairs of Uniform (0,1) random variables that follow the specified Copula. Then we can use the inverse distribution method to obtain $\{w_s, u_s\}$ and the series $\{z_s = w_s - u_s\}$.

[6] For the Gaussian Copula, the methodology is different, since the conditional Copula remains an integral, and it will be presented next.

Draws under the Clayton Copula

Our goal is to produce a single bivariate pair of Uniform random numbers (ν_1, ν_2) which follows the chosen Copula $C(\nu_1, \nu_2)$.[7]

Step 1 Generate two *independent* Uniform random numbers (ν_0, ν_2), i.e., two independent draws from the $U(0, 1)$ distribution.

Step 2 Calculate the conditional Copula $C(\nu_1|\nu_2)$ which is the derivative of the Copula (not the copula density), with respect to ν_2. For the Clayton Copula, we have

$$C(\nu_1|\nu_2)=\frac{\partial C(\nu_1,\nu_2)}{\partial \nu_2}=\frac{\partial}{\partial \nu_2}\left[\left(\nu_1^{-\theta}+\nu_2^{-\theta}-1\right)^{-1/\theta}\right]=\frac{\left(\nu_1^{-\theta}+\nu_2^{-\theta}-1\right)^{-1-1/\theta}}{\nu_2^{1+\theta}}. \tag{10.53}$$

Step 3 Set the conditional Copula equal to ν_0 and then solve for ν_1. Doing this for the Clayton Copula results in

$$C(\nu_1|\nu_2)=\nu_0 \implies \nu_1=\left[1+\nu_2^{-\theta}\left(\nu_0^{-\theta/(1+\theta)}-1\right)\right]^{-1/\theta}. \tag{10.54}$$

The pair (ν_1, ν_2) is a draw from the Clayton Copula with parameter θ (the value of θ will have been obtained in the estimation stage). Note that we do not use further ν_0.

Step 4 Compute $w = F_w^{-1}(\nu_1)$ and $u = F_u^{-1}(\nu_2)$. The parameters of the distribution functions F_w and F_u will have been obtained in the estimation stage. The pair (w, u) thus computed is a draw from a correlated bivariate random vector with marginal distributions F_w and F_u, and whose dependence is represented by the Clayton Copula with parameter θ.

Step 5 Create $z = w - u$ using the two values obtained previously.

Step 6 Repeat S times to obtain a series $\{z_s\}$, and use this series to compute the unconditional expected values.

Draws under the Frank Copula

Step 1 Generate two *independent* Uniform random numbers (ν_0, ν_2), i.e., two independent draws from the $U(0, 1)$ distribution.

Step 2 For the Frank Copula, the Copula itself can be written:

$$C(\nu_1,\nu_2)=-\frac{1}{\theta}\cdot\ln\left(e^{-\theta}-e^{-\theta\nu_1}-e^{-\theta\nu_2}+e^{-\theta(\nu_1+\nu_2)}\right)+\frac{1}{\theta}\ln\left(e^{\theta}-1\right)$$

[7] For simplicity we ignore the indicator function here.

and so

$$C(\nu_1|\nu_2) = \frac{\partial C(\nu_1,\nu_2)}{\partial \nu_2} = -\frac{1}{\theta}\cdot\frac{\theta e^{-\theta\nu_2} - \theta e^{-\theta(\nu_1+\nu_2)}}{e^{-\theta} - e^{-\theta\nu_1} - e^{-\theta\nu_2} + e^{-\theta(\nu_1+\nu_2)}}$$

$$= \frac{e^{-\theta(\nu_1+\nu_2)} - e^{-\theta\nu_2}}{e^{-\theta} - e^{-\theta\nu_1} - e^{-\theta\nu_2} + e^{-\theta(\nu_1+\nu_2)}} \tag{10.55}$$

Step 3 Following the previous steps and after a little algebra, we obtain

$$\nu_0 = \frac{\partial C(\nu_1,\nu_2)}{\partial \nu_2} \implies \nu_1 = -\frac{1}{\theta}\cdot \ln\left(\frac{1-\nu_0+\nu_0\exp\{\theta\nu_2-\theta\}}{1-\nu_0+\nu_0\exp\{\theta\nu_2\}}\right). \tag{10.56}$$

Steps 4–6 are identical to those of the Clayton Copula.

Draws under the Gaussian Copula
Finally, if we have used the Gaussian Copula and its density, the algorithm is as follows:

Step 1 Draw three standard Normal random variates x_1, x_2, x_3.

Step 2 Create two correlated standard Normals (y_1, y_2) with correlation coefficient ρ by

$$y_1 = \sqrt{1-|\rho|}x_1 + \sqrt{|\rho|}x_3,$$

$$y_2 = \sqrt{1-|\rho|}x_2 + \text{sgn}(\rho)\sqrt{|\rho|}x_3.$$

Here, $\text{sgn}(\rho)$ is the sign function taking the value 1 if $\rho > 0$ and the value -1 if $\rho < 0$ (the value of ρ will have already been estimated).

Step 3 Create two Uniform (0,1) variates by $\nu_1 = \Phi(y_1)$ and $\nu_2 = \Phi(y_2)$. These now represent draws from the Gaussian Copula with dependence parameter ρ.

Step 4 Create the correlated pair (w, u) by $w = F_w^{-1}(\nu_1)$, $u = F_u^{-1}(\nu_2)$, and obtain $z = w - u$.

Step 5 Repeat S times to obtain the series $\{z_s\}$.

10.4 Conditioning Out the Source of Dependence

In all three previous models, the two latent forces (w, u) were treated as the primary entities, and we postulated the existence of statistical dependence between them. In Chap. 4 where we presented bargaining 2TSF models, in the Nash bargaining case, each of these two variables was decomposed multiplicatively, as $w = \eta g$ and $u = (1-\eta)d$, where g was the "own evaluation premium" of the seller in a transaction, d was the "prudential discount" of the buyer, and $\eta \in (0,1)$ was the relative bargaining power of the seller. It was η that created the statistical dependence between (w, u). Note that we assume (reasonably) that η is independent from (w, u).

For the Nash bargaining model specifically, but also more generally, if in our model the latent forces have such an explicit structure, we can apply a conditioning technique that will allow us not only to estimate the model but also to obtain measures for η. We present it next based on the 2TSF Nash bargaining model.

Most of the available 2TSF composed error densities spring from the convolution of three single-parameter random variables, $f_\varepsilon(\varepsilon; \sigma_v, \sigma_w, \sigma_u)$ for $\varepsilon = v+w-u$. The distributions of w and u have support in $(0, \infty)$, and this makes them "closed under scaling," in the sense that, for a distribution D and θ the distribution parameter, we have

$$X \sim D_X(\theta) \implies Z = aX \sim D_X(a\theta).$$

This property is not universal. Multiplying a random variable by a scalar does not always mean that we can just insert the scaled parameters in the exact same density expression. For example, if the support of the distribution has a finite *non-zero* boundary, it is not closed under scaling. But when the property holds, the conditional density $f_{\varepsilon|\eta}(\varepsilon|\eta)$ will have the exact same form as the marginal over (v, w, u), where η and $(1-\eta)$ will operate as scaling factors of g and d to arrive correspondingly at w and u, multiplying the parameters of their distributions:

$$f_{\varepsilon|\eta}(\varepsilon|\eta; \sigma_v, \sigma_g, \sigma_d) = f_\varepsilon(\varepsilon; \sigma_v, \eta\sigma_g, (1-\eta)\sigma_d).$$

Essentially, here we assume that g and d have Exponential or Half-Normal marginal distributions, and so $w = \eta g$ will have a *conditional* Exponential distribution, given η (i.e., treating η as a fixed number).[8] Analogous results hold for u. So, the formula for $f_{\varepsilon|\eta}$ will be the familiar Normal-Exponential or Normal-Half-Normal 2TSF composite error density, with the only difference that in place of the scale parameters σ_w and σ_u, we will have $\eta\sigma_g$ and $(1-\eta)\sigma_d$, respectively.

Then the composite error marginal density is obtained by integrating out η from the joint density:

$$\begin{aligned} f_\varepsilon(\varepsilon) &= \int_0^1 f(\varepsilon, \eta)d\eta = \int_0^1 f_{\varepsilon|\eta}(\varepsilon|\eta)\, f_\eta(\eta)d\eta \\ &= \int_0^1 f_\varepsilon(\varepsilon; \sigma_v, \eta\sigma_g, (1-\eta)\sigma_d) f_\eta(\eta)d\eta. \end{aligned}$$

Now, since the relative bargaining power η ranges in $(0, 1)$, we assume that it follows a Beta distribution with density

[8] This has the consequence that the *marginal* distribution of w will not be Exponential.

$$f_\eta(\eta;\gamma,\delta) = \frac{\eta^{\gamma-1}(1-\eta)^{\delta-1}}{\mathrm{B}(\gamma,\delta)},$$

where $\mathrm{B}(\gamma,\delta)$ is the Beta function. For computational purposes, we use its relation to the Gamma function:

$$\mathrm{B}(\gamma,\delta) = \frac{\Gamma(\gamma)\Gamma(\delta)}{\Gamma(\gamma+\delta)}.$$

The great advantage of the Beta distribution is that with only two parameters, it is extremely flexible: it nests "standard-looking" unimodal densities with or without skewness, the Uniform with its flat density but also the family of the arcsine distributions (symmetric or asymmetric) that have a minimum in the middle values and rising probability mass at the extremes. It can also become monotonic (increasing or decreasing). In fact the only "inflexibility" of the Beta distribution is that it cannot represent populations with multiple finite modes (but then again, distributions with multiple modes are always mixtures anyway).

It follows that the sample log-likelihood based on the marginal distribution of the composite error term can be written:

$$\begin{aligned}
\mathcal{L}_n &= \sum_{i=1}^n \ln\left(\int_0^1 f_\varepsilon(y_i - \boldsymbol{x}_i'\boldsymbol{\beta};\sigma_v,\eta\sigma_g,(1-\eta)\sigma_d)f_\eta(\eta)d\eta\right)\\
&= -n\ln\mathrm{B}(\gamma,\delta) + \sum_{i=1}^n \ln\left(\int_0^1 f_\varepsilon(y_i - \boldsymbol{x}_i'\boldsymbol{\beta};\sigma_v,\eta\sigma_g,(1-\eta)\sigma_d)\right.\\
&\qquad\left.\times\eta^{\gamma-1}(1-\eta)^{\delta-1}d\eta\right).
\end{aligned}$$

We could here too apply simulated maximum likelihood, but we present a numerical alternative: we can use numerical integration by Gauss-Legendre quadrature, with weights adjusted for the [0, 1] limits, that give, for some function h,

$$\int_0^1 h(\zeta)d\zeta \approx \frac{1}{2}\sum_{j=1}^m \xi_j h(\zeta_j),$$

where (ζ_j, ξ_j) are Gauss-Legendre nodes and weights for the interval (0, 1). So, in practice, we will work with the sample log-likelihood

$$\begin{aligned}
\mathcal{L}_n &\approx -n\ln 2 - n\ln\Gamma(\gamma) - n\ln\Gamma(\delta) + n\ln\Gamma(\gamma+\delta)\\
&+\sum_{i=1}^n \ln\left(\sum_{j=1}^m \xi_j f_\varepsilon(y_i - \boldsymbol{x}_i'\boldsymbol{\beta};\sigma_v,\zeta_j\sigma_g,(1-\zeta_j)\sigma_d)\cdot\zeta_j^{\gamma-1}(1-\zeta_j)^{\delta-1}\right).
\end{aligned} \tag{10.57}$$

The parameters under estimation are $(\boldsymbol{\beta}, \sigma_v, \sigma_g, \sigma_d, \gamma, \delta)$.

10.4.1 Prediction and Conditional Expectations of Latent Forces

Sample-level metrics are easy to obtain, since for g, d and η, we have specified distributions and have just estimated their parameters, while for w and u, we can use the independence between η and (w, u), for example,

$$E(w) = E(\eta g) = E(\eta)E(g), \;\; \mathrm{Var}(w) = \mathrm{Var}(\eta g) = E(\eta^2)E(g^2) - [E(\eta)]^2[E(g)]^2.$$

And so on. More complicated is the issue of obtaining observation-specific measures, based as usual on the obtained residuals. In fact, while we will be able to obtain individual measures for $w - u$ and for η, there is a price to pay: individual measures conditional on ε, either for g and d, or even for w and u appear intractable, mainly because we have exhausted our easy-to-use distributions for g and d, meaning that w and u and consequently ε have unmanageable marginal distributions, for theoretical derivations.

What we can have is the following: as in the previous section,

$$E(e^{w-u}) = \int_{-\infty}^{\infty} e^z f_z(z)dz = E(e^z),$$

and,

$$E(e^{w-u}|\varepsilon) = \frac{1}{f_\varepsilon(\varepsilon)} \int_{-\infty}^{\infty} f_z(z)e^z f_v(\varepsilon - z)dz = \frac{1}{f_\varepsilon(\varepsilon)} E_Z\big(e^z f_v(\varepsilon - z)\big).$$

We can obtain draws from $z = \eta g - (1 - \eta)d$, since (η, g, d) are independent, they have specified marginal distributions, and we have estimated their parameters. Moreover, $f_\varepsilon(\varepsilon)$ can be computed numerically from

$$f_\varepsilon(\varepsilon) = \int_0^1 f_\varepsilon\,(\varepsilon|\eta)\, f(\eta)d\eta.$$

So, we have, for a set of draws $\{z_s\}$ of length S,

$$E(e^{w-u}) \approx S^{-1} \sum_{s=1}^{S} e^{z_s}, \tag{10.58}$$

$$E(e^{w_i-u_i}|\varepsilon_i) \approx \frac{1}{f_\varepsilon(\widehat{\varepsilon}_i)} S^{-1} \sum_{s=1}^{S} e^{z_s} f_v(\widehat{\varepsilon}_i - z_s). \tag{10.59}$$

As before, for computing $E(e^{-z})$, we simply use

$$E(e^{u-w}) \approx S^{-1}\sum_{s=1}^{S} e^{-z_s}, \tag{10.60}$$

$$E(e^{u_i-w_i}|\varepsilon_i) \approx \frac{1}{f_\varepsilon(\widehat{\varepsilon}_i)} S^{-1}\sum_{s=1}^{S} e^{-z_s} f_v(\widehat{\varepsilon}_i - z_s). \tag{10.61}$$

We can also obtain individual measures for the relative bargaining power, η, apart from characterizing its marginal distribution through the estimates of γ and δ. By Bayes' theorem for densities,

$$\begin{aligned} E(\eta_i|\varepsilon_i) &= \int_0^1 \eta f_{\eta|\varepsilon}(\eta|\varepsilon_i)d\eta = \int_0^1 \eta \frac{f_{\varepsilon|\eta}(\varepsilon_i|\eta)\cdot f_\eta(\eta)}{f_\varepsilon(\varepsilon_i)}d\eta \\ &= \frac{\int_0^1 \eta \cdot f_\varepsilon\,(\varepsilon_i|\eta)\, f(\eta)d\eta}{\int_0^1 f_\varepsilon\,(\varepsilon_i|\eta)\, f(\eta)d\eta}. \end{aligned} \tag{10.62}$$

We can apply the same numerical integration as before, separately for the numerator and the denominator, to compute this conditional expectation predictor, using the estimated parameters and the value of the residual $\widehat{\varepsilon}_i$ for each observation. Note that the $1/2$ factor from the Gauss-Legendre quadrature as well as the normalizing Beta function cancel out here.

Contrast this with, for example, the model of Harding et al. (2003) where, in a standard hedonic-pricing regression for the housing market, the authors include linearly a vector of buyer and seller characteristics as proxies for idiosyncratic bargaining power, and they have to assume a priori symmetric bargaining power to identify the effects of these characteristics on the observed price. Therefore, our approach opens a window and sheds light into a previously dark room.

10.4.2 Application: Bilateral Wage Bargaining in Ghana

For this application, we use a different part of the African matched employer-employee earnings dataset from Bigsten et al. (2000) that was used in the 2TSF Nash bargaining application in Chap. 4. Here the sample relates to Ghana. The data was collected in three waves, for the years 1992–1994. We also used the same regressor set (plus squared Potential Experience), as in the aforementioned application. The usable sample size was $n = 1{,}910$. We ignore the endogeneity issue in this case, and we focus on illustrating only the results from the estimation of relative bargaining power over the agreed upon wage, in a 2TSF Nash bargaining framework.

We will employ the Normal-Exponential 2TSF composite error density, where in $\varepsilon = v + w - u$, v is assumed $N(0, \sigma_v)$, $w|\eta = \eta g$ and $u|\eta = (1-\eta)d$ are

Table 10.1 Composite error parameters from the Ghana sample

Parameter	Estimate	Std. dev.	Moments	
σ_v	0.4315	0.0551		
σ_g	0.3735	0.0728	$E(w)$	0.1926
σ_d	0.6437	0.0650	$E(u)$	0.3117
γ	1.3046	0.0406	$E(\eta)$	0.5158
δ	1.2246	0.0169	$SD(\eta)$	0.2660

Exponential with scale parameters $\sigma_w = \eta\sigma_g$, $\sigma_u = (1-\eta)\sigma_d$, so g and d are assumed to have Exponential marginals. Then, the density of ε conditional on η has the familiar form:

$$f_\varepsilon(\varepsilon|\eta) = \frac{\exp\{a_1\}\,\Phi(b_1) \;+\; \exp\{a_2\}\,\Phi(b_2)}{\sigma_w + \sigma_u},$$

$$a_1 = \frac{\sigma_v^2}{2\sigma_u^2} + \frac{\varepsilon}{\sigma_u}, \quad b_1 = -\left(\frac{\varepsilon}{\sigma_v} + \frac{\sigma_v}{\sigma_u}\right), \quad a_2 = \frac{\sigma_v^2}{2\sigma_w^2} - \frac{\varepsilon}{\sigma_w}, \quad b_2 = \frac{\varepsilon}{\sigma_v} - \frac{\sigma_v}{\sigma_w}.$$

The conditioning variable η lives inside the scale parameters σ_w and σ_u. The log-likelihood for this particular specification is

$$\mathcal{L}_n = -n\ln \mathrm{B}(\gamma,\delta) + \sum_{i=1}^{n} \ln\left(\int_0^1 \frac{\exp\{a_1\}\,\Phi(b_1) \;+\; \exp\{a_2\}\,\Phi(b_2)}{\eta\sigma_g + (1-\eta)\sigma_d} \cdot \eta^{\gamma-1}(1-\eta)^{\delta-1} d\eta\right),$$

to be maximized over $(\boldsymbol{\beta}, \sigma_v, \sigma_g, \sigma_d, \gamma, \delta)$. Table 10.1 has the estimated distribution parameters of the composite error and the resulting moments.

The average prudential discount from the part of the firms, $\sigma_d = E(d)$, is much stronger than the self-evaluation premium of the workers $\sigma_g = E(g)$. As we have announced in Chap. 4, in this model, we can compute surplus extraction measures free of the relative bargaining power. Since we have assumed g and d to have Exponential marginal densities, we obtain average measures:

$$\text{Worker Surplus} = 1 - E\left(\exp\{-g\}\right) = 1 - \frac{1}{1+\sigma_g} = 0.2719,$$

$$\text{Employer Surplus} = E\left(\exp\{d\}\right) - 1 = \frac{1}{1-\sigma_d} - 1 = 1.8066.$$

It is important to communicate these results in a non-misleading and useful way. For convenience, we replicate here from Chap. 4 the mathematical definition of "surplus" in a bargaining framework:

$$\text{Seller/worker Surplus (SS)} \equiv \frac{\omega - \omega_f^T}{\omega} = 1 - \exp\{-g\},$$

$$\text{Buyer/employer Surplus (BS)} \equiv \frac{\omega_e^T - \omega}{\omega} = \exp\{d\} - 1.$$

Looking first at the worker surplus, we indirectly learn that the firm's target-wage ω_f^T was 0.7281 of the actual wage (the ratio $\omega_f^T/\omega = \exp\{-g\}$). The distance from unity is the "worker surplus," $1 - \exp\{-g\}$, namely, the proportion of the actual wage that was beyond the firm's target. Note that "worker surplus" is *not* a markup over the firm's target wage, and it is not "the actual wage is XX% higher than the firm's target wage." A correct description of the result would be "on average, the amount that the wage was above the employer's target was equal to 27% of the actual wage." The reason we use as base the actual wage and not the firm's target wage is because the sellers can have a much better assessment of their performance when various distances are measured with respect to the actual outcome. And the goal of the estimation here is *not* to provide "tools for bragging" ("I forced you to pay me XX% more than you wanted"), but rather the tangible benefits for the seller from their own-evaluation premium ("how much of what I actually got can be attributed to my bargaining drive").

Turning to the employer surplus, we indirectly learn that the worker target wage was ≈2.8 times the actual wage (the ratio $\omega_e^T - \omega = \exp\{d\}$). Employer surplus expresses what the employers *avoided to pay*, in the same spirit as how we measure the producer surplus in standard microeconomics. So, since they did pay the wage, we subtract one unit and we arrive at the magnitude 1.80. A way to communicate this result is to say "on average, the amount by which the wage was below the workers' target was 180% of actual wage." Put another way, the workers targeted almost three times (2.8) the amount they got. For those with experience in Human Resources and Personnel matters, this estimate should not come as a surprise.

Finally, we discuss the estimation of the distribution of relative bargaining power η. The parameters γ and δ of the Beta distribution of η are both higher than unity and so assign declining probability mass to the (0, 1) extremes. Being small in value, they signal that there is no significant concentration of probability mass near the middle. They are also close, indicating a roughly symmetric distribution around the 0.5 value.

Combining the expected values of g, d, and η, we obtain the expected values of the bargaining performance (w, u). We observe that $E(u) > E(w)$, and the bargaining performance of firms is stronger (where "performance" is the combined effect of relative bargaining power and negotiation skills).

Turning to individual measures, we compute $E(\eta|\varepsilon)$ using 7-point Gauss-Legendre quadrature. We compare the postulated marginal distribution with the empirical distribution of $E(\eta|\varepsilon)$ in Table 10.2.

That the distribution of $E(\eta|\varepsilon)$ is more concentrated than the marginal distribution should not come as a surprise: the conditional expectation is a shrinkage

Table 10.2 Distribution of η and of $E(\eta|\varepsilon)$

			Quantiles		
	η	$E(\eta\|\varepsilon)$		η	$E(\eta\|\varepsilon)$
Mean	0.516	0.516	0.1	0.1445	0.3629
median	0.521	0.523	0.3	0.3433	0.4737
Std. dev.	0.266	0.111	0.5	0.5205	0.5229
Minimum	0.000	0.192	0.7	0.6941	0.5705
Maximum	1.000	0.877	0.9	0.8789	0.6460

estimator. It does not go below 0.2 or above 0.88, and 80% of the predictions for worker relative bargaining power range close to the middle (0.35–0.65).

With this result, we close this short empirical illustration and indeed the chapter on modeling intra-error dependence in the 2TSF model.

Chapter 11
Software for 2TSF Analysis

This chapter touches at the heart of implementing two-tier stochastic frontier analysis: after all is said, analyzed, and described, data must be uploaded to a software package and software commands must be run over this data to produce estimation and computation results, which will then be used for inference.

The chapter presents how three well-known software packages, the open-source `R` and `gretl` and the commercial Stata, can be put to use for purposes of 2TSF analysis.

When we present a code, it can be copied and used by readers for their own research (with the appropriate tweaking of course, in order to match the specifics of their data set). When the use of the software is intermediated by a "command" or a "function package," where the user has to load the data set and make certain choices, and the rest is up to the functionality of said packed routines, we describe what this command/function package can do and what output it produces.

We must clarify from the start that this is *not* a tutorial in using these software platforms; a certain degree of familiarity with them is assumed.[1]

11.1 2TSF Estimation in R

11.1.1 Basic Implementation

Here we will discuss implementing many of the core aspects of the 2TSF model, illustrating the writing of the likelihood function, calculation of the metrics, and conducting inference between the Normal-Exponential and Normal-Half Normal

[1] A recent publication with an educational focus that introduces `R` and `gretl` in parallel is Kivedal (2024).

A. Papadopoulos, C. F. Parmeter, *Two-Tier Stochastic Frontier Analysis for the Social Sciences*, https://doi.org/10.1007/978-3-031-81513-3_11

2TSF models. We begin by writing a function in R that will calculate the individual likelihood values.

First, we consider the Normal-Exponential 2TSF likelihood. Our base for this is the `function()` command, which is quite accommodating. As it turns out for the majority of optimizers in R, the first element in your function **must** be the parameter vector over which you will optimize.

```
twotier.ne <- function(p, y, xx){

nr  <- ncol(xx) ## Calculate number of regressors in regression

sigv <- p[nr+1] ## Assume homoskedastic two sided component
sigw <- p[nr+2]
sigu <- p[nr+3]

## Construct residuals
e <- y - xx%*%p[1:nr]

a1 <- e/sigu + (sigv^2)/(2*sigu^2)
b1 <- -e/sigv - sigv/sigu

a2 <- (sigv^2)/(2*sigw^2) - e/sigw
b2 <- e/sigv - sigv/sigw

## Calculate individual likelihood values
ll <- -log(sigw+sigu)+log(pnorm(b2)*exp(a2)+pnorm(b1)*exp(a1))

## return will send the individual likelihood values
return(ll)

}
```

A few notes about the construction of this function. Aside from the parameter vector (`p`), we are also including the dependent variable (`y`) and the vector of covariates that form the regression function (`xx`). The way that we are constructing the likelihood function is with the knowledge that the first `nr` elements of `p` refer to the regression specification and the remaining elements are allocated to the three variance components.[2]

[2] Note that we could have used the Exponential transformation to construct the variances. In essence, this would allow us to avoid implementing constrained optimization and guarantee that our variance parameters will be non-negative. One downside of this is that when we report estimates of our variances, if we wish to include standard errors, then we would need to invoke the Delta method since the parameters we are optimizing over are not the parameters we are interested in.

After all of the parameters have been defined, we simply construct the terms as listed in Chap. 7 and then create the vector of log-likelihoods. We could also sum these values up to obtain the full log-likelihood, but to conduct the Vuong (1989) test, we need the individual values.

We now repeat this for the Normal-Half-Normal 2TSF likelihood. One additional thing that we need to do to construct the log-likelihood function is to access the VGAM library (Yee, 2015) so that we can quickly and efficiently call the bivariate standard Normal distribution function (pbinorm()).

```
library(VGAM)

twotier.hn <- function(p, y, xx){

nr  <- ncol(xx) ## Calculate number of regressors in regression

sigv <- p[nr+1] ## Assume homoskedastic two sided component
sigw <- p[nr+2]
sigu <- p[nr+3]

## Calculate reparameterizations
theta1  <- sigw/sigv
theta2  <- sigu/sigv
s       <- sqrt(sigv^2+sigw^2+sigu^2)
omega1  <- s*sqrt(1+theta2^2)/theta1
omega2  <- s*sqrt(1+theta1^2)/theta2
lambda1 <- (theta2/theta1)*sqrt(1+theta1^2+theta2^2)
lambda2 <- (theta1/theta2)*sqrt(1+theta1^2+theta2^2)

## Construct residuals
e <- y - xx%*%p[1:nr]

D <- pbinorm(e/omega1, 0, cov12=lambda1/sqrt(1+lambda1^2))-
pbinorm(e/omega2, 0, cov12=-lambda2/sqrt(1+lambda2^2))

## Calculate individual likelihood values
ll <- log(2*sqrt(2)/sqrt(pi))-log(s)-e^2/(2*s^2)+log(D)

## Return individual likelihood values
return(ll)

}
```

Notice that the structure of the function is identical. We input the parameter vector first, followed by the other pieces of information. Next we construct the relevant pieces for the likelihood function and then we calculate the likelihood values and return them. As a sidenote, the call to pbinorm contains the entry cov12, whereas we have used ρ, the correlation coefficient. pbinorm is for a Standard Bivariate Normal distribution, so the variance parameters are both equal to 1, making cov12 interpretable as a correlation coefficient.

To see how to put this in action, let us consider the incomplete information example of Chap. 3 that involved estimating a hedonic model using data from the house listing service in Windsor, Ontario, Canada. This dataset, as was previously stated, is freely available in R through the AER package (Kleiber and Zeileis, 2008) and was originally studied in Anglin and Gençay (1996). We also grab this opportunity to highlight another aspect of empirical work: how different computational procedures may result in different results—and then, *how* different becomes a robustness diagnostic. In Chap. 3, we initially used OLS starting values for the slope coefficients (the regression betas) and the values 0.123, 0.10, and 0.140 for the variance components ($\sigma_v, \sigma_w, \sigma_u$). We then applied Powell's box constraint method. The optimum that this found was then used as the vector of starting values for the call to MaxLik in R.

Here, we are optimizing using the Nelder-Mead simplex method, which is a non-gradient-based method. MaxLik offers a range of optimizers, and we caution users that they should use a variety of starting values and optimization methods to ensure that their estimates can be classified as a global optima. But our work here is purely for pedagogical purposes. We also note that our starting values were found by repeated invocation of optimization (multistarts).

```
## Load AER library
library(AER)

## Load in data
data("HousePrices")

attach(HousePrices)

## Set up variables as needed
## dependent variable
y <- log(price)

## independent variables
llot <- log(lotsize)
lbd  <- bedrooms
lsty <- stories
lfb  <- bathrooms

## Now convert factors to {0,1}
drv  <- ifelse(driveway=="yes", 1, 0)
rec  <- ifelse(recreation=="yes", 1, 0)
ffin <- ifelse(fullbase=="yes", 1, 0)
ghw  <- ifelse(gasheat=="yes", 1, 0)
```

```
ca    <- ifelse(aircon=="yes", 1, 0)
reg   <- ifelse(prefer=="yes", 1, 0)

## Garage places is a pure count
gar <- garage

## Now put everything together
xx <- cbind(1, drv, rec, ffin, ghw, ca, gar, reg, llot, lbd,
lfb, lsty)
```

We have loaded in the data and reconfigured all of the variables so that they appear as needed prior to maximization of the likelihood function. Now we estimate our two different 2TSF likelihood functions.

```
## Load maxLik package
  library(maxLik)

## Construct A and B matrices to impose variance constraints
## We have a total of 15 parameters, 3 of which have lower
bounds
A <- matrix(0, 3, 15)
B <- matrix(0, 3, 1)

A[1, 13] <- 1
A[2, 14] <- 1
A[3, 15] <- 1

## Starting parameter vectors for optimization
p.start.ne  <- c(8.22776379, 0.12584321, 0.05928428, 0.10201268,
0.16453450, 0.16135696, 0.04976582, 0.12769348,
0.24858275, 0.03091735, 0.16894982, 0.09524657,
0.11188304, 0.10960680, 0.14151329)

p.start.hn <- c(8.11413483, 0.14204995, 0.04975210, 0.10058137,
0.16038626, 0.16518561, 0.04948550, 0.12470841,
0.25554172, 0.04944552, 0.19352065, 0.07941103,
0.06192610, 0.20287579, 0.26141331)

## Estimate both NE and HN 2TSF models
opt.ml.ne <- maxLik(twotier.ne, grad=NULL, hess=NULL,
method="nm", control=list(iterlim=20000),
constraints=list(ineqA=A, ineqB=B),
start=p.start.ne, y=y, xx=xx)

opt.ml.hn <- maxLik(twotier.hn, grad=NULL, hess=NULL,
```

```
method="nm", control=list(iterlim=20000),
constraints=list(ineqA=A, ineqB=B),
start=p.start.hn, y=y, xx=xx)
```

We can use the `summary()` command to print out to the screen the coefficient estimates and standard errors for each of the two models. Table 11.1 presents the maximum likelihood estimates and corresponding standard errors, along with estimates from OLS.[3]

Table 11.1 OLS and 2TSF estimates for the data sample of Chap. 3

	OLS	NE	NHN
(Intercept)	7.745	8.217	8.130
	(0.216)	(0.208)	(0.214)
drv	0.110	0.121	0.120
	(0.028)	(0.027)	(0.027)
rec	0.058	0.062	0.059
	(0.026)	(0.025)	(0.025)
ffin	0.104	0.100	0.105
	(0.022)	(0.021)	(0.021)
ghw	0.179	0.160	0.175
	(0.044)	(0.046)	(0.045)
ca	0.166	0.157	0.157
	(0.021)	(0.020)	(0.021)
gar	0.048	0.052	0.052
	(0.011)	(0.011)	(0.011)
reg	0.132	0.128	0.126
	(0.023)	(0.021)	(0.022)
ln lot	0.303	0.251	0.262
	(0.027)	(0.026)	(0.026)
bd	0.034	0.032	0.032
	(0.014)	(0.014)	(0.015)
fb	0.166	0.168	0.166
	(0.020)	(0.019)	(0.020)
sty	0.092	0.094	0.094
	(0.013)	(0.011)	(0.012)
SD(ε)	0.208	0.212	0.208
σ_v	–	0.110	0.059
		(0.023)	(0.050)
σ_w	–	0.111	0.206
		(0.017)	(0.026)
σ_u	–	0.143	0.260
		(0.017)	(0.023)

[3] We note that the tables presented do not look like the raw screen output of the R software.

We see little difference in the parameter estimates between the two specifications. This is not surprising as it is commonly the case in the single-tier frontier literature as well. Recall from Chap. 7 that we cannot directly compare σ_w and σ_u across the two specifications since they do not have the same meaning. For the Normal-Exponential model, they represent the mean and the standard deviation of the w and u variables. But in the Normal-Half-Normal, they are just scale parameters (and higher than both the mean and the standard deviation). Note also that $\mathrm{SD}(\varepsilon)$, the standard error of the regression, is estimated in OLS but computed and not estimated in the 2TSF models, by the square root of the formula $\mathrm{Var}(\varepsilon) = \mathrm{Var}(v) + \mathrm{Var}(w) + \mathrm{Var}(u)$. One interesting feature between the two specifications is the size difference in σ_v. For the NHN specification, we see that the two-sided noise standard deviation is almost half what it is for the NE setting. The composite error standard deviation $\mathrm{SD}(\varepsilon)$ is very close in the three models, indicating that what the 2TSF models do here is to decompose it and reveal its inner structure.

Before moving on, readers are invited to compare Table 11.1 with Table 3.1 (which uses the same data set), as regards the OLS column but also the NE column. They will find that the OLS columns are identical but the NE columns are not. This is due to the different computational routines used, as mentioned earlier.[4]

Next, we execute the Vuong (1989) test of specification between the two models. The discussion in Chap. 7 suggests that as the Vuong test statistic diverges to $\pm\infty$, we favor one distribution over the other.

```
## Need sample size
n <- dim(HousePrices)[1]

## Calculate Vuong (1989) test between NE and NHN 2TSF models
li.NE <- twotier.ne(coefficients(opt.ml.ne), y, xx)
li.HN <- twotier.hn(coefficients(opt.ml.hn), y, xx)

## Difference of likelihood contributions
li.diff <- li.NE-li.HN

LRn          <- sum(li.diff)
omegahat2.n <- mean(li.diff^2)-mean(li.diff)^2

## Calculate Vuong statistic
Tn <- sqrt(1/n)*LRn/sqrt(omegahat2.n)
```

[4] Another use of this comparison is for one to see clearly how cryptic raw estimation results may look as regards the labels of the variables (Table 11.1), compared to how they look in tables curated for publication (Table 3.1). Table 11.1 has been treated with formatting commands, but we have left the variable labels intact.

```
## Calculate c and -c from standard Normal for given
significance level
alpha <- 0.05
c.max <- qnorm(1-alpha/2)
c.min <- qnorm(alpha/2)
```

What this chunk of code does first is to calculate the vector of likelihood contributions for each of our two specifications, using the maximum likelihood estimates just found (notice we use the `coefficients()` call to extract those). From there, we add up the differences in these contributions and compute the variance of our test statistic (LRn). We then calculate our Vuong statistic and compare this to the upper and lower cutoffs for our prescribed confidence level (`c.min` and `c.max`). If the two models cannot be discriminated between themselves, then `Tn` should lie between `c.min` and `c.max`. Here we get the value 4.181, which is decisively bigger than 1.96 (the 0.975 Normal quantile), and so we reject the null hypothesis that the two distributions are equally informative, at 95% confidence level, favoring the NE model.

11.1.2 Calculating Individual Measures

We can use the housing example to construct various metrics of interest as discussed in Chap. 2. Here for pedagogical purposes, we include an example to construct the M_7 metric from our initial estimates of the Normal-Exponential hedonic information deficiency example. We will construct the measure both assuming independence and accounting for the induced dependence that arises when we condition on ε. The following code computes all conditional expectations, although not all of them are used for the M_7 metric.

```
## Strip off parameter estimates and create epsilon series.
par.hat <- coefficients(opt.ml.ne)
ep.hat  <- y-xx%*%par.hat[1:12]

## Pull distributional parameters
sig.v <- par.hat[13]
sig.w <- par.hat[14]
sig.u <- par.hat[15]

## Use 7.23 and 7.27 to construct metrics
## Setup necessary parameters needed
lambda <- 1/sig.w+1/sig.u
a1     <- sig.v^2/(2*sig.u^2)+ep.hat/sig.u
b1     <- -(ep.hat/sig.v+sig.v/sig.u)
a2     <- sig.v^2/(2*sig.w^2)-ep.hat/sig.w
```

```
b2       <- ep.hat/sig.v-sig.v/sig.w
chi1     <- pnorm(b2)+exp(a1-a2)*pnorm(b1)
chi2     <- exp(a2-a1)*chi1

Eew.cond <- (lambda/(chi2*(lambda-1)))*
 (pnorm(b1)+exp(0.5*((b2+sig.v)^2-b1^2))*pnorm(b2+sig.v))

Eemw.cond <- (lambda/(chi2*(1+lambda)))*
 (pnorm(b1)+exp(a2-a1-b2*sig.v+0.5*sig.v^2)*pnorm(b2-sig.v))

Eeu.cond <- (lambda/(chi1*(lambda-1)))*
 (pnorm(b2)+exp(0.5*((b1+sig.v)^2-b2^2))*pnorm(b1+sig.v))

Eemu.cond <- (lambda/(chi1*(1+lambda)))*
 (pnorm(b2)+exp(a1-a2-b1*sig.v+0.5*sig.v^2)*pnorm(b1-sig.v))

## Now 7.29 to contrast with dependence
Eewmu.cond <- exp((1+sig.u)*(a1+(sig.v^2/(2*sig.u))))
*pnorm(b1-sig.v)+
 exp((1-sig.w)*(a2-(sig.v^2/(2*sig.w))))*pnorm(b2+sig.v)

Eewmu.cond <- Eewmu.cond/(exp(a1)*pnorm(b1)+exp(a2)*pnorm(b2))

## M7 Metric
M7.I <- Eew.cond*Eemu.cond-1
M7.D <- Eewmu.cond-1
```

The mean of M_7 presuming independence is 0.003, while the quartiles are −0.065, 0.002, and 0.065, respectively. Alternatively, acknowledging that the series are dependent under the Exponential transformation, we have that the mean of M_7 is 0.000, while the quartiles are −0.001, −0.014, and 0.061, respectively.[5] The means are equal up to three decimal places, while the quartiles vary minimally. Figure 11.1 plots out the kernel density estimate of the full M_7 series, both assuming independence and accounting for dependence of the series. As is clear, there is almost no discernible difference in the two series. But this will not necessarily always be the case, especially if the two latent forces w and u are more imbalanced than in this specific data set. Readers are advised to use the proper formula that takes into account the dependence between the predictors of w and u when based on the common information contained in ε.[6]

[5] This last result should also be compared to the corresponding results for the same data set and metric in the empirical application of Chap. 3.

[6] This does not contradict the assumption that w and u themselves are independent, since conditional dependence and unconditional independence can coexist.

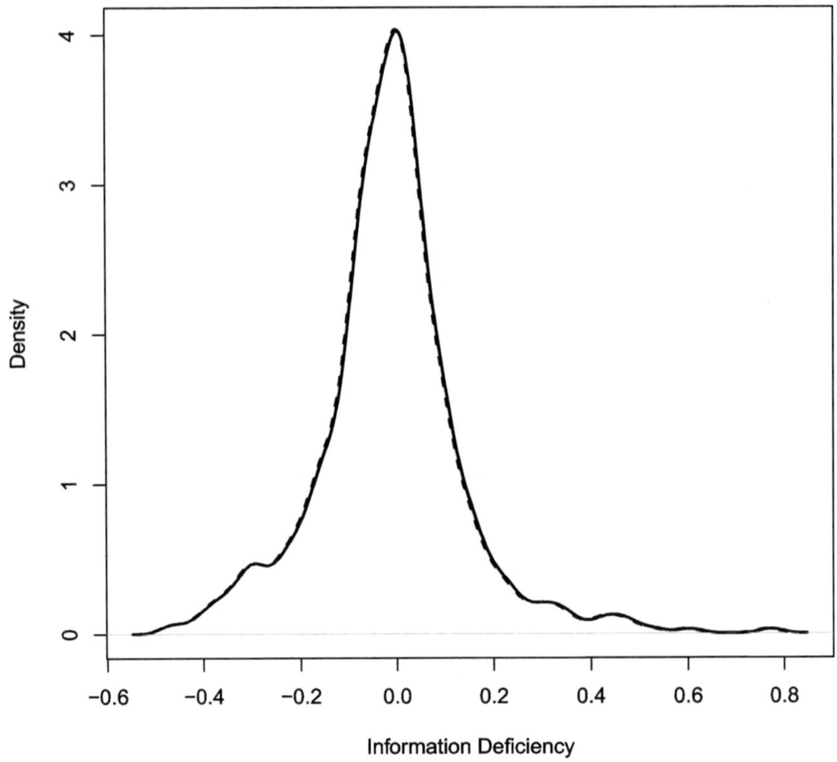

Fig. 11.1 Kernel density plot of two M_7 series for hedonic information deficiency example under the Normal-Exponential specification

11.1.3 Heterogeneity

The previous discussion presented a basic overview of implementing the two core specifications of the 2TSF model. However, as we have mentioned at various points of the book, we may also be interested in explaining what drives w and u to be larger/smaller in expectation. This necessitates some changes to the framework in order to operationalize the code. We consider two distinct cases, modeling the distributional parameters directly and invoking the scaling property, and we use the data sample from Chap. 8.

11.1.3.1 Modeling the Distributional Parameters Directly

To allow the distributions of w and u to change based on observable data, we consider access to data z_w and z_u (which may or may not overlap with $\boldsymbol{x}$). To then include this requires only changing a few pieces of our likelihood function at the beginning (but no changes otherwise). Consider the Normal-Exponential 2TSF likelihood function, `likelihood.ne()`, presented at the start of this chapter. To allow determinants, we would modify the beginning of it as

```
twotier.ne <- function(p, y, xx, zw, zu){

nr  <- ncol(xx) ## Calculate number of regressors in regression
nzw <- ncol(zw) ## Calculate number of determinants for u
component
nzu <- ncol(zu) ## Calculate number of determinants for w
component

sigv <- exp(p[nr+1]) ## Assume homoskedastic two sided component
sigw <- exp((zw%*%p[(nr+2):(nr+nzw+1)]))
sigu <- exp((zu%*%p[(nr+nzw+2):(nr+nzu+nzw+1)]))

## Construct residuals
e <- y - xx%*%p[1:nr]

a1 <- e/sigu + (sigv^2)/(2*sigu^2)
b1 <- -e/sigv - sigv/sigu

a2 <- (sigv^2)/(2*sigw^2) - e/sigw
b2 <- e/sigv - sigv/sigw

## Calculate individual likelihood values
ll <- -log(sigw+sigu)+log(pnorm(b2)*exp(a2)+pnorm(b1)*exp(a1))

## return will send the individual likelihood values
return(ll)

}
```

The above code formulates the functional dependence of σ_w and σ_u using the Exponential function. Other mathematical formulations are also possible. You should verify that after we construct `sigv`, `sigw`, and `sigu`, the code is identical to what was presented previously. The main change is adding in more inputs to the function itself, `zw` and `zu`, and then tailoring the code to allow for subvectors of parameters that are used to construct the unknown scale parameters for the two Exponential distributions. We have also changed the specification of `sigv` to follow suit with the Exponential transformation. This is not needed but it provides us with an alternative to constructing variance parameters. A consequence of this transformation is that constrained optimization is no longer required when one calls `maxLik()`. A similar set of changes could be deployed to modify `likelihood.hn()`.

Table 11.2 Determinants of deficiency on housing prices across specifications

	Seller $\sigma_u(z_u)$			Buyer $\sigma_w(z_w)$		
	NE	NHN	NLS	NE	NHN	NLS
μ_u and μ_w	0.672	1.042	0.855	0.197	0.111	0.106
	(0.111)	(0.063)		(0.067)		(0.081)
Out of town	–	–	–	0.043	0.066	0.112
				(0.057)	(0.079)	(0.097)
First time	–	–	–	0.008	0.003	0.139
				(0.052)	(0.033)	(0.163)
Income	−0.266	−0.293	−0.225	0.239	0.288	0.268
	(0.037)	(0.027)		(0.038)	(0.017)	(0.066)
Own business	−0.264	−0.207	−0.249	0.067	0.121	0.157
	(0.068)	(0.042)		(0.056)	(0.072)	(0.091)
Age	−0.273	−0.256	−0.28	0.461	0.718	0.695
	(0.121)	(0.048)		(0.112)		(0.308)
Black	0.349	0.241	0.288	−0.56	−3.996	−0.655
	(0.104)	(0.078)	(0.036)	(0.151)	(0.565)	(0.339)
Married	−0.16	−0.139	−0.109	−0.045	−0.008	−0.058
	(0.054)	(0.041)	(0.014)	(0.059)	(0.087)	(0.184)
Single female	−0.029	−0.06	−0.05	0.002	0.127	0.077
	(0.064)	(0.047)	(0.051)	(0.073)	(0.105)	(0.132)
Education	−0.37	−0.307	−0.279	0.301	0.458	0.409
	(0.067)	(0.035)		(0.064)	(0.065)	(0.137)
Kids	−0.028	−0.022	0.01	−0.092	−0.128	−0.094
	(0.055)	(0.034)	(0.043)	(0.043)	(0.059)	(0.086)

Table 11.2 presents the coefficients for z_w and z_u that stem from optimization of the 2TSF NE and NHN specifications using the housing data analyzed in Kumbhakar and Parmeter (2010) and that was discussed in Chap. 8 (along with the NLS estimates that we will discuss next).[7] As has also been pointed out in Chap. 8, two of the NHN parameter estimates do not have corresponding standard errors. This was due to negative values along the diagonal of the Hessian provided by `maxLik`. An alternative would be to construct the variance-covariance matrix of the maximum likelihood estimator using the outer product of gradients as discussed in Chap. 7. Since our aim here is purely illustrative, we do not pursue this further, but note that these are issues likely to be encountered in applied practice.

Looking at the NE and NHN estimates (i.e., comparing columns 1–2 and 4–5), we observe total agreement between the two models as regards the direction of influence (signs). On the other hand, as for the sigma parameters in the homoskedastic model we saw earlier, here too it holds in an analogous manner that the delta estimates, the coefficients on the z-variables, are not in general directly

[7] In fact, Table 11.2 is identical to Table 8.2.

comparable between specifications, as we have cautioned the readers in various instances throughout the book. The essence is that we are interested in comparing the moments (means, standard deviations) of our latent forces, because it is moments that are meaningful to interpret behaviorally and structurally, rather than primary parameters that may not directly correspond to such moments, or be related to them but in different ways across specifications.

Comparing the effects between Sellers and Buyers now, we observe that they all have the *opposite* sign, except for being married and having kids. While for all other influences it holds that, what benefits the Seller also independently handicaps the Buyer and vice versa, being married and having kids exerts a negative influence on both Sellers and Buyers, as regards the effects on price. This indicates reduced price variability, especially when couples/families are engaged in both sides of such a transaction, which makes sense, since couples and families are collective entities, leaving less room for idiosyncratic individuality and promoting rational decision-making, which in turn intensifies attempts to gather information.

11.1.3.2 Invoking the Scaling Property

As was discussed in Chap. 8, when we have access to determinants of deficiency, we no longer need to rely on distributional assumptions to identify the model. Rather, the scaling property is enough to allow for a nonlinear least squares solution. In R, this can be undertaken via the `nls()` command that is part of the `stats` package installed with base R. As an example, suppose for our 2TSF framework, we have three x variables (`x1`, `x2`, and `x3`), and z_w and z_u each had two elements (`zw1` and `zw2` and `zu1` and `zu2`, respectively), then the Exponential scaling model is estimated using

```
mod.nls <- nls(y-b0-b1*x1-b2*x2-b3*x3-exp(dw0+dw1*zw1+dw2*zw2)+
  exp(du0+du1*zu1+du2*zu2),
data=data, start=list(b0=value, b1=value, b2=value, b3=value,
dw0=value, dw1=value, dw2=value,
du0=value, du1=value, du2=value))
```

where `value` is a generic place holder for whichever value we wish to start the optimization procedure at.

What makes the call to `nls()` onerous is that with many unknown parameters, each parameter must be written out *explicitly* with its own name and this name must also appear in the list for the starting point of the optimization (`start`). We mention here that the use of `nls` is also different than standard calls that rely on the `Formula` environment in R. We are writing the form of the residual that is to be squared in our call to `nls`, hence the "−" sign on `exp(dw0+dw1zw1+dw2zw2)` and the "+" sign on `exp(du0+du1zu1+du2zu2)`. As an alternative, we can construct our own function and pass it to a nonlinear optimizer (similar to our construction of the likelihood function and then invoking `maxLik`). In R, there are a wide variety of optimizers, with optim() the base optimizer (with a wide variety of techniques) available from stock installation of R.

Here our function is

```
twotier.nls <- function(p, y, xx, zu, zw){

nr  <- ncol(xx) ## Calculate number of regressors in regression
nzw <- ncol(zw) ## Calculate number of determinants for u
component
nzu <- ncol(zu) ## Calculate number of determinants for w
component

sigw <- exp((zw%*%p[(nr+1):(nr+nzw)]))
sigu <- exp((zu%*%p[(nr+nzw+1):(nr+nzu+nzw)]))

## Construct residuals
e <- y - xx%*%p[1:nr]-sigw+sigu

## We minimize the sum of squared residuals
return(sum(e^2))

}
```

In this function, it is presumed that the user would pass along a column of ones inside of both zw and zu to capture the unknown means of the deficiency components. For clarity, we mention here that there is no need to define or estimate a sigv parameter; hence, it is not defined in our function twotier.nls(). This is due to the fact that as our discussion in Chap. 8 made clear, the error term in a two-tier stochastic frontier model, which adopts the scaling property, is not equal to v of the original 2TSF model. If one is using optim() to optimize, you can turn on the option hessian=TRUE so that the Hessian of the objective function is calculated and standard errors can be obtained.

Table 11.2 (columns 3 and 6) presents the coefficient estimates for z_w and z_u that stem from NLS optimization of the 2TSF specification with the scaling property using the same data as previously. This model also agrees with the other two as regards the sign of the coefficients (i.e., comparing column 3 to 2 and to 1 and column 6 to 5 and to 4). Moreover, it so happens that the NLS estimates *are* directly comparable to the NE estimates. This comes from the property of the Exponential distribution that its scale parameter equals its mean value, while in the nonparametric NLS model, the estimates of the constants μ_u and μ_w represent an estimation of the mean value. So for the NE model, μ_u is the "base sigma" parameter, equal to its mean if all deltas are zero, while for the NLS model, it is the mean value of the base random variable u^*, which would be the only latent force if the deltas were zero. This makes them and the deltas comparable across the two models. And we see that the estimates are broadly in agreement. For both the Seller and the Buyer side, a notable difference is the effect of education, visibly lower in strength per the NLS model for the Sellers, visibly higher for the Buyers.

11.1.4 Allowing Dependence: Closed-Form Specification

We return to the data sample of Chap. 3 and repeat our analysis of the incomplete information setting, but now we allow for dependence between the two latent forces. To begin, we have to create our likelihood functions that we will optimize over. This is no different than our earlier construction, except that we now have an additional distributional parameter to account for when setting up everything (and of course the distribution is different). First, the code for the Correlated Exponential 2TSF model (Sect. 10.1):

```
twotier.nce <- function(p, y, xx){

nr  <- ncol(xx) ## Calculate number of regressors in regression

sigv    <- p[nr+1]  ## Assume homoskedastic two sided component
a.prime <- p[nr+2]
b.prime <- p[nr+3]
m       <- p[nr+4]

## Create residuals
e <- y - xx%*%p[1:nr]

## Define omega parameters as above equation (10.8)
omega2 <- (e/sigv)+b.prime*sigv
omega3 <- (e/sigv)-a.prime*sigv

term1 <- m*b.prime*exp(0.5*omega2^2)*pnorm(-omega2)
term2 <- (1-m)*a.prime*exp(0.5*omega3^2)*pnorm(omega3)

## Construct Density
f.e <- sqrt(2*pi)*dnorm(e/sigv)*(term1+term2)

## Calculate individual likelihood values
ll <- log(f.e)

## return will send the individual likelihood values
return(ll)

}
```

As should be clear, there is little to add on the coding of the Correlated Exponential 2TSF likelihood function. All that has changed is an additional parameter (and that we renamed some of the parameters), which is what controls for the dependence between w and u.

We will also construct the function to return the likelihood values for the Correlated Truncated Normal 2TSF specification, or BTN model (Sect. 10.2):

```
twotier.nctn <- function(p, y, xx){

nr  <- ncol(xx) ## Calculate number of regressors in regression

sigv <- p[nr+1] ## Assume homoskedastic two sided component
sigw <- p[nr+2]
sigu <- p[nr+3]
rho  <- p[nr+4]

## Construct residuals
e <- y - xx%*%p[1:nr]

## Define shorthands
sige <- sqrt(sigv^2+sigw^2+sigu^2-2*rho*sigw*sigu)
psiw <- sqrt(sigv^2+(1-rho^2)*sigw^2)
psiu <- sqrt(sigv^2+(1-rho^2)*sigu^2)
r    <- (rho*sigv^2+(1-rho^2)*sigw*sigu)/(psiw*psiu)
aw   <- (rho*sigw-sigu)/(sige*psiw)
au   <- (rho*sigu-sigw)/(sige*psiu)

## Density as defined in Equation 10.23
f.e    <- (2*pi/acos(-rho))*dnorm(e/sige)*pbinorm(aw*e, -au*e,
cov12=r)/sige
## Calculate individual likelihood values
ll <- log(f.e)

## return will send the log of the density of the composite
error
return(ll)

}
```

Now, prior to estimation, as in the independent component setting, we have to place box constraints on our parameters. This is almost identical to our earlier setup, except that for the Correlated Exponential specification, we have $m \in [0, 1]$ and in the BTN model we have $\rho \in [-1, 1]$. This means we have to place *both* upper and lower bounds on these parameters. For the estimation of these two models, the box constraints look different, so we will estimate each model separately.

```
A <- matrix(0, 5, 16)
B <- matrix(0, 5, 1)

A[1, 13] <- 1 ## sigma_v>0
A[2, 14] <- 1 ## a^prime>0
A[3, 15] <- 1 ## b^prime>0
A[4, 16] <- 1 ## m>0

## m<1
A[5, 16] <- -1
B[5]     <- 1

p.start.nce <- c(7.82477633, 0.11087207, 0.05757073, 0.10282750,
0.15951663, 0.15413679, 0.04375504, 0.12354117,
0.30143483, 0.03092422, 0.16379650, 0.09652393,
0.11538181, 9.07547700, 7.16551247, 0.64971669)

opt.ml.nce <- maxLik(twotier.nce, grad=NULL, hess=NULL,
method="nm", control=list(iterlim=20000),
constraints=list(ineqA=A, ineqB=B),
start=p.start.nce, y=y, xx=xx)
```

Here the start values we selected were found through several iterations of starting values and are used here for pedagogical concerns. As we stated earlier, in serious applied settings, a range of starting values should be tried to assess the robustness of the estimates to the specific set of starting values. We now turn our attention to optimization of the Correlated Truncated Normal specification.

```
A <- matrix(0, 5, 16)
B <- matrix(0, 5, 1)

A[1, 13] <- 1 ## sigma_v>0
A[2, 14] <- 1 ## sigma_u>0
A[3, 15] <- 1 ## sigma_w>0

## rho>-1
A[4, 16] <- 1
B[4]     <- 1

## rho<1
A[5, 16] <- -1
B[5]     <- 1

p.start.nctn <- c(7.96949933, 0.12758384, 0.05329819,
```

```
0.11283846, 0.18143522, 0.16622376, 0.04746548, 0.12184049,
0.28011132, 0.04260589, 0.15475334, 0.09307778,
0.04183599, 0.34719496, 0.41250827, 0.81821724)

opt.ml.nctn <- maxLik(twotier.nctn, grad=NULL, hess=NULL,
method="nm", control=list(iterlim=20000),
constraints=list(ineqA=A, ineqB=B),
start=p.start.nctn, y=y, xx=xx)
```

We can use the `summary()` command to print out the coefficient estimates and standard errors for each of the two models to the screen. Table 11.3 presents the maximum likelihood estimates and corresponding standard errors.

There are several interesting features that can be gleaned from the table. First, the coefficient estimates for the hedonic model themselves are comparable both in sign and magnitude across the specifications. Second, the estimate of σ_v does appear to hinge on the specification deployed (in both models, it represents the standard deviation of the Normal noise component). For the Correlated Exponential, we have an estimate of σ_v that is almost three times larger than the corresponding estimate in the Correlated Truncated Normal specification. It is not possible to compare the other parameters between the two specifications. Moreover, the lack of identification in the Correlated Exponential does not allow us to obtain marginal moments of w and u, which would be another dimension along which we could compare the results between models.

However, we can again use the Vuong (1989) test of specification to help adjudicate between the two. In this case, we have a statistic of 0.697. Compared to the upper and lower cutoffs at the 5% level, (1.96, −1.96), we see that we cannot reject the null of equivalent specification. In this case, we may decide to use the BTN specification since it is fully identified and the moments of u and w can be easily assessed. We have that $E(w) = 0.371$ and $E(u) = 0.312$, both of which are similar to the first moments when we did not model dependence. Further, the variances are $Var(w) = 0.063$ and $Var(u) = 0.045$. Quite close to the results from our earlier analysis of this data. Further, the estimated value of ρ, 0.819, is quite high and statistically different from 0, suggesting dependence between w and u.

11.1.5 Allowing Dependence: Copulas and Simulated Methods

Next we turn our attention to allowing dependence where we cannot rely on a closed form for the density of the composite error. Here we will present the steps to engage in simulated maximum likelihood estimation, as discussed in Sect. 10.3. To take Halton draws, we access the `randtoolbox` package (Dutang and Savicky, 2023) in R. The command we need is `halton()`. We first deploy the Gaussian Copula.

Table 11.3 Two-tier stochastic frontier estimates allowing dependence between w and u

	Correlated Exponential		Correlated Truncated Normal
(Intercept)	7.852 (0.266)		7.984 (0.215)
drv	0.112 (0.027)		0.126 (0.028)
rec	0.057 (0.025)		0.058 (0.026)
ffin	0.102 (0.021)		0.108 (0.021)
ghw	0.155 (0.046)		0.183 (0.044)
ca	0.154 (0.020)		0.167 (0.022)
gar	0.045 (0.011)		0.048 (0.011)
reg	0.124 (0.021)		0.122 (0.023)
ln lot	0.298 (0.026)		0.280 (0.027)
bd	0.033 (0.014)		0.039 (0.015)
fb	0.165 (0.019)		0.158 (0.020)
sty	0.095 (0.011)		0.092 (0.012)
σ_v	0.117 (0.036)		0.042 (0.066)
a'	9.146 (1.770)	σ_u	0.412 (0.281)
b'	7.179 (0.850)	σ_w	0.346 (0.256)
m	0.650 (0.670)	ρ	0.819 (0.316)

```
library(randtoolbox)

## Number of Halton Draws
S <- 200

## Construct Halton Draws
H.draw <- halton(S, dim=2, init=FALSE)
```

```
twotier.copula <- function(p, y, xx, H.draw){

nr  <- ncol(xx) ## Calculate number of regressors in regression

sigv <- p[nr+1] ## Assume homoskedastic two sided component
sigw <- p[nr+2] ## Parameter for w marginal
sigu <- p[nr+3] ## Parameter for u marginal
rho  <- p[nr+4] ## Correlation for Normal copula

## Calculate residuals
e <- y - xx%*%p[1:nr]

## Length of vector to take draws over
n <- length(e)

f.e <- as.numeric()

for (i in 1:n){

## Create 'observations' for u and w
w.si <- qexp(H.draw[ ,1], rate=1/sigw)
u.si <- qexp(H.draw[, 2], rate=1/sigu)

## Calculate distribution
Fw.si <- pexp(w.si, rate=1/sigw)
Fu.si <- pexp(u.si, rate=1/sigu)

## Construct summands of likelihood
## Noise density
fv <- dnorm(e[i]-w.si+u.si, sd=sigv)

## Gaussian copula density
w.star <- qnorm(Fw.si)
u.star <- qnorm(Fu.si)

cdens <- (1/sqrt(1-rho^2))*exp((-(rho^2)*(w.star^2+u.star^2)+
 2*rho*w.star*u.star)/(2*(1-rho^2)))

## Average over S draws
f.e[i] <- mean(fv*cdens)

}

## Take logarithm
```

```
ll <- log(f.e)

## return will send the log of the density of the composite
error
return(ll)

}
```

This function looks quite similar to our earlier functions. A key difference is the use of a loop to construct the average for each observation. Note that because we will be optimizing this function, we create the Halton draws outside of the function. This will ensure that every time an optimizer calls `twotier.copula`, the same Halton draws are used. Further, as pointed out in Sect. 10.3, we use the same Halton draws for all observations. The setup here is for the Gaussian Copula with Exponential marginals for both w and u. If we wanted to use, say, the Clayton or Frank Copulas, we would keep everything the same in the function call except the chunk of code that starts `##Gaussian copula density`, which we could replace with

```
## Frank copula density
num   <- theta*(1-exp(-theta))*exp(-theta*(Fw.si+Fu.si))
denom <- exp(-theta)-1+(exp(-theta*Fw.si)-1)*
   (exp(-theta*Fu.si)-1)
cdens <- num/(denom^2)
```

or

```
## Clayton copula density
num   <- (Fw.si^(-theta)+Fu.si^(-theta)-1)^(-2-1/theta)
denom <- (Fw.si*Fu.si)^(1+theta)
cdens <- (1+theta)*num/denom
```

For consistency with the text, we have written the `rho` parameter that appears in the Gaussian copula density with `theta` for the Frank and Clayton Copulas. Naturally, this would need to be changed in the beginning of the code for `twotier.copula()`; otherwise, R would throw an error. We also note that the first three parameters signifying the scale parameters for v, w, and u all still need to be restricted to be greater than 0, but the copula parameter needs to be restricted in the Gaussian framework since $\rho \in [-1, 1]$ and in the Clayton framework since $\theta > -1$[8]. For the Frank Copula, θ is unbounded (aside from being restricted to differ from 0). Finally, we mention that our code here is for pedagogical purposes as it relates to the discussion in the text. There are a variety of more code friendly changes that could be made to the construction of the Halton draws that we are eschewing here in favor of parsimony with our discussion in the book.

[8] For completeness, the Clayton code could include the indicator function that may become effective when we allow θ to be negative. The command would be "cdens<-ifelse(cdens>0,cdens,0)".

Table 11.4 presents the estimates using simulated maximum likelihood with 200 Halton draws for the Gaussian, Frank, and Clayton Copulas. There are various interesting features that we can draw from this set of results. First, the coefficient estimates for the hedonic function are quite similar across all three specifications. Second, the variance parameters for both w and u (with Exponential marginals) are also similar between the three different copulas. Third, the variance of v does appear to be sensitive to the choice of copula. Both the Gaussian and Clayton Copulas produce estimates of σ_v that are an order of magnitude larger than the Frank Copula. Fourth, the direction and level of dependence vary considerably between the three different copulas. With the Gaussian Copula, we have positive dependence (though the ρ parameter estimate is very inaccurately estimated and could be positive or negative at conventional levels of statistical confidence), while the estimate of θ from the Frank Copula is negative and statistically significant and the Clayton Copula produces a $\widehat{\theta}$ that is also negative but again, imprecisely estimated.

11.1.6 Calculation of Individual Metrics via Simulation

Following the discussion in Sect. 10.3.3, here we provide a simple illustration to calculate $E(e^w|\varepsilon)$ and $E(e^u|\varepsilon)$ (for w and u following Exponential marginal distributions). For this example, we will assume that our maximum simulated likelihood estimates using a Gaussian Copula were saved in the object `opt.ml.ncop.g`. Further, we will use 200 Halton draws to average over.

```
## Strip off parameter estimates and create epsilon series.
par.hat <- coefficients(opt.ml.ncop.g)

## Number of regressors
nr  <- ncol(xx)

## Calculate residuals
ep.hat  <- y-xx%*%par.hat[1:nr]

## length of residual vector to loop over
n <- length(ep.hat)

## Assign parameter estimates
sigv <- par.hat[nr+1] ## Assume homoskedastic two sided
component
sigw <- par.hat[nr+2] ## Parameter for u marginal
sigu <- par.hat[nr+3] ## Parameter for w marginal
rho  <- par.hat[nr+4] ## Correlation for Normal Copula

## Construct density using Halton draws
f.e <- as.numeric()

for (i in 1:n){
```

Table 11.4 Two-tier stochastic frontier estimates allowing dependence between w and u using copulas

	Gaussian	Frank	Clayton
(Intercept)	8.183	8.214	8.220
	(0.208)	(0.116)	(0.208)
drv	0.121	0.106	0.120
	(0.027)		(0.026)
rec	0.056	0.061	0.062
	(0.027)	(0.008)	(0.025)
ffin	0.099	0.110	0.098
	(0.021)	(0.008)	(0.021)
ghw	0.169	0.220	0.167
	(0.047)	(0.017)	(0.048)
ca	0.158	0.171	0.162
	(0.021)	(0.007)	(0.021)
gar	0.050	0.041	0.049
	(0.011)	(0.006)	(0.012)
reg	0.132	0.121	0.132
	(0.022)	(0.009)	(0.021)
ln lot	0.255	0.255	0.250
	(0.026)	(0.016)	(0.026)
bd	0.031	0.016	0.033
	(0.014)	(0.003)	(0.016)
fb	0.171	0.174	0.165
	(0.020)	(0.006)	(0.020)
sty	0.093	0.093	0.095
	(0.011)	(0.003)	(0.011)
σ_v	0.121	0.017	0.104
	(0.033)	(0.003)	(0.054)
σ_w	0.123	0.139	0.116
	(0.036)	(0.001)	(0.026)
σ_u	0.157	0.159	0.148
	(0.030)	(0.002)	(0.019)
ρ	0.274	–	–
	(0.516)		
θ	–	−1.825	−0.042
		(0.600)	(0.388)

```
w.si <- qexp(H.draw[ ,1], rate=1/sigw)
u.si <- qexp(H.draw[, 2], rate=1/sigu)

## Construct summands
## Noise density
fv <- dnorm(ep.hat[i]-w.si+u.si, sd=sigv)
```

```
## Copula density
Fw.si <- pexp(w.si, rate=1/sigw)
Fu.si <- pexp(u.si, rate=1/sigu)

w.star <- qnorm(Fw.si)
u.star <- qnorm(Fu.si)

cdens <- (1/sqrt(1-rho^2))*exp((-(rho^2)*(w.star^2+u.star^2)+
 2*rho*w.star*u.star)/(2*(1-rho^2)))

f.e[i] <- mean(fv*cdens)

}

## Now calculate desired expectations
Eew.sim <- as.numeric()
Eeu.sim <- as.numeric()

for (i in 1:n){

w.si <- qexp(H.draw[ ,1], rate=1/sigw)
u.si <- qexp(H.draw[, 2], rate=1/sigu)

## Construct summands
## Noise density
fv <- dnorm(ep.hat[i]-w.si+u.si, sd=sigv)

## Copula density
Fw.si <- pexp(w.si, rate=1/sigw)
Fu.si <- pexp(u.si, rate=1/sigu)

w.star <- qnorm(Fw.si)
u.star <- qnorm(Fu.si)

cdens <- (1/sqrt(1-rho^2))*exp((-(rho^2)*(w.star^2+u.star^2)+
 2*rho*w.star*u.star)/(2*(1-rho^2)))

## Create expectations
Eew.sim[i] <- mean(exp(w.si)*fv*cdens)/f.e[i]
Eeu.sim[i] <- mean(exp(u.si)*fv*cdens)/f.e[i]

}
```

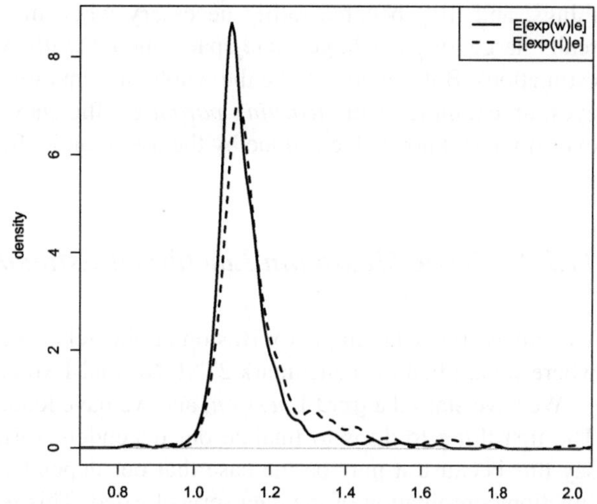

Fig. 11.2 Kernel density plots of $E(e^w|\varepsilon)$ and $E(e^u|\varepsilon)$ for the hedonic information deficiency example under Normal-Exponential specification allowing dependence using the Gaussian copula density

Figure 11.2 displays kernel density estimates for the two series created through simulation via 200 Halton draws. The density for $E(e^u|\varepsilon)$ has a slightly smaller peak and a thicker right-hand tail than the density for $E(e^w|\varepsilon)$.

11.2 2TSF Estimation in `gretl`

While the `gretl` open-source community is nowhere near as vast as the R universe has become, it is close to the econometrician's heart because it is focused on econometric work ("gretl" is the acronym of "**G**nu **R**egression, **E**conometrics and Time-series **L**ibrary").

Moreover, `gretl` has passed third-party rigorous tests for reliability, and it is packed with features and capabilities that continue to increase courtesy of its core supporting team but also through contributions of `gretl` users.[9] To certain individuals, scripting in `hansl` (the language of `gretl`) feels more natural than scripting in R. In addition, comparing scripts increases efficiency in using both platforms. Therefore, some of the material in this section covers the same ground as the previous one.

Finally, we clarify that the scripts that we will present belong to the "beginners" category, on purpose. For example, many things could be wrapped up in modular user-defined functions, leading to a procedure "load the data set, invoke the user-made function, execute," and even make all this into convenient graphical user interfaces (GUIs). `gretl` and its `hansl` scripting language are capable of all these

[9] Its User Guide runs 500 pages, and it is one out of the ten supporting documents that are part of the software installation.

things, and they become rather necessary when the estimation project becomes really large (e.g., a large state-space model with Markov Chain Monte Carlo estimation). But it would make the whole presentation unfriendly for newly arrived eyes, an example of the *learning paradox*: efficiency in learning requires that the exposition of knowledge is done by the use of an inefficient structure.

11.2.1 Basic Maximum Likelihood Estimation

We will use the data sample for Boston Public Schools that we presented in Sect. 5.4, where we applied the benchmark 2TSF Normal-Exponential specification.

We have started a `gretl` *session*, and we have loaded our data set in it. The first thing to do is to finalize our dependent variable and our regressors. We say this because it may be the case that our dependent variable, for example, is a function/combination of the data series loaded. This is indeed the case here, where as dependent variable, we used test scores from the Massachusetts Comprehensive Assessment System (MCAS) in Mathematics and in English Language Arts (ELA), specifically, we took the natural logarithm of their average from Spring Semester 2001 (the original data series were these two series of scores, not in logarithms). Using the GUI, we created our dependent variable and we named it `DepY`.

Analogous data manipulations may be needed also for our regressors, depending on the model we have in mind and what we want to study. Eventually, the data set will have been enhanced with these created series.

Then we create a *named list* that will include the regressors. This is easily done through the GUI. In our case, this list was given the name "RegrListEdRelInc" (don't ask why).

Next, to make our code a little modular and transferable to other data sets, we will use a generic name for the dependent variable, say `Depvar`, and we will also use a generic name for the regressor list, say `Xlist` that will go into the estimation. It is now time to open a "new script" `gretl` window, where we will write our code. The first lines are[10]

```
series Depvar = DepY
list Xlist = RegrListEdRelInc
smpl --no-missing DepY Xlist
```

The third line of the code that uses the *command* `smpl` makes sure that the sample size is restricted to those observations for which there are no missing values for any variable involved. We then proceed to some preliminary actions.

```
ols Depvar Xlist --quiet
matrix bcoef = $coeff
scalar sv = .5
```

[10] The colors used to differentiate commands, objects, functions, and comments will appear automatically on the screen.

```
scalar sw = .5
scalar su = .5

   set max_verbose full
#The next three lines make the names of the variables appear in the
estimation output
string varblnames = varname(Xlist)
string errnames = ,"sigmav,sigmaw,sigmau"
string allcoefnames = varblnames ~ errnames
```

The first line uses the command ols, executes ordinary least squares, and keeps quiet (does not print any output). The second line creates a vector (a vector is a matrix) with the regression betas from the OLS estimation. The next three lines assign starting values to the three scale parameters of the NE specification. The next line, which uses the all-purpose command set, is a reporting command so that we can see the iterations of maximum likelihood in detail (this may be helpful when the estimator does not converge and we want to get an idea when, how, and on account of which estimated coefficients did maximum likelihood fail). The last three lines prepare the situation so that in the estimation output appear the names of the variables whose coefficients we estimate.

Next comes the maximum likelihood command block.

```
mle logl = check ? likely : NA
series res = Depvar - lincomb(Xlist, bcoef)

   series a1 = sv^2/(2*su^2) + res/su
series b1 = -res/sv - sv/su
series a2 = sv^2/(2*sw^2) - res/sw
series b2 = res/sv - sv/sw
series likely = -log(sw + su) + log(exp(a1)*cnorm(b1) +
exp(a2)*cnorm(b2))
   scalar check = (sv>0) && (sw>0) && (su>0)
params bcoef sv sw su
param_names allcoefnames
end mle --verbose --robust
```

A few notes on the above command block: We write the expression for the *likelihood contribution* ℓ_i, i.e., the likelihood for a single observation, not the sample likelihood, and we define it as a series. This is the likely series inside the mle command. The syntax in the first line involving "check," "?," ": NA" relates to the constraint that the scale parameters must be strictly positive (see the line that starts with "scalar check = ..."). It tells gretl not to abort if any one of the constraints is not satisfied, but to continue trying with a different set of candidate values.

The second line creates the residuals using the gretl *function* lincomb. The next lines are the components of the NE composite error likelihood, which we repeat here for convenience of comparison and visual mapping:

$$a_1 = \frac{\sigma_v^2}{2\sigma_u^2} + \frac{\varepsilon}{\sigma_u}\,, \quad b_1 = -\left(\frac{\varepsilon}{\sigma_v} + \frac{\sigma_v}{\sigma_u}\right),$$

$$a_2 = \frac{\sigma_v^2}{2\sigma_w^2} - \frac{\varepsilon}{\sigma_w}, \qquad b_2 = \frac{\varepsilon}{\sigma_v} - \frac{\sigma_v}{\sigma_w}.$$

while the density/likelihood is

$$f_\varepsilon(\varepsilon) = \frac{\exp\{a_1\}\Phi(b_1) + \exp\{a_2\}\Phi(b_2)}{\sigma_w + \sigma_u}.$$

The likelihood expression itself uses `gretl` functions for the natural logarithm, another to exponentiate a magnitude, and `cnorm` that computes the standard Normal CDF.

Also, note the inclusion in the command block of the `allcoefnames` string we created earlier. Moreover, this is optimization using numerical derivatives. As a consequence, we have to state explicitly what are the entities under estimation, hence the `params` command. Carefully note that the order of writing down the coefficients to estimate must be the same as the order in which they appear in the name-string `allcoefnames`. Alternatively, we would have included the exact expression for the derivatives of ℓ_i with respect to the parameters, and then the `params` command would have been omitted (but with 2TSF likelihoods, these derivatives are usually tedious to derive). In this case too, the order of writing the derivatives must follow the order of `allcoefnames`. To the right of the last line, `verbose` indicates that we do want to see the iterations (to different extent of details, regulated by what we have decided through the `set max_verbose` command earlier), and `robust` makes the software compute robust (or "QML") standard errors. The raw output we get is shown in Table 11.5.

Note that we get the final values of the gradient of the (sample) log-likelihood so that we know how close we really are to the "zero-gradient" condition for a maximum. Table 5.11 in Chap. 5 is essentially the lower part of this raw output, with rounded/truncated figures and more explicit names for the variables (and without the dreadful significance asterisks).

11.2.2 Individual Metrics

What is created *inside* the `mle` command is destroyed when the command exits. So the series `a1`, `a2`, etc., which will be needed in order to compute the individual metrics of Chap. 2, are not there to wait for us (but the final values of the `bcoef` vector with the regression betas, and of the sigma coefficients are available, because they have been declared outside and before the `mle` command). If we want to avoid this, we should declare *before* the `mle` command the series we want to have available after, as zero series, namely, include additional code before `mle`.

```
...
series a1 = 0
series b1 = 0
```

Table 11.5 gretl output from MLE estimation

```
--- FINAL VALUES:
loglikelihood = 4961.70565409 (steplength = 1.024e-007)
 Parameters:        5.3734   -0.0093403      0.010553      0.065203          0.070710
                 -0.032529     0.029342     -0.017403     0.0052065          0.045793
                 -0.049718     0.031795      0.032899      0.024823
 Gradients:     -0.0018190   0.00031832    0.00018190   -0.00059117       -9.0949e-005
                0.00036380  -0.00072760   -0.00054570    -0.0011823       -9.0949e-005
               -0.00068212  -0.00036380    0.00054570    -0.0023192  (norm 2.67e-002)
Tolerance = 1.81899e-012

Function evaluations: 524
Evaluations of gradient: 71

Model 7: ML, using observations 1-3194
logl = check ? -log(sw + su) + log(exp(a1)*cnorm(b1) +
exp(a2)*cnorm(b2)):NA
QML standard errors
                        estimate   std. error       z     p-value

 const                 5.37341    0.00615777    872.6     0.0000   ***
 sex2                -0.00934030  0.00180841   -5.165   2.41e-07   ***
 Drace_2_Hispanic     0.0105525   0.00497982    2.119     0.0341   **
 Drace_3_White        0.0652029   0.00469303    13.89   6.93e-044  ***
 Drace_4_Asian        0.0707103   0.00699238    10.11   4.86e-024  ***
 bil2                -0.0325292   0.00380107   -8.558   1.15e-017  ***
 income               0.0293416   0.00203669    14.41   4.71e-047  ***
 sped                -0.0174027   0.000737761  -23.59   5.05e-123  ***
 AlignRace            0.00520652  0.00496943    1.048     0.2948
 DRelIncome_0         0.0457932   0.00402500    11.38   5.43e-030  ***
 DRelIncome_2        -0.0497178   0.00382369   -13.00   1.18e-038  ***
 sigmav               0.0317955   0.00333015    9.548   1.33e-021  ***
 sigmaw               0.0328988   0.00254792    12.91   3.85e-038  ***
 sigmau               0.0248232   0.00222041    11.18   5.14e-029  ***

 Log-likelihood        4961.706   Akaike criterion      -9895.411
 Schwarz criterion    -9810.445   Hannan-Quinn          -9864.946
```

```
series a2 = 0
series b2 = 0
series res = 0
```

These series would be filled by `mle`, and when it exits, they will contain the series that are based on the final estimates.

But if we have not included this in our initial presentation code, we should create these series after the mle command block, by, say

```
series resmle = Depvar - lincomb(Xlist, bcoef)
series a1 = sv^2/(2*su^2) + resmle/su
series b1 = -resmle/sv - sv/su
series a2 = sv^2/(2*sw^2) - resmle/sw
series b2 = resmle/sv - sv/sw
```

We could use any name we want of course (to the left of the = sign).[11] Then we just write out the formulas for the conditional expectations and then the expressions for the M-metrics of Chap. 2. For example, and since the model has the dependent variable in logarithms,

```
series chi1 = cnorm(b2) + exp(a1-a2)*cnorm(b1)
series chi2 = exp(a2-a1)*chi1
scalar lambda = 1/sw + 1/su
series Exp_w_neg = (lambda/(1+lambda))*(1/chi2)*
                  (cnorm(b1)+exp(a2 - a1 - b2*sv + 0.5*(sv^2))*
                     cnorm(b2-sv))
series M1 = 1-Exp_w_neg
```

11.2.3 Heterogeneity

In relation to the material in Chap. 8, suppose we have data series on covariates (z_1, z_2) that we believe influence the positive latent force w, specifically, its scale parameter σ_w. How would we incorporate them in the code?

Suppose that we postulate a quadratic relation:

$$\sigma_{wi} = \delta_1 z_{1i}^2 + \delta_2 z_{2i}^2.$$

The following modifications of the code are required: first, after the `ols` command, instead of providing a starting value for σ_w (`sw`), we should provide starting values for δ_1 and δ_2, say,

```
...
scalar sv = .5
scalar d1 = .2
scalar d2 = .3
scalar su = .5
```

Second, we should modify accordingly the string with the names of the parameters under estimation.

```
...
string errnames = ,"sigmav,d1,d2,sigmau"
string allcoefnames = varblnames ~ errnames.
```

Third, the `mle` command block will now become

```
mle logl = check ? likely : NA
series res = Depvar - lincomb(Xlist, bcoef)

   series sw = d1*z1^2 + d2*z2^2
```

[11] Analogously, if we want to obtain the likelihood contribution series to perform a Vuong specification test, we should write the expression for the likelihood after and outside the mle command block.

```
series a1 = sv^2/(2*su^2) + res/su
series b1 = -res/sv - sv/su
series a2 = sv^2/(2*sw^2) - res/sw
series b2 = res/sv - sv/sw
series likely = -log(sw + su) + log(exp(a1)*cnorm(b1) + exp(a2)*
        cnorm(b2))
series swneg = (sw<=0)
scalar swposcheck = sum(swneg)
scalar check = (sv>0) && (swposcheck=0) && (su>0)
params bcoef sv d1 d2 su
param_names allcoefnames
end mle --verbose --robust
```

Note that we did not replace all instances of σ_w(`sw`) with the expression that includes the z-covariates, but we defined it as a *series* inside the command, and then we used `sw` as a convenient temporary shorthand. Note also that in the `params` line, we now declare that we estimate `bcoef sv d1 d2 su`.

See also how things change so that we still guarantee that the scale parameters will be positive. Here, we want *every single* σ_{wi} to be positive. To check this, we first create the binary series `swneg` that takes the value "1" whenever an `sw` becomes zero or negative. Then we sum this series into the scalar `swposcheck`. If all `sw` are positive, then `swposcheck` must be equal to zero, and the `check` condition is modified accordingly. We should not put non-negativity constraints directly on the deltas, because what we need is for the σ_w series to have only positive elements, and this can happen even if one of (z_1, z_2) influence it negatively.

11.2.4 COLS and GMM Estimation

In Chap. 7, we presented Corrected OLS (COLS) and generalized method of moments (GMM) as alternative estimation methods for the 2TSF model. In COLS, we run first OLS, and then use the residuals to form a system of equations reflecting the moments/cumulants of the assumed distributional specification, through which we estimate the composite error distribution parameters. In GMM, we estimate jointly the regression betas and the distribution parameters.

11.2.4.1 Combined COLS/GMM Estimation

What is lacking in the COLS approach is standard errors for the distribution parameters. To remedy this, we can implement a combined COLS/GMM approach, where the second stage of COLS is not a deterministic solution to a system of equations, but a GMM estimation of the said system, to obtain standard errors. We start with this implementation, modifying accordingly the presentation in Sect. 7.1.4. One detail to keep in mind is that here, since we are estimating only the distribution parameters, we are going to use the OLS residuals from the first stage. *And these residuals have zero mean by construction.* So, while the general presentation in Sect. 7.1.4 discusses, and uses, residuals centered on their mean, here the mean is by construction zero.

We will use the same sample from Cornwell and Rupert (1988) that was estimated in the application of Chap. 7. The first stage is an OLS estimation plus some preparatory actions, where among other things, we declare the series that will hold the orthogonality conditions.

```
series depvar = LWAGE
scalar obsn = $nobs
ols depvar Xlist --quiet
matrix bcoef = $coeff
series res = $uhat
scalar qv = .4
scalar qw = .3
scalar qu = .3
matrix W = I(4)
series eq2 = 0
series eq3 = 0
series eq4 = 0
series eq5 = 0
```

Notes: Here we named the distribution parameters, the sigmas, by `qv,qw,qu`—for fun. The $\boldsymbol{W}$ matrix is the weight matrix for GMM, specified as being the Identity matrix. If we iterate the estimator (and we will), $\boldsymbol{W}$ will be replaced in subsequent iterations by a matrix automatically computed by the software.

We will implement the NHN 2TSF model, so we take the expressions for the cumulant equations from Sect. 7.2.2. We estimate three parameters, but we will use four equations to have an overidentifying system (this is why $\boldsymbol{W}$ is specified as 4×4). The GMM command block is

```
gmm
series res = depvar - lincomb(Xlist,bcoef)
series res2 = res^2
series res3 = res^3
series res4 = res^4
series res5 = res^5
scalar Avres2sq = ((1/obsn)*sum(res2))^2
scalar Avres2 = (1/obsn)*sum(res2)
scalar Avres3 = (1/obsn)*sum(res3)
series eq2 = res2 - sqrt(qv)^4 - (1-2/$pi)*(sqrt(qw)^4 + sqrt(qu)^4)
series eq3 = res3 - ((4-$pi*sqrt(2)/($pi^(3/2)))*(qw^3 - qu^3)
series eq4 = (res4 - 3*Avres2sq) - (8*($pi-3)/($pi^2))*(qw^4 + qu^4)
series eq5 = (res5-10*Avres2*Avres3)
                    -(qw^5-qu^5)*(sqrt(2)/($pi^5/2))*(96-40*$pi+3*$pi^2)

orthog eq2 ; const
orthog eq3 ; const
orthog eq4 ; const
orthog eq5 ; const
weights W
params qv qw qu
end gmm --verbose --iterate
```

Table 11.6 gretl output from COLS/GMM estimation of scale parameters

```
--- FINAL VALUES:
GMM criterion 43.9095234726 (step 4.096e-009, norm 7.29e-003)
--- Iteration
1:  GMM criterion 43.9095234726 (norm 1.14e-002)
--- FINAL VALUES:
GMM criterion 43.9095234726 (step 8.192e-010, norm 1.14e-002)

Orthogonality conditions - descriptive statistics
 OC       mean  std.  dev

  0  0.000196    0.006500
  1  0.001585    0.006245
  2  0.000346    0.005647
  3  0.001586    0.005838
Tolerance = 1.81899e-012

Function evaluations:  337
Evaluations of gradient:  58

Model 7:  Iterated GMM, using observations 1-595
     estimate  std.  error       z  p-value

 qv  0.129051       0.260674  0.4951   0.6206
 qw  0.331614       0.154647   2.144   0.0320  **
 qu  0.359107       0.138187   2.599   0.0094  ***
GMM criterion:  Q = 0.000124029 (TQ = 0.0737975)
J test:  Chi-square(1) = 0.0737975 [0.7859]
```

In the syntax for the orthogonality conditions, say for "`orthog eq2; const`," the `orthog` command implies that the expression is set equal to zero, the series `const` is a series of ones, and the ";" symbol stands here for inner product (or sumproduct). So in practice, the software sums the elements of the `eq2` series, and set this sum equal to zero. Each such equation is the difference of a sample cumulant from its theoretical expression given the NHN specification; hence, the sum of its elements represents indeed an orthogonality condition when it is set equal to zero. The output we get is shown in Table 11.6.

As with the gradient in maximum likelihood estimation, here the software prints out the attained value for each orthogonality condition, so we can see how close we are to the ideal zero value. Compared to maximum likelihood estimation of the NHN specification (see the rightmost column of Table 7.1), GMM has trouble estimating σ_v, while ML appears more assertive on the matter. We also note that the J-test does not reject the NHN specification.

11.2.4.2 Full GMM Estimation

To avoid any confusion, we will provide the full script for joint GMM estimation of the regression betas and of the distribution parameters. Although we use OLS estimation to obtain starting values, we re-estimate the betas by GMM.

```
series depvar = LWAGE
scalar obsn = $nobs
ols depvar Xlist --quiet
matrix bcoef = $coeff
series res = $uhat
scalar qv = .4
scalar qw = .3
scalar qu = .3
matrix W = I(17)
series eqb0 = 0
series eqb1 = 0
series eqb2 = 0
series eqb3 = 0
series eqb4 = 0
series eqb5 = 0
series eqb6 = 0
series eqb7 = 0
series eqb8 = 0
series eqb9 = 0
series eqb10 = 0
series eqb11 = 0
series eqb12 = 0
series eq2 = 0
series eq3 = 0
series eq4 = 0
series eq5 = 0

string varblnames = varname(Xlist)
string errnames = ,"sigmav,sigmaw,sigmau"
string allcoefnames = varblnames ~ errnames
```

The model includes 12 regressors plus a constant term, and we use four orthogonality conditions for the three error distribution parameters; hence, the dimension of $\boldsymbol{W}$ is 17×17. We also use the same command as we used in the maximum likelihood script, so that the names of the variables appear in the output. The GMM command block here is

```
gmm
series resc = depvar - lincomb(Xlist,bcoef) - sqrt(2/$pi)*(qw - qu)
series resc2 = resc^2
series resc3 = resc^3
series resc4 = resc^4
series resc5 = resc^5
scalar Avres2sq = ((1/obsn)*sum(resc2))^2
scalar Avres2 = (1/obsn)*sum(resc2)
scalar Avres3 = (1/obsn)*sum(resc3)
series eqb0 = resc*Xlist[1]
```

```
series eqb1 = resc*Xlist[2]
series eqb2 = resc*Xlist[3]
series eqb3 = resc*Xlist[4]
series eqb4 = resc*Xlist[5]
series eqb5 = resc*Xlist[6]
series eqb6 = resc*Xlist[7]
series eqb7 = resc*Xlist[8]
series eqb8 = resc*Xlist[9]
series eqb9 = resc*Xlist[10]
series eqb10 = resc*Xlist[11]
series eqb11 = resc*Xlist[12]
series eqb12 = resc*Xlist[13]
series eq2 = resc2 - sqrt(qv)^4 - (1-2/$pi)*(sqrt(qw)^4 + sqrt(qu)^4)
series eq3 = resc3 - ((4-$pi*sqrt(2)/($pi^(3/2)))*(qw^3 - qu^3)
series eq4 = (resc4 - 3*Avres2sq) - (8*($pi-3)/($pi^2))*(qw^4 + qu^4)
series eq5 = (resc5-10*Avres2*Avres3)
                -(qw^5-qu^5)*(sqrt(2)/($pi^5/2))*(96-40*$pi+3*$pi^2)

orthog eqb0 ; const
orthog eqb1 ; const
orthog eqb2 ; const
orthog eqb3 ; const
orthog eqb4 ; const
orthog eqb5 ; const
orthog eqb6 ; const
orthog eqb7 ; const
orthog eqb8 ; const
orthog eqb9 ; const
orthog eqb10 ; const
orthog eqb11 ; const
orthog eqb12 ; const
orthog eq2 ; const
orthog eq3 ; const
orthog eq4 ; const
orthog eq5 ; const
weights W
params bcoef qv qw qu
param_names allcoefnames
end gmm --verbose --iterate
```

In the first line of the code, we have defined the centered residual proper by subtracting from the right-hand side its postulated mean, which is $(\sqrt{2/\pi})(\sigma_w - \sigma_u)$. We get the output shown in Tables 11.7 and 11.8.

One can verify that the estimates for the regression coefficients are virtually identical with the NHN maximum likelihood estimates (Table 7.1) and also that the estimates of the scale parameters are very similar to what we obtained by the COLS/GMM method previously. We close the presentation of GMM estimation in gretl by noting that the gmm command supports also upper and lower bounds on the estimated parameters (using a different syntax than the maximum likelihood command). The lower bound is certainly useful for estimating the scale parameters of the distribution of the error term that are always expressed as positive numbers,

Table 11.7 `gretl` output from full GMM estimation. Convergence and orthogonality conditions

```
--- FINAL VALUES:
GMM criterion 32.1333514764 (step 2.09715e-015, norm 1.96e-001)

--- Iteration
1:  GMM criterion 32.1333851205 (norm 1.78e-001)
--- FINAL VALUES:
GMM criterion 32.1333851205 (step 1.04858e-014, norm 1.78e-001)

Orthogonality conditions - descriptive statistics
 OC       mean  std. dev

  0  0.002171   0.013200
  1  0.049890   0.341310
  2  1.383267  11.359279
  3  0.103072   0.622567
  4  0.000674   0.008682
  5  0.000439   0.007351
  6  0.000772   0.007755
  7  0.001509   0.010508
  8  0.001871   0.012097
  9  0.000123   0.003442
 10  0.000499   0.007194
 11  0.029368   0.176149
 12  0.000101   0.003369
 13  0.000178   0.006502
 14  0.001419   0.006253
 15  0.000303   0.005666
 16  0.001309   0.005871
Tolerance = 1.81899e-012

Function evaluations:  1217
Evaluations of gradient:  147
```

while upper and lower bounds will be needed if, for example, we estimate a correlation coefficient.

11.2.5 The Tug-of-War Model (with Copula)

In this section, we will present `gretl` code that relates to Chap. 12, where we develop "Tug-of-War" (ToW) 2TSF models (readers may want to first consult that chapter). In short, in a ToW model, the composite error term does not contain a zero-mean noise component, but only the two competing latent forces. In our empirical adventures, we have encountered many cases, especially with small samples, where the initial estimation of a standard 2TSF model returns an estimate of σ_v that is

Table 11.8 gretl output from full GMM estimation. Estimates and specification test

```
Model 9:  Iterated GMM, using observations 1-595
                estimate  std. error         z      p-value

 const         5.60243      0.196501    28.51  8.58e-179  ***
 EXP           0.0292552    0.00648568   4.511  6.46e-06  ***
 EXP2       -0.000483857  0.000124175  -3.897  9.76e-05  ***
 WKS           0.00334485   0.00281680   1.187    0.2350
 OCC          -0.160174     0.0371742   -4.309  1.64e-05  ***
 IND           0.0863627    0.0278118    3.105    0.0019  ***
 SOUTH        -0.0593699    0.0316593   -1.875    0.0608  *
 SMSA          0.165732     0.0290675    5.702  1.19e-08  ***
 MS            0.0950581    0.0503953    1.886    0.0593  *
 FEM          -0.322948     0.0555928   -5.809  6.28e-09  ***
 UNION         0.106800     0.0305798    3.493    0.0005  ***
 ED            0.0572514    0.00705754   8.112  4.98e-016 ***
 BLK          -0.189894     0.0523276   -3.629    0.0003  ***
 sigmav        0.125347     0.261203    0.4799    0.6313
 sigmaw        0.340104     0.149133     2.281    0.0226  **
 sigmau        0.354849     0.135699     2.615    0.0089  ***
GMM criterion:  Q = 9.07659e-005 (TQ = 0.0540057)
J test:  Chi-square(1) = 0.0540057 [0.8162]
```

virtually zero, which is equivalent to the noise component being nonexistent. But as we explain in Chap. 12, we cannot just keep the standard model and report that the noise component does not exist; we have to re-estimate using a ToW model.

In general, the `mle` code will not be different as regards its structure, although it will contain a different log-likelihood. But what really differentiates matters in the example that we will provide is the inclusion of a copula density in order to account for the possible endogeneity of a regressor (see Sect. 14.1.4 for a quick discussion and references to fuller treatments).

11.2.5.1 A Copula to Forgive Your Endogeneity Sins

We consider one specific regressor, say x, as endogenous. We decide to use the Gaussian copula density. The Gaussian copula density takes as arguments variables of the form $x_* = \Phi^{-1}(F_x(x))$, namely, the variable transformed twice: we first pass it through its own distribution function and then through the inverse Normal distribution function (see also Sect. 10.3.2). In order to construct x_* from a given sample, we need to obtain its observations ranked; we could use the ready-made `gretl` function `ranking` for this purpose, but this function subtracts half of ties, something that, for the intended use of the function here, could be consequential with small samples and a possibly quasi-discrete behavior of the variable in question. So first we define our own `rankingalt` function (a "user-defined" function) that provides the rankings without subtracting half of ties.

```
function series rankingalt(series X)
matrix a = aggregate(const, X)
matrix b = cum(a[,2])
return replace(X, a[,1], b)
end function
```

Then we create the twice-transformed variable, and two auxiliary series called `lowbound` and `highbound`, which we will discuss in a while.

```
series x_star = invcdf(N,(1/(obsn+1))*rankingalt(x1))
series lowbound = 0.000001
series highbound = const - 0.000001
```

We now move to familiar territory,

```
ols Depvar Xlist --quiet
matrix bcoef = $coeff
scalar sw = .6
scalar su = .7
scalar rhole = -0.5

set max_verbose full
#The next three lines make the names of the variables appear in the
estimation output
string varblnames = varname(Xlist)
string errnames = ,"sigmaw,sigmau,rhole"
string allcoefnames = varblnames ~ errnames
```

The change here relates to which distribution parameters we declare: we have again σ_w and σ_u (`sw` and `su`), we do not have σ_v, but now we have `rhole`, which is the dependence parameter that characterizes the bivariate Normal copula density, essentially a correlation coefficient ρ between the (transformed) endogenous regressor and the (transformed) error term, ranging in $(-1,\ 1)$. The string with the parameter names changes accordingly.

Next, the mle command block is

```
mle logl = check ? likely : NA
series res = Depvar - lincomb(Xlist, bcoef)

series logdensAL = -ln(sw+su) + (res<=0)*(res/su) - (res>0)*(res/sw)
series likely = -0.5*ldet(R2) -0.5*qRq + 0.5*((invcdf(N,CDF))^2) +
logdensAL

# the following relate to the copula density
series CDFraw = (res<=0)*(su/(sw+su))*exp(res/su)
                + (res>0)* (1-(sw/(sw+su))*exp(-res/sw))
list cdflist = lowbound CDFraw highbound
series CDF = median(cdflist)
matrix R2 = {1,rhole; rhole, 1}
matrix q = {x_star, invcdf(N,CDF)}
matrix qR = q*inv(R2)
series qRq = sumr(q.*qR)

scalar check = (abs(rhole)<1) && (sw>0) && (su>0)
params bcoef sw su rhole
param_names allcoefnames
```

```
end mle
```

Notes: the `logdensAL` series is the log-density of the Asymmetric Double Exponential distribution, which is the 2TSF ToW error density when the two opposing latent forces each follow an Exponential distribution (see Sect. 12.1.1). All other components of the log-likelihood `likely` come from the Gaussian (Normal) copula density.[12] The CDF of the error density is the series `CDFraw`. But this is not what we input in the copula density; we input the series `CDF=median{lowbound, CDFraw, highbound}`. At each observation, this series will be equal either to the value of `CDFraw`, if it is in-between the two bounds, or to the bound closer to `CDFraw`, if it is below (or above) it, respectively.[13] We do this to avoid crashing the software if it so happens that for even just one observation, `CDFraw` evaluates so close to 1 that the $\Phi^{-1}(F(\varepsilon))$ will evaluate to "infinity," or so close to 0 that $\Phi^{-1}(F(\varepsilon))$ will evaluate to "minus infinity." This is the functionality of the series `lowbound` and `highbound`—a hedge against outliers that will inflate or deflate greatly the value of the residual at an observation forcing the software to abort.

The output from this maximum likelihood estimation will be similar to the one shown previously.

11.2.6 Using the Conditional Mode as Predictor of Latent Forces

The ultimate goal of estimation using a 2TSF model is to obtain sample-level estimates and predictions at the observation level for the two latent forces (then, the metrics of Chap. 2 take over to support with inference). Throughout this book, we have used exclusively the expected value and the conditional expectation respectively for this purpose. But as we discuss in Sect. 14.2.1, there exists an alternative predictor, the conditional *mode*, which may be a more appropriate predictor when we examine setups with one-off decisions or outcomes, rather than recurring ones. So here we will provide `gretl` code to find the conditional mode series, i.e., the argmax of the density of a latent force, say u, conditional on the error term:

$$\text{mode} : f_{u|\varepsilon}(\text{mode}) = \max .$$

If we want the mode of the conditional density of an exponentiated variable, say $\exp\{-u\}$, we can readily obtain its density by a change of variables as

[12] So if one wants to use the ToW model without the endogeneity aspect, one does not need them.

[13] Note the special functionality of the "median" `gretl` function here. If we had inputted a single series, we would get the median of the series. But when we input more than one series through a list, say our three, we do not get the three individual medians of these series, but a *series* of the median values *across* the three series per observation.

$$k \equiv e^{-u} \implies f_{k|\varepsilon}(k) = \frac{1}{k} f_{u|\varepsilon}(-\ln k).$$

Suppose we have successfully estimated a 2TSF NE model. We now have available estimates for the regression betas, and hence we can compute the residuals that stand in for the composite error term. We also have estimates for the three scale parameters. The first step is to write a user-defined function that computes the value of the density whose mode we want to find. The function will take as arguments ε, σ_v, σ_w, and σ_u (`res, sv, sw, su`), and it will return the density value.

For this example, we will find the conditional mode of $f_{u|\varepsilon}(u)$. Looking up its formula in Chap. 7, we have

$$f(u_i|\varepsilon_i) = \frac{\lambda \exp\{-\lambda u_i\}\, \Phi(u_i/\sigma_v + b_{2i})}{\chi_{1i}},$$

with

$$\lambda = \sigma_w^{-1} + \sigma_u^{-1}, \quad \chi_{1i} = \Phi(b_{2i}) + \exp\{a_{1i} - a_{2i}\}\, \Phi(b_{1i}).$$

Since the logarithmic transformation is strictly monotonic, it is always advisable to perform maximization at the logarithmic level. The script for creating the function is

```
function scalar ulogdensity (const scalar sv, const scalar sw,
                    const scalar su, series res, matrix mod)
scalar k = $t2
scalar a1 = sv^2/(2*su^2) + res[k] /su
scalar b1 = -res[k] /sv - sv/su
scalar a2 = sv^2/(2*sw^2) - res[k] /sw
scalar b2 = res[k] /sv - sv/sw
scalar chi1 = cnorm(b2) + exp(a1-a2)*cnorm(b1)
scalar lambda = 1/sw + 1/su
logdensu = log(lambda) -lambda*mod + log(cnorm(mod/sv + b2) ) -
         log(chi1)
return logdensu
end function
```

Some notes: the "mod" argument is a moniker, a stand-in label for u, the argument of the conditional density over which we will eventually maximize (in the next stage). But also, the conditioning variable ε (`res`) appears. Realize that we want to find the mode given the conditioning variable. So in practice, we will find n modes (for n the sample size), one for each observation in the sample, one for each value of the conditioning variable. So what the `ulogdensity` function does is to compute the value of the density (with respect to its argument, `mod`), given the value of `res` that corresponds to a specific observation of the sample. This is why the function is defined as a "scalar" (it returns a scalar value), and this is why the familiar $a_1, b_1, a_2, b_2, \chi_1$ are here defined as scalars and not as series, as they were before.

How do we control at which observation we are? Acknowledge the parameter k that takes the strange value $t2. This is hansl-speak for k to take on the index value of "the last observation in the *current* sample" (so if the current sample has 382 observations, k=382), and keep firmly in mind the word "current." We see that k indexes the residual series, res[k], and this is how we will be able to compute the density at a specific observation.

The next step is to prepare some things for the "BFGScmax" maximizer with constraints.

```
matrix modeu = 0.5*ones(1,1) #initialize the object we seek to compute.
matrix bounds = {1,1.0e-20,0.9999} #argmax must be between 0 and 1.
The bounds are the second and third values.
series modeuvals = 0 #declare the series to hold the results
```

The first two lines create tools needed by the maximizer, and the last line declares the series that will hold the modes of the conditional density per observation. Then we create a loop construct.

```
loop m=1..obsn --quiet
smpl 1 m
matrix modeu[1,1] = 0.5 #initial value for the argmax we seek
densval = BFGScmax(&modeu, bounds, ulogdensity(sv, sw, su, resmle,
modeu))
smpl 1 obsn
series modeuvals[m] = modeu[1,1]
endloop
```

We run a loop over the whole sample, observation per observation (obsn is the size of the sample). In each iteration, we first restrict the sample size to not exceed the *current* index of the loop. So if we have obsn=88 and we are currently at m=66, after the first smpl command, we work with the subsample $(1, \ldots, 66)$. Ah, but then the scalar k hiding inside the logudensity function takes the value 66, in this specific iteration. The BFGScmax maximizer finds the mode of the density for observation 66 and stores it in modeu (which is a scalar disguised for syntax purposes as a 1×1 matrix—this is so because this maximizer can perform maximization in multiple dimensions and therefore requires a matrix object).

We then reinstate the sample size and record the result in the series modeuvals at position m. The loop moves on to the next observation, and the cycle is repeated. The full series includes the "most likely" predictions of the value of the latent force u per observation (because these values maximize the conditional density).

A final note: observe that while we have constructed the function ulogdensity using arguments res,mod, in its actual implementation, we used arguments resmle,modeu. This is standard fare in gretl user-defined functions, and very useful, because the function can thus be reusable in other settings and name-calling. The names we use in the construction of the function are generic stand-ins; what matters is to place at each position when calling the function the same type of object (so if the second argument is a scalar, when calling the function, one should put the name of an actual scalar in the second position, etc.). Apart from reusability of the code, using labels not used in actual estimation when building functions guards also against inadvertently changing the values of estimates from previous steps when using the function, without wanting to do so. At the same time, to remember what to place in which position, the labels used in function-building could be very close to the labels that will be used in function-usage. So, for example, if we want to include in a function the parameter σ_w that goes by the label `sw` in our estimations, in building the function, we could use something like `fsw`.

11.3 The `sftt` Command in Stata for 2TSF Estimation

A "blackbox" implementation of the Normal-Exponential and the Normal-Half-Normal specifications of the 2TSF model (with and without determinants of latent forces) is available in Stata (Lian et al., 2023) through the `sftt` command. `sftt` also fits 2TSF models with the scaling property using nonlinear least squares. In addition, there are two subcommands, `sftt sigs` and `sftt eff`, to assist in post-estimation analysis and construction of the various metrics discussed here.

Since it is likely that a not negligible number of readers may feel elated to learn that they can eschew writing code and just use this ready-made Stata command, we decided to describe in some detail its capabilities, so that these readers are not subject to information deficiencies but have full information while deciding what to do.

11.3.1 Basic Maximum Likelihood Estimation

The `sftt` command estimates both the NE and the NHN 2TSF specifications by maximum likelihood. One can also include determinants of inefficiency in the basic "heteroskedasticity" method, i.e., by making the scale parameters of the distributions of the latent forces functions of covariates (see the relevant section of Chap. 8). There is not much to comment here: this being the Stata environment, all the software functionality surrounding benchmark maximum likelihood estimation is available (like robust standard errors).

11.3.2 NLS Estimation

The command can also implement nonlinear least squares (NLS), where the latent forces *themselves* (not the scale parameters of their distributions) are modeled as functions of covariates. This is the same method as the one we use in Chap. 8. Namely, the `sftt` command unmodified executes a standard NLS estimation routine estimating jointly the regression constant and the mean levels of the latent forces (but the user can fix any one of them to an arbitrary value, like 0 or 1). Lian et al. (2023) are aware of the possible identification problems that may arise, and so in their simulated example for the NLS model, they do not include a constant term in the regression (i.e., they fix it to zero), in order to be able to estimate efficiently the mean levels of the latent forces. But this is an intervention that will have consequences in real-world data samples, since, if the constant term of the regression exists in the data but is absent from the estimating equation, we will get a biased and inconsistent estimator. If researchers encounter problems in estimating these three parameters jointly, any attempted remedy should not alter the structure of the model in an ad hoc way; otherwise, the results will be questionable.

11.3.3 Presenting Estimation Results

In the Stata environment, it is common to obtain as automatic output the *logarithm* of the scale parameters, and this is what the `sftt` command does. This is a coding trick to force the estimator to go away from negative values, but other tricks exist (see the previous sections), and we say this because this trick has two undesirable side effects: first, while it is clearly stated in the software printout that what we look at is the logarithm of the scale parameter, seeing negative numbers in relation to scale parameters creates unnecessary informational noise, and it is not immediately interpretable in any case. Second, the standard errors produced relate to this logarithm, not to the original scale parameter, and one has to convert them by the Delta method, which is an approximation.

The software prints out the actual estimated scale parameters with the use of the `sftt sigs` subcommand (but without standard errors). This subcommand prints also a "variance analysis" where the relevant magnitudes of the scale parameters are computed, for example, the ratio

$$\frac{\sigma_w^2 + \sigma_u^2}{\sigma_v^2 + \sigma_w^2 + \sigma_u^2}.$$

The reader may have realized by now that we have avoided such kind of metrics in this book, and this is because the variances are not in the same scale as the data, and we consider such analytics not very useful, *apart* from assessing from a technical point of view the "strength" of the various signals that exist in the data.

Also, note that such ratios represent a *variance* analysis *only* for the Normal-Exponential 2TSF model, where the scale parameters of the latent forces correspond to the standard deviation, and hence their squares to their variances. But if one has implemented the NHN specification that the `sftt` command offers, the squared scale parameters do not equal the variance—in fact, they visibly overestimate it. In such a case, the user of the `sftt` command should compute the proper "variance analysis" outside the command.

Finally, we mention something minor but which can prove consequential under stress: the command output lists everywhere *first* the u component and *then* the w, while in this book, we followed throughout the order "$v/w/u$" in writing equations and presenting results.

11.3.4 Individual Metrics

The `sftt` command, through the subcommand `sftt eff`, is sophisticated enough to provide the series for certain conditional expectations needed for serious inference in the 2TSF model, as well as printing out summary statistics of them. We have two comments here: first, not all such conditional expectations are provided, as the reader may verify by comparing the related list of variables in Lian et al. (2023) with the conditional expectation expressions we provide and use in this book (see, e.g., Chaps. 2 and 7). Second, and perhaps more importantly, the *interpretation* given in Lian et al. (2023) of what these conditional expectations measure is not fully in line with the interpretations we have discussed and put forth in Chap. 2, the reason being that the Lian et al. (2023) viewpoints come from a less mature period in developing and using the 2TSF framework.

This is where the capabilities of the `sftt` command end, at the time of writing this book. In particular, the command does not allow dependence between w and u, cannot include Copulas to handle regressor endogeneity, neither it is able to estimate a 2TSF panel data model or execute simulated maximum likelihood on its own.

It is very fortunate that this command exists in a popular environment like Stata, facilitating the familiarization with and use of the 2TSF model in applied research, but one must keep in mind the comments discussed above. Eventually, one is advised to take the coding route to 2TSF analysis. It is in order to support this advice that we have written extensive sections for 2TSF coding in `R` and in `gretl`, but also the reason why we have set up a website companion to this book, https://github.com/Parms23/2TSF, where even more detailed code can be freely downloaded and used.

Part IV
Extensions

Chapter 12
Extensions I: 2TSF Tug-of-War Models

Tug-of-War (ToW) is an ancient universal game competition, still found in many countries and civilizations, where two teams pull a rope toward opposite directions until one side is swept by the force of the other's pull, tumbles, and falls.[1] This is exactly what our two latent forces w and u do, (almost but not fully) figuratively speaking. But up to now, their rivalry in the 2TSF model has been intermediated by the existence of the "noise" component v, leading to the 2TSF composite error term $\varepsilon = v + w - u$. In this chapter, we dispense with v and we present 2TSF models where only w and u appear in the error term: here, $\varepsilon = w - u$.

The reason for this is not modeling variety for variety's shake: extensive experience with applying 2TSF models to data has shown that, especially with small samples sometimes even up to several hundreds of observations, it is often the case that the scale parameter of the noise component σ_v is estimated as practically zero, telling us that a v component does not exist in the data under the 2TSF model. But why is it then necessary to develop a distinct model without v? Could we not keep the full 2TSF model and just report that σ_v is estimated as zero?

The answer is "no" for at least three reasons: first, the asymptotic properties of the maximum likelihood estimator when the true value of one coefficient lies at the boundary of the parameter space (and 0 is a boundary for a scale parameter) change and make subsequent inference more complicated. Second, in such cases, usually the estimator fails to provide standard errors for *any* estimated coefficient. Third, the various mathematical expressions characterizing the three-error-component 2TSF model (like the composite error density, or the conditional expectations for individual predictions) include σ_v in ways that do not allow straightforward insertion of the value 0 to obtain an arithmetic result. In theoretical derivations, we may be able to avoid this issue through algebraic manipulations and resolve a limiting expression with σ_v tending to zero, but computational software is not designed to

[1] There are also variations: a wooden stake instead of a rope, two persons rather two teams, or even a "sitting" version, where one contender has the goal to drag the other toward their side.

A. Papadopoulos, C. F. Parmeter, *Two-Tier Stochastic Frontier Analysis for the Social Sciences*, https://doi.org/10.1007/978-3-031-81513-3_12

handle such limits. In practice, researchers are advised to start with the full 2TSF model, and if they see non-convergence or $\widehat{\sigma_v} \approx 0$, to implement a ToW model.

We present 2TSF-ToW models under the assumption that w and u are independent, as in Chap. 7, but also, we draw from the techniques of Chap. 10 and develop a Copula model where they are assumed to be dependent. We include an application where we apply the first variant on sectors of the Greek economy using data before and after the depression of 2009–2018, in an attempt to determine the effects of the depression on management and overall efficiency of firms.

A note on notation: we have encountered this pure difference of w minus u in previous chapters, and there we used the symbol z for it, $z = w - u$. Here, to keep clearly visible that this is the full composite error term, we will use $\varepsilon = w - u$.

12.1 Tug-of-War Distributions of Independent Latent Forces

12.1.1 The Asymmetric Double Exponential Specification

The well-known Laplace distribution can be obtained as the difference of two independently and identically distributed (iid) Exponential random variables, and this is why its alternative name is "Double Exponential." For our case, ε is the difference of two Exponential random variables, $w \sim \text{Exp}(\sigma_w)$ and $u \sim \text{Exp}(\sigma_u)$, with the scale parameters being different in principle. This makes the distribution a version of an *Asymmetric* Laplace distribution, but we use the "Double Exponential" naming convention to stress that we are distinctly interested in its two generating components. Its density function is

$$f_\varepsilon(\varepsilon) = \frac{1}{\sigma_w + \sigma_u} \cdot \begin{cases} e^{\varepsilon/\sigma_u} & \varepsilon \leq 0 \\ e^{-\varepsilon/\sigma_w} & \varepsilon \geq 0. \end{cases} \tag{12.1}$$

This density is a continuous function, but not differentiable at zero. The (also continuous) distribution function of this composite error is

$$F_\varepsilon(\varepsilon) = \begin{cases} \frac{\sigma_u}{\sigma_w+\sigma_u} e^{\varepsilon/\sigma_u} & \varepsilon \leq 0 \\ 1 - \frac{\sigma_w}{\sigma_w+\sigma_u} e^{-\varepsilon/\sigma_w} & \varepsilon \geq 0. \end{cases} \tag{12.2}$$

Maximum Likelihood Estimation For estimation purposes, we can write the density without branches using the indicator function $I_\varepsilon \equiv I\{\varepsilon \leq 0\}$, as

$$f_\varepsilon(\varepsilon) = \frac{I_\varepsilon \cdot e^{\varepsilon/\sigma_u} + (1 - I_\varepsilon) \cdot e^{-\varepsilon/\sigma_w}}{\sigma_w + \sigma_u}. \tag{12.3}$$

The log-likelihood function for observation i is

$$\ell_i = -\ln(\sigma_w + \sigma_u) + \ln\left(I_{\varepsilon_i} \cdot e^{\varepsilon_i/\sigma_u} + (1 - I_{\varepsilon_i}) \cdot e^{-\varepsilon_i/\sigma_w}\right).$$

Because I_{ε_i} is binary, the above can be equivalently written:

$$\ell_i = -\ln(\sigma_w + \sigma_u) + I_{\varepsilon_i} \cdot \frac{\varepsilon_i}{\sigma_u} - (1 - I_{\varepsilon_i})\frac{\varepsilon_i}{\sigma_w}. \tag{12.4}$$

The log-likelihood of the sample is

$$\mathcal{L} = -n \cdot \ln(\sigma_w + \sigma_u) + \sum_{i=1}^{n} \ln\left(I_{\varepsilon_i} \cdot e^{\varepsilon_i/\sigma_u} + (1 - I_{\varepsilon_i}) \cdot e^{-\varepsilon_i/\sigma_w}\right),$$

or alternatively,

$$\mathcal{L} = -n \cdot \ln(\sigma_w + \sigma_u) + \sum_{i=1}^{n} I_{\varepsilon_i}\frac{\varepsilon_i}{\sigma_u} - \sum_{i=1}^{n}(1 - I_{\varepsilon_i})\frac{\varepsilon_i}{\sigma_w}. \tag{12.5}$$

There are only two distribution parameters to estimate, alongside the regression coefficients $\boldsymbol{\beta}$, which appear in the log-likelihood through $\varepsilon_i = y_i - \mathbf{x}_i'\boldsymbol{\beta}$.

COLS, CNLS, and GMM Estimation For these estimation methods, everything is the same as for the benchmark 2TSF NE specification (see Sect. 7.2.1), except the second cumulant, which is here

$$\varphi_2 = \kappa_2(\varepsilon) = \sigma_w^2 + \sigma_u^2.$$

Moreover, strictly speaking, we only need two cumulant equations since we estimate only two distribution parameters. In a GMM setting, one could decide to keep the additional cumulant equations in order to form an over-identified system in pursuit of estimator efficiency.

Metrics The sample-level metrics are exactly the same as those presented in Sect. 7.2.1, but the formulas for the individual metrics change. We will exploit here the fundamental property of a conditional expectation, $E(\varepsilon_i|\varepsilon_i) = \varepsilon_i$. This should be intuitive: if we know (i.e., "conditional on") that ε_i has been realized equal to, say, 0.53, it is evident that our "best prediction" for it should be 0.53: $E(\varepsilon_i|\varepsilon_i = 0.53) = 0.53$. But this implies, by the additivity property of conditional expectation, that

$$\varepsilon_i = E(w_i - u_i|\varepsilon_i) = E(w_i|\varepsilon_i) - E(u_i|\varepsilon_i) \implies E(u_i|\varepsilon_i) = E(w_i|\varepsilon_i) - \varepsilon_i.$$

In words, we do not need to compute separately $E(w_i|\varepsilon_i)$ and $E(u_i|\varepsilon_i)$, only, say, the first, since we will have available the residuals from the estimation and then obtain $E(u_i|\varepsilon_i)$ through

$$E(u_i|\varepsilon_i) = E(w_i|\varepsilon_i) - \varepsilon_i. \tag{12.6}$$

This property extends beneficially to some of the exponentiated formulas:

$$E(e^{\varepsilon_i}|\varepsilon_i) = e^{\varepsilon_i} \implies E(e^{w_i-u_i}|\varepsilon_i) = e^{\varepsilon_i} \tag{12.7}$$

$$\implies E(e^{u_i-w_i}|\varepsilon_i) = e^{-\varepsilon_i}. \tag{12.8}$$

However, the individual conditional expectations each require their own formulas. So in all, the formulas we need for the individual metrics are

$$E(w_i|\varepsilon_i), \quad E(e^{w_i}|\varepsilon_i), \quad E(e^{-w_i}|\varepsilon_i), \quad E(e^{u_i}|\varepsilon_i), \quad E(e^{-u_i}|\varepsilon_i).$$

These are, using as in Chap. 7 the composite $\lambda = \sigma_w^{-1} + \sigma_u^{-1}$, for which here too we require $\lambda > 1$, for these conditional expectations to exist:

$$E(w_i|\varepsilon_i) = \begin{cases} \frac{1}{\lambda} & \varepsilon_i \leq 0 \\ \frac{1}{\lambda} + \varepsilon_i & \varepsilon_i \geq 0. \end{cases} \implies E(u_i|\varepsilon_i) = \begin{cases} \frac{1}{\lambda} - \varepsilon_i & \varepsilon_i \leq 0 \\ \frac{1}{\lambda} & \varepsilon_i \geq 0. \end{cases} \tag{12.9}$$

We see that for half of the domain of ε_i (the opposite in each case), these metrics become constant, i.e., uninformative as regards the per observation strength of the corresponding latent force. These results align with what Horrace and Parmeter (2018) found in developing the single-tier Laplace stochastic frontier model.

Turning to the exponentiated measures we have,

$$E(e^{w_i}|\varepsilon_i) = \begin{cases} \frac{\lambda}{\lambda-1} & \varepsilon_i \leq 0 \\ \frac{\lambda}{\lambda-1}e^{\varepsilon_i} & \varepsilon_i \geq 0, \end{cases} \qquad E(e^{-w_i}|\varepsilon_i) = \begin{cases} \frac{\lambda}{\lambda+1} & \varepsilon_i \leq 0 \\ \frac{\lambda}{\lambda+1}e^{-\varepsilon_i} & \varepsilon_i \geq 0. \end{cases} \tag{12.10}$$

and

$$E(e^{u_i}|\varepsilon_i) = \begin{cases} \frac{\lambda}{\lambda-1}e^{-\varepsilon_i} & \varepsilon_i \leq 0 \\ \frac{\lambda}{\lambda-1} & \varepsilon_i \geq 0, \end{cases} \qquad E(e^{-u_i}|\varepsilon_i) = \begin{cases} \frac{\lambda}{\lambda+1}e^{\varepsilon_i} & \varepsilon_i \leq 0 \\ \frac{\lambda}{\lambda+1} & \varepsilon_i \geq 0. \end{cases} \tag{12.11}$$

For computational purposes, we combine the two branches into one expression using indicator functions. This completes the presentation of the Tug-of-War 2TSF model with independent and Exponential latent forces.

12.1.2 The Asymmetric Double Half-Normal Specification

For this specification, we assume that $\varepsilon = w - u$ is the difference of two independent Half-Normal random variables, which are allowed to have different scale parameters: $w \sim \text{HN}(\sigma_w)$, $u \sim \text{HN}(\sigma_u)$. The density of the composite error ε is

$$f_\varepsilon(\varepsilon) = \begin{cases} (4/s)\phi(\varepsilon/s)\,\Phi(\lambda\varepsilon/s) & \varepsilon \le 0 \\ (4/s)\phi(\varepsilon/s)\,\Phi\left(-\lambda^{-1}\varepsilon/s\right) & \varepsilon \ge 0 \end{cases} \quad \lambda = \sigma_w/\sigma_u,\ s = \sqrt{\sigma_w^2 + \sigma_u^2}. \tag{12.12}$$

The symbols ϕ and Φ represent the density and distribution function of the standard Normal, respectively, as always. This distribution has been called the "Asymmetric Double Half-Normal" (ADHN) by Papadopoulos (2015a), and we keep the name based on our comments earlier. Like the Laplace, it has a kink at the point where its two branches meet, and it can be considered the "Laplace analogue" in the Normal family. Its mean equals the difference of the means, and its variance the sum of the variances of w and u. Its skewness and excess kurtosis range in $(-0.995, 0.995)$ and $(0.434, 0.869)$, respectively. This distribution can be seen as a tool to model situations where the properties of the data are qualitatively congruent to the Laplace distribution (fatter tails than the Normal and concentration of values around zero), but not so strongly realized.

The distribution function of an ADHN random variable is

$$F_\varepsilon(\varepsilon) = \begin{cases} 4\Phi_2(\varepsilon/s\,,\ 0;\ -\delta) & \varepsilon \le 0 \\ 4\Phi_2\left(\varepsilon/s\,,\ 0;\ \lambda^{-1}\delta\right) - 1 & \varepsilon \ge 0. \end{cases} \tag{12.13}$$

Here, $\Phi_2(\cdot, \cdot; \delta)$ is the bivariate standard Normal integral, evaluated at the vector $(\varepsilon/s, 0)$ with correlation coefficient

$$\delta = \lambda/\sqrt{1+\lambda^2}.$$

Maximum Likelihood Estimation For estimation purposes, we can write the density using the indicator function as before:

$$f_\varepsilon(\varepsilon) = (4/s)\phi(\varepsilon/s) \cdot \left[I_\varepsilon \cdot \Phi(\lambda\varepsilon/s) + (1 - I_\varepsilon) \cdot \Phi\left(-\lambda^{-1}\varepsilon/s\right)\right]. \tag{12.14}$$

The log-likelihood function for observation i can be written:

$$\ell_i = \ln\left(\frac{4}{\sqrt{2\pi}}\right) - \ln s - \frac{1}{2}\frac{\varepsilon_i^2}{s^2} + I_{\varepsilon_i} \cdot \ln \Phi\left(\lambda\varepsilon_i/s\right) + (1 - I_{\varepsilon_i}) \cdot \ln \Phi\left(-\lambda^{-1}\varepsilon_i/s\right). \tag{12.15}$$

COLS, CNLS, and GMM Estimation As with the Double Exponential specification, the only thing that changes in the structure of these essentially Method-of-Moments estimators is the second cumulant, which now includes only the variances of w and u:

$$\varphi_2 = \kappa_2(\varepsilon) = \left(1 - \frac{2}{\pi}\right) \cdot (\sigma_w^2 + \sigma_u^2).$$

Metrics The sample-level metrics are exactly the same as those presented in Sect. 7.2.2, since w and u are Half-Normals. For the individual metrics, we have here too the convenient results:

$$\begin{aligned} E(u_i|\varepsilon_i) &= E(w_i|\varepsilon_i) - \varepsilon_i, \\ E(e^{w_i - u_i}|\varepsilon_i) &= e^{\varepsilon_i} \\ E(e^{u_i - w_i}|\varepsilon_i) &= e^{-\varepsilon_i}. \end{aligned}$$

As regards the rest of the conditional expectation formulas that we need to compute the individual metrics, they are

$$E(w_i|\varepsilon_i)$$
$$= \begin{cases} [\Phi\left(\lambda\varepsilon_i/s\right)]^{-1} \cdot \left[\sigma^*\phi(\mu_i^*/\sigma^*) + \mu_i^*\Phi(\mu_i^*/\sigma^*)\right] & \varepsilon_i \le 0 \\ \left[\Phi\left(-\lambda^{-1}\varepsilon_i/s\right)\right]^{-1} \cdot \left[\sigma^*\phi((\mu_i^* - \varepsilon_i)/\sigma^*) + \mu_i^*\Phi((\mu_i^* - \varepsilon_i)/\sigma^*)\right] & \varepsilon_i > 0. \end{cases} \tag{12.16}$$

with

$$\sigma^* = \frac{\sigma_w\sigma_u}{s}, \qquad \mu_i^* = \frac{\sigma_w^2}{s^2}\varepsilon_i.$$

Also

$$E(e^{w_i}|\varepsilon_i)$$
$$= \begin{cases} [\Phi\left(\lambda\varepsilon_i/s\right)]^{-1} \cdot \Phi(\sigma^* + \lambda\varepsilon_i/s) \cdot \exp\left\{\frac{\sigma_w^2}{s^2}\left(\frac{\sigma_u^2}{2} + \varepsilon_i\right)\right\} & \varepsilon_i \le 0 \\ \left[\Phi\left(-\lambda^{-1}\varepsilon_i/s\right)\right]^{-1} \cdot \Phi(\sigma^* - \lambda^{-1}\varepsilon_i/s) \cdot \exp\left\{\frac{\sigma_w^2}{s^2}\left(\frac{\sigma_u^2}{2} + \varepsilon_i\right)\right\} & \varepsilon_i > 0. \end{cases} \tag{12.17}$$

$$E(e^{-w_i}|\varepsilon_i) = \begin{cases} [\Phi(\lambda\varepsilon_i/s)]^{-1} \cdot \Phi(-\sigma^* + \lambda\varepsilon_i/s) \cdot \exp\left\{\frac{\sigma_w^2}{s^2}\left(\frac{\sigma_u^2}{2} - \varepsilon_i\right)\right\} & \varepsilon_i \le 0 \\ \left[\Phi\left(-\lambda^{-1}\varepsilon_i/s\right)\right]^{-1} \cdot \Phi(-\sigma^* - \lambda^{-1}\varepsilon_i/s) \cdot \exp\left\{\frac{\sigma_w^2}{s^2}\left(\frac{\sigma_u^2}{2} - \varepsilon_i\right)\right\} & \varepsilon_i > 0. \end{cases} \tag{12.18}$$

For the u latent force, we have

$$E(e^{u_i}|\varepsilon_i) = \begin{cases} [\Phi(\lambda\varepsilon_i/s)]^{-1} \cdot \Phi(\sigma^* + \lambda\varepsilon_i/s) \cdot \exp\left\{\frac{\sigma_u^2}{s^2}\left(\frac{\sigma_w^2}{2} - \varepsilon_i\right)\right\} & \varepsilon_i \le 0 \\ \left[\Phi\left(-\lambda^{-1}\varepsilon_i/s\right)\right]^{-1} \cdot \Phi(\sigma^* - \lambda^{-1}\varepsilon_i/s) \cdot \exp\left\{\frac{\sigma_u^2}{s^2}\left(\frac{\sigma_w^2}{2} - \varepsilon_i\right)\right\} & \varepsilon_i > 0. \end{cases} \tag{12.19}$$

$$E(e^{-u_i}|\varepsilon_i) = \begin{cases} [\Phi(\lambda\varepsilon_i/s)]^{-1} \cdot \Phi(-\sigma^* + \lambda\varepsilon_i/s) \cdot \exp\left\{\frac{\sigma_u^2}{s^2}\left(\frac{\sigma_w^2}{2} + \varepsilon_i\right)\right\} & \varepsilon_i \le 0 \\ \left[\Phi\left(-\lambda^{-1}\varepsilon_i/s\right)\right]^{-1} \cdot \Phi(-\sigma^* - \lambda^{-1}\varepsilon_i/s) \cdot \exp\left\{\frac{\sigma_u^2}{s^2}\left(\frac{\sigma_w^2}{2} + \varepsilon_i\right)\right\} & \varepsilon_i > 0. \end{cases} \tag{12.20}$$

With these, all necessary formulas for the ToW ADHN specification have been provided. We turn now to present a ToW specification that incorporates dependence between w and u through the use of Copulas.

12.2 A Copula Tug-of-War Model for Dependent Latent Forces

We have already encountered *two* Tug-of-War distributions with intra-error dependence, in Chap. 10, where we explored the case of statistical dependence between the two latent forces. There, it was convenient to derive the distribution of the difference $w - u$ as an intermediate step to arrive at the more complicated expressions for the three-component error term. In a ToW setting, these distributions could be used as-is, incorporating also dependence.

But there are issues: for the case of Freund's Bivariate Exponential extension (which is a component in the Correlated 2TSF Exponential model of Sect. 10.1), a little algebra reveals that if $\varepsilon = w - u$, then the conditional expectations $E(w|\varepsilon)$ and the like, which we need for our metrics, include not the identifiable parameters but the non-identified ones, in a way that renders the specification useless for empirical work. As regards the bivariate Truncated Normal (see Sect. 10.2), it is rather restrictive in its stochastic assumptions, since it requires that the marginal distributions of w and u have the same skewness parameter.

For this reason, we take the Copula pathway to introduce dependence. While the spirit is the same as in Sect. 10.3, the absence of the noise component v creates certain differences that must not be overlooked.

12.2.1 The Likelihood Function

To model dependence between variables using Copulas, ultimately at the density level in order to apply maximum likelihood, we need their joint density, which incorporates dependence through the multiplicative presence of the copula density. For variables w and u under dependence that is represented by the Copula $C\,(F_w(w), F_u(u))$ with copula density $c\,(F_w(w), F_u(u))$, we have

$$f_{w,u}(w, u) = c\,(F_w(w), F_u(u)) \cdot f_w(w) \cdot f_u(u).$$

Further, we want to arrive at the density of the composite error $\varepsilon = w - u$. We can use the transformation $\varepsilon = w - u \implies w = \varepsilon + u \geq 0 \implies u \geq -\varepsilon$, through which we have

$$f_{\varepsilon,u}(\varepsilon, u) = f_{w,u}(\varepsilon + u, u) = c\,(F_w(\varepsilon + u), F_u(u)) \cdot f_w(\varepsilon + u) \cdot f_u(u).$$

Then we have, using the indicator function $I\{u \geq -\varepsilon\}$,

$$f_\varepsilon^{(u)}(\varepsilon) = \int_{-\varepsilon}^{\infty} f_{\varepsilon,u}(\varepsilon, u)\, du$$

$$\implies f_\varepsilon^{(u)}(\varepsilon) = \int_0^{\infty} I\{u \geq -\varepsilon\} \cdot c\,(F_w(\varepsilon + u), F_u(u))) \cdot f_w(\varepsilon + u) \cdot f_u(u)\, du.$$

This can be simulated by using draws from an Exponential random variable, i.e., based on the density $f_u(u)$, but draws from this density require the estimated parameter σ_u each time. To avoid this complication, we insert in the integrand the neutral term $e^u e^{-u} = e^u f_{\text{Exp}}(u; 1)$, and we have

$$f_\varepsilon^{(u)}(\varepsilon) =$$

$$\int_0^{\infty} f_{\text{Exp}}(u; 1) \cdot \left[e^u I\{u \geq -\varepsilon\} \cdot c\,(F_w(\varepsilon + u), F_u(u)) \cdot f_w(\varepsilon + u) \cdot f_u(u) \right] du,$$

which can be computed using S draws from a standard Exponential $t \sim \text{Exp}(1)$ by

$$f_{\varepsilon}^{(u)}(\varepsilon_i) \approx S^{-1} \sum_{s=1}^{S} \left[e^{t_s} I\{t_s \geq -\varepsilon_i\} \cdot c\left(F_w(\varepsilon_i + t_s), F_u(t_s)\right) \cdot f_w(\varepsilon_i + t_s) \cdot f_u(t_s) \right]. \tag{12.21}$$

Notice that "our" u variable does not appear anymore in the expression, since it has become the dummy variable of integration and then the random draw for summation and averaging. This is why we have switched notation from u to t for the running variable, in order to make this fully clear. The u variable is of course still very much *structurally* present, because the specification for its density and the corresponding distribution function do not disappear. For example, the expression $f_u(t)$ means that we take a random draw t from a standard Exponential distribution and we insert it into the assumed density function f_u of u whatever that density is—it has nothing to do with what the density of t is.

But also notice that we have indexed the composite error density with superscript (u) to indicate which is the last variable that we are integrating out to get to it. The reason we did this is that in the process, we will also need the alternative equivalent expression for the composite density. This alternative starts with the transformation $\varepsilon = w - u \implies u = w - \varepsilon \geq 0 \implies w \geq \varepsilon$, through which we have

$$f_{\varepsilon,w}(\varepsilon, w) = f_{w,u}(w, w - \varepsilon) = c\left(F_w(w), F_u(w - \varepsilon)\right) \cdot f_w(w) \cdot f_u(w - \varepsilon).$$

Using here the indicator function $I\{w \geq \varepsilon\}$, we have in an analogous manner

$$f_{\varepsilon}^{(w)}(\varepsilon) = \int_{\varepsilon}^{\infty} f_{\varepsilon,w}(\varepsilon, w)\, dw$$

$$\implies f_{\varepsilon}^{(w)}(\varepsilon) = \int_{0}^{\infty} I\{w \geq \varepsilon\} \cdot c\left(F_w(w), F_u(w - \varepsilon)\right) \cdot f_w(w) \cdot f_u(w - \varepsilon)\, dw.$$

By the same manipulations as before and drawing again from a standard Exponential $t \sim \text{Exp}(1)$, we arrive at

$$f_{\varepsilon}^{(w)}(\varepsilon_i) \approx S^{-1} \sum_{s=1}^{S} \left[e^{t_s} I\{t_s \geq \varepsilon_i\} \cdot c\left(F_w(t_s), F_u(t_s - \varepsilon_i)\right) \cdot f_w(t_s) f_u(t_s - \varepsilon_i) \right]. \tag{12.22}$$

Here we have substituted t for our w variable.

Estimation is carried out by simulated maximum likelihood, along the lines we discussed in Chap. 10. Any one of the two alternative forms of the composite error density can be used for estimation purposes. But we will need both for the derivation and computation of the various conditional expectations.

12.2.2 Individual Metrics

To compute individual metrics, we need the related conditional expectations. Looking first at the w variable, and setting generically $g(w) = w, \exp\{\pm w\}$, we have in general

$$E(g(w)|\varepsilon) = \int_{\varepsilon}^{\infty} g(w) f(w|\varepsilon)\, dw = \int_{0}^{\infty} g(w) I\{w \geq \varepsilon\} \frac{f(w, \varepsilon)}{f^{(w)}(\varepsilon)} f(w|\varepsilon)\, dw.$$

It is here that it becomes preferable to use the expression for the density of ε that is associated with each variable. This can be computed by simulation as

$$E(g(w_i)|\varepsilon_i) \approx \frac{S^{-1} \sum_{s=1}^{S} \left[g(t_s) e^{t_s} I\{t_s \geq \varepsilon_i\} \cdot c\left(F_w(t_s), F_u(t_s - \varepsilon_i)\right) \cdot f_w(t_s) \cdot f_u(t_s - \varepsilon_i) \right]}{S^{-1} \sum_{s=1}^{S} \left[e^{t_s} I\{t_s \geq \varepsilon_i\} \cdot c\left(F_w(t_s), F_u(t_s - \varepsilon_i)\right) \cdot f_w(t_s) \cdot f_u(t_s - \varepsilon_i) \right]}. \tag{12.23}$$

For $g(u) = \exp\{\pm u\}$ (since we do not need to compute $E(u|\varepsilon)$ separately), we have similarly

$$E(g(u)|\varepsilon) = \int_{-\varepsilon}^{\infty} g(u) f(u|\varepsilon)\, du = \int_{0}^{\infty} g(u) I\{u \geq -\varepsilon\} \frac{f(u, \varepsilon)}{f^{(u)}(\varepsilon)} f(u|\varepsilon)\, du,$$

and

$$E(g(u_i)|\varepsilon_i) \approx \frac{S^{-1} \sum_{s=1}^{S} \left[g(t_s) e^{t_s} I\{t_s \geq -\varepsilon_i\} \cdot c\left(F_w(\varepsilon_i + t_s), F_u(t_s)\right) \cdot f_w(\varepsilon_i + t_s) \cdot f_u(t_s) \right]}{S^{-1} \sum_{s=1}^{S} \left[e^{t_s} I\{t_s \geq -\varepsilon_i\} \cdot c\left(F_w(\varepsilon_i + t_s), F_u(t_s)\right) \cdot f_w(\varepsilon_i + t_s) \cdot f_u(t_s) \right]}. \tag{12.24}$$

In both cases, t_s comes from a standard Exponential distribution.

12.2.3 Choosing the Marginal Distributions and the Copula

The marginal distributions can be chosen separately from the Copula, and they can be different (on the matter of distributional heterogeneity see some discussion on chapter 14). The sample-level marginal moments then will depend on these choices, and they are not affected by the presence of dependence between w and u.

The choice of Copula can be guided by the causal factors that we believe are responsible for the statistical dependence: do we expect the variables to be negatively or positively correlated? Do we expect similar dependence along the supports of w and u, or do we expect perhaps stronger correlation between their extreme values compared to their central realizations? Each Copula endows the variable with different forms of dependence, in a known and well-studied way. The Copulas we have presented in Chap. 10 cover some of these properties, and they are certainly among the first candidates to be used.

With these remarks, we conclude the theoretical part and we turn to the empirical application of the chapter.

12.3 Application: The Greek Depression 2009–2018

Due to its severity and length, the Greek economic depression of 2009–2018 is rather more widely known than what the size and the geopolitical presence of the country would have us anticipate: gross domestic product eventually fell by approximately 26% compared to 2008, wages fell by even more in most industries, and it took a decade for the economy to start getting out of the slump. Clear macroeconomic indications of a recession were already present in 2009 (with GDP falling by 4.5% and unemployment rising by 2.5 percentage points), although at the time, this was not the prevailing mood in the country. Three years later, in 2012, the crisis and depression were in full force, with unemployment at 25%, i.e., three times its 8% pre-depression long-term rate. We set the final year of the depression at 2018, because 2019 was the third year in a row where GDP grew (although still 23% away from its pre-depression levels), and employment had gained back half of its losses (unemployment in 2019 stood at 17%).

A study was undertaken, in collaboration with the Foundation for Economic and Industrial Research (IOBE), in order to examine if, how, and how much the structure of production changed due to the depression. The study was conducted at the three-digit sectoral level based on the NACE rev2 classification.[2] Due to data limitations, we were able to estimate eventually only 40 such sectors, but many different economic activities are covered (except agriculture). For each sector, we estimated two cross-sectional samples: one from 2008 (pre-depression) and one from 2019 (after the depression). So in all, we estimated 80 separate samples. This is the "before-and-after" methodology, where the cumulative change over time if any, appears clearly, and also, it had the time to settle into its new status/level.

---

[2] Alecos Papadopoulos thanks professor Nikos Vettas, Director General of the Foundation for Economic and Industrial Research (IOBE, www.iobe.gr) for granting access to the data, and also IOBE researcher Alexandros Louka who prepared the data set.

We estimated production functions where the production inputs were Fixed Assets, Current Assets, and Labor (head count). We were able to separate Assets into Fixed and Current, because we used balance sheet data. The functional form of the production function was the translog, allowing for interaction terms and varying output elasticities of inputs per observation. Control variables were the age of each firm, whether or not it was headquartered in the two big urban centers of Greece (the country's capital Athens and the "capital of the North" Thessaloniki), and an index of long-term financing (long-term funds—equity or liabilities—as a percentage of total funds in the Balance Sheet). Our dependent variable was gross value added (labor costs plus EBITDA for gross capital returns ("Earnings Before Interest, Tax, Depreciation, and Amortization"). The dependent variable and the production inputs were used in logarithms.

We implemented maximum likelihood estimation. To handle endogeneity of regressors, for capital we used the beginning-of-period values in each sample, while for labor we applied a copula methodology using the Gaussian copula density.

Apart from the traditional scope and goals of production function estimation (output elasticities, complementarity/substitutability of inputs), we also aimed at recovering the contribution of management, the level of external inefficiency, and overall efficiency, applying the two-tier stochastic frontier model presented in Sect. 5.1.

The total number of observations from 2008 was $n = 6{,}824$, while for 2019, it was $n = 11{,}960$. There were 2,792 firms appearing on both years—these were the survivor firms in our database, operating before, during, and after the depression.

But because we estimated each sector and each year separately, sample sizes where small: for 2008, we had samples as small as $n = 53$ and only in two sectors was the sample size higher than $n = 500$ (average sample size was $n = 171$, median was $n = 123$). For 2019, the smallest sample size was $n = 99$ and the largest $n = 1{,}275$. But only in five sectors do we have a sample with size larger than $n = 500$ (average sample size here was $n = 299$, median $n = 186$). Table 12.1 provides the full list of sectors and sample sizes per year.

In all cases, except for the sample of the highest size, we encountered the issue discussed in the introduction of this chapter: the scale parameter σ_v of the noise error component was estimated as essentially zero. So, in order to be able to proceed with the study, we applied the Tug-of-War model with Exponential marginal distributions.

12.3.1 Results Related to Management and Efficiency

Small sample sizes do not allow to drill down in each sector individually with much confidence, so we did not compute individual metrics per sample. But the sample-level metrics and the overall picture from the 40 sectors provide a reliable aggregate assessment.

Table 12.1 Three-digit NACE rev2 sectors of the Greek economy and sample sizes

SecID	Description	2008	2019
101	Processing and preserving of meat and production of meat products	76	123
103	Processing and preserving of fruits and vegetables	114	224
105	Manufacture of dairy products	63	112
107	Manufacture of bakery and farinaceous products	87	165
108	Manufacture of other food products	71	156
110	Manufacture of beverages	76	150
141	Manufacture of wearing apparel, except fur apparel	120	139
172	Manufacture of articles of paper and paperboard	84	136
181	Printing and service activities related to printing	105	145
222	Manufacture of plastic products	151	223
236	Manufacture of articles of concrete, cement, and plaster	113	133
251	Manufacture of structural metal products	134	182
259	Manufacture of other fabricated metal products	64	114
310	Manufacture of furniture	110	99
412	Construction of residential and nonresidential buildings	303	435
421	Construction of roads and railways	60	104
432	Electrical, plumbing, and other construction installation activities	72	135
451	Sale of motor vehicles	224	248
453	Sale of motor vehicle parts and accessories	92	190
461	Wholesale on a fee or contract basis	72	99
463	Wholesale of food, beverages, and tobacco	345	987
464	Wholesale of household goods	828	1,275
465	Wholesale of information and communication equipment	176	235
466	Wholesale of other machinery, equipment, and supplies	396	740
467	Other specialized wholesale	544	893
469	Nonspecialized wholesale trade	127	233
471	Retail sale in nonspecialized stores	78	141
475	Retail sale of other household equipment in specialized stores	204	271
477	Retail sale of other goods in specialized stores	222	370
494	Freight transport by road and removal services	53	192
522	Support activities for transportation	174	404
551	Hotels and similar accommodation	277	998
561	Restaurants and mobile food service activities	156	178
581	Publishing of books, periodicals, and other publishing activities	136	137
620	Computer programming, consultancy, and related activities	227	405
682	Rental and operating of own or leased real estate	150	226
702	Management consultancy activities	212	366
711	Architectural and engineering activities and related technical consultancy	89	174
731	Advertising	125	138
791	Travel agency and tour operator activities	114	285

Table 12.2 Management factor, external inefficiency, and managed efficiency scores. Basic statistics

Metric	Variable	Mean	Median	S.D.	Min	Max
$M_5 + 1$	Management factor 2008 (mng2008)	1.40	1.35	0.322	1.10	2.92
	Management factor 2019 (mng2019)	1.51	1.44	0.254	1.25	2.50
M_6	External inefficiency 2008 (inef2008)	0.287	0.264	0.0766	0.185	0.506
	External inefficiency 2019 (inef2019)	0.346	0.351	0.0794	0.200	0.515
$M_7 + 1$	Managed Efficiency 2008 (eff2008)	0.992	0.968	0.210	0.573	1.95
	Managed efficiency 2019 (eff2019)	0.982	0.956	0.158	0.666	1.42

We discuss only the results pertaining to the 2TSF aspect of the study. Was management efficiency higher in 2019 compared to 2008? What happened with external inefficiency and overall efficiency?

Here we collect and compare the 40 points per sample-level metric per year, each representing a specific sector. Table 12.2 contains the basic statistics of these series of 40 points.

From Table 12.2, we see that the distribution of management gross output shifters (metric $M_5 + 1$) has moved to the right in 2019 compared to where it was in 2008. This is an indication that harsh environments like a depression do perform an evolutionary adaptation/weeding-out act. In fact in 33 out of the 40 sectors, the average management factor has increased. But external inefficiency (metric M_6) has done the same—the environment in 2019 was not as friendly as in 2008. In 32 out of the 40 sectors (not the same 32 as before), average external inefficiency was higher in 2019. The overall effect, on managed efficiency (metric $M_7 + 1$), is negative, being a bit more below unity than in 2008, while its distribution appears more concentrated. Managed efficiency has increased in only 18 sectors out of the 40.

Note that "managed efficiency" being close to 1 does not translate into "almost full efficiency" with the meaning that this concept has in the traditional single-tier frontier analysis of production/technical efficiency. It only means that management just manages to counter and offset external inefficiency, and inputs are producing "as if" they were unmanaged while external inefficiency was zero. This is a rather uninspiring level of efficiency, since with the management factor present (and its associated costs), we would want to see managed efficiency visibly higher than unity.

In Fig. 12.1, we present the scatter plots of the three pairs. We have put the year 2019 in the horizontal axis, so points *below* the 45°-degree line indicate an *increase* in the magnitude of each variable, and the further away the point is from it, the bigger the change. These figures verify visually what we have discussed: both the management factor and external inefficiency have mostly increased (first two figures), while the end effect is mixed (third figure).

While the 10-year-long Greek depression appears to have had some beneficial side effects as regards the efficiency level of management in Greek firms across sectors, the more harsher economic environment did not allow for these gains to be

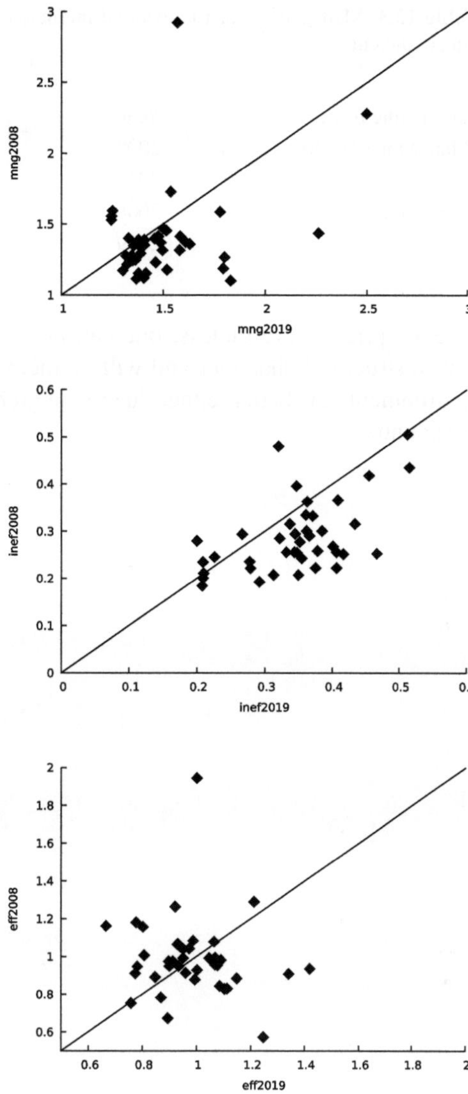

Fig. 12.1 Management factor, external inefficiency, and managed efficiency scores Scatterplots

materialized in overall efficiency of firms. We can further illuminate the situation by looking at Kendall's tau coefficients, a nonparametric measure of ordinal association ("concordance"). These are presented in Table 12.3.

The degree of ordinal association of qualitative co-movement of managed efficiency has decreased in relation to management and has increased (in absolute terms) in relation to external inefficiency, going from 2008 to 2019. This indicates that, in 2019 compared to 2008, managed efficiency tends to be determined more by external inefficiency than by management efficiency. This does not make the management factor irrelevant, but it indicates that external inefficiency extracts

Table 12.3 Management factor, external inefficiency, and managed efficiency scores. Kendall's tau coefficients

		Managed Efficiency	
Competing forces	Year	2008	2019
Management factor	2008	0.5328	
	2019		0.4554
External inefficiency	2008	– 0.2577	
	2019		– 0.4287

a heavy price. Nevertheless, one can hope that the management efficiency gains have a structural character and will be there to perform when (or if) the economic environment gets better, either due to the global tides or due to structural reforms in the country.

Chapter 13
Extensions II: 2TSF Binary Response Models

As promised in Chap. 1, we develop in this penultimate chapter of the book two 2TSF binary response (BR) models, the first one to handle unobservable latent forces that silently affect the binary responses we observe, and the second one, where we decompose and measure separately the possibly conflicting effects that a binary Treatment has on a binary outcome. This is completely new territory for two-tier stochastic frontier analysis, and we believe that it augments importantly its reach and ability to support scientists in their research projects. Applications of both models have been presented in Chap. 6.

13.1 Unobservable Effects on Binary Outcomes

Conditional probability models are among the most widely used statistical tools around the globe. Among them, the "Logit" model appears to be the dominant one. Already at the turn of the millennium, Hosmer and Lemeshow (2000) started their preface in their very popular book on Logistic regression analysis by stating "The use of Logistic regression modeling has exploded during the past decade"—and we have lived another quarter of a century since then. They motivate the use of Logistic regression (i.e., the Logit model) by plotting a binary outcome (having or not coronary heart disease) against age groups. Then, commenting on the emerging shape of the curve, they write:

> The curve is said to be S-shaped. It resembles a plot of a cumulative distribution of a random variable. Many distribution functions have been proposed for use in the analysis of a dichotomous outcome variable. (...) There are two primary reasons for choosing the logistic distribution. First, from a mathematical point of view, it is an extremely flexible and easily used function, and second, it lends itself to a clinically meaningful interpretation.

Conditional probability models are more often than not motivated in such a way, as a "convenient tool" to model the probabilities related to binary (or more generally

A. Papadopoulos, C. F. Parmeter, *Two-Tier Stochastic Frontier Analysis for the Social Sciences*, https://doi.org/10.1007/978-3-031-81513-3_13

dichotomous) outcomes: we model probabilities as a function of covariates, we need a function that is by design constrained in the interval [0, 1], and many distribution functions have this property for *any* value of their argument, *voilá*.[1]

What is not so frequently presented is how the conditional probability model can arise from an underlying causal regression specification, which leads to the BR model. This foundational approach is usual in econometric settings, and we will also adopt it here, not for tribal reasons, but because it is in this underlying model that the unobservables are hiding.

13.1.1 Derivation of the Binary Response Model

As mentioned in Chap. 1, discrete choices/binary responses often depend on whether a state variable (or a combination of state variables) has reached a critical threshold or not. We mentioned there three examples: war breaks out when the state variable "animosity" exceeds some threshold value; hospital admission happens when the state variable "health damage" exceeds some threshold value; a decision to buy an expensive durable good like a car depends on a combination of state variables that relate to family income, family size, and family age. These state variables, say y_*^ℓ, are not directly observable. But we have data on a binary variable $y \in \{0, 1\}$ that takes the value 1 when the action was taken (or the status has changed), indicating that y_*^ℓ has exceeded some (also unknown) threshold c, and it takes the value 0 when the action was not taken, indicating that y_*^ℓ has not exceeded the critical threshold. So the two are linked by the equivalence between events:

$$\{y = 1\} \Leftrightarrow \{y_*^\ell \geq c\}, \qquad \{y = 0\} \Leftrightarrow \{y_*^\ell < c\}.$$

Since both y_*^ℓ and the threshold c are unobservable (and will remain so), we can define $y^\ell \equiv y_*^\ell - c$, where we can think of y^ℓ as the *excess* state variable (since it represents the value of the original state in excess of the threshold), and write the equivalence relation as

$$\{y = 1\} \Leftrightarrow \{y^\ell \geq 0\}, \qquad \{y = 0\} \Leftrightarrow \{y^\ell < 0\}.$$

When events are equivalent, they have the same probability distribution. So we can form a probabilistic equality statement, conditional on the regressors $\boldsymbol{x}$ since we have data on them:

$$\Pr(y = 1|\boldsymbol{x}) = \Pr(y^\ell > 0|\boldsymbol{x}).$$

[1] Many, but not all. For example, the distribution function of a Uniform random variable is $F_U(u) = u$ and it depends on $u \in [0, 1]$ to automatically satisfy this.

The regressors represent our ideas about which factors may affect the (unknown) value of y^ℓ. We can then model the situation by assuming a relation between the unobservable excess state variable and these observable factors:

$$y^\ell = g(\boldsymbol{x}; \boldsymbol{\beta}) + \varepsilon.$$

The coefficients $\boldsymbol{\beta}$ characterize the function g and the relation between $\boldsymbol{x}$ and y^ℓ. The (continuous) random variable ε represents all other forces affecting the unobservable y^ℓ. At this point, an identification issue arises: suppose that ε has standard deviation σ_ε. Let $\gamma > 0$ be some scalar. Suppose somebody appeared and argued that the correct specification of the latent regression is

$$y^\ell = g(\boldsymbol{x}; \boldsymbol{\beta}) + \gamma\varepsilon.$$

Can we discriminate between the two specifications, effectively determining the scale of the equation? The answer is negative. First, note that

$$y^\ell = g(\boldsymbol{x}; \boldsymbol{\beta}) + \gamma\varepsilon \implies \frac{y^\ell}{\gamma} = \frac{g(\boldsymbol{x}; \boldsymbol{\beta})}{\gamma} + \varepsilon.$$

So the right-hand side equation, which has dependent variable y^ℓ/γ, has the same error term as our original equation. Second, remember that the only thing we know about y^ℓ is when it is positive and when it is negative, through knowledge of $y \in \{1, 0\}$. But when $y = 1$ *both* y^ℓ *and* y^ℓ/γ are positive, and when $y = 0$ both y^ℓ and y^ℓ/γ are negative. This implies that what is available as data does not contain discriminating information as regards the scale of the underlying latent regression, and so we cannot identify and estimate that scale level. It is for this reason that in conditional probability models, the standard deviation of the underlying error term is fixed to a known value.

So we cannot estimate the scale, and we cannot estimate the regression equation either (because we do not have data on the dependent variable). What we can do is to combine the probability statement with the regression specification:

$$\begin{aligned}
\Pr(y = 1|\boldsymbol{x}) &= \Pr(y^\ell > 0|\boldsymbol{x}) \\
&= \Pr(g(\boldsymbol{x}; \boldsymbol{\beta}) + \varepsilon > 0|\boldsymbol{x}) \\
&= 1 - \Pr(\varepsilon \le -g(\boldsymbol{x}; \boldsymbol{\beta})|\boldsymbol{x}).
\end{aligned}$$

The probability of a random variable being smaller or equal than some value is the definition of its distribution function (CDF). Here we have the distribution conditional on $\boldsymbol{x}$. Denote this distribution function:

$$\varepsilon|\boldsymbol{x} \sim F_{\varepsilon|\boldsymbol{x}}(\varepsilon|\boldsymbol{x}).$$

It is this distribution function that must have a fixed and known standard deviation. We arrive at

$$\Pr(y = 1|\boldsymbol{x}) = 1 - F_{\varepsilon|\boldsymbol{x}}(-g(\boldsymbol{x}; \boldsymbol{\beta})|\boldsymbol{x}).$$

To obtain the full probability distribution of the binary variable y, we write

$$\Pr(y|\boldsymbol{x}) = \left[1 - F_{\varepsilon|\boldsymbol{x}}(-g(\boldsymbol{x}; \boldsymbol{\beta})|\boldsymbol{x})\right]^{y} \cdot \left[F_{\varepsilon|\boldsymbol{x}}(-g(\boldsymbol{x}; \boldsymbol{\beta})|\boldsymbol{x})\right]^{1-y}. \tag{13.1}$$

Equation (13.1) can be used to compute the probability for both events $\{y = 1\}$ and $\{y = 0\}$, so it is a proper probability mass function (pmf) for the binary y variable. And we can estimate it by maximum likelihood because we have data both on $\boldsymbol{x}$ and also on the binary y.

This is how the abstract binary response model arises from a latent regression specification. *Then*, we can obtain any number of variants by assuming a specific distribution for the error term conditional on the regressors, and several have been used in the literature and in applied studies. In the benchmark Logit and "Probit" models, the error term is assumed to follow a zero-mean Logistic or a zero-mean Normal distribution, respectively. This, among other things, allows one to exploit the symmetry properties of these distributions, due to which the above expression can be simplified, because under symmetry around zero, we have for a CDF that $1 - F(-z) = F(z)$.[2] This is how we arrive at the Logistic (Logit) model where $\Pr(y = 1|\boldsymbol{x}) = \Lambda(\boldsymbol{x}'\boldsymbol{\beta})$ (with Λ representing the Logistic CDF), or, in the Probit model where $\Pr(y = 1|\boldsymbol{x}) = \Phi(\boldsymbol{x}'\boldsymbol{\beta})$.

The final modeling step is to assume a functional form for $g(\boldsymbol{x}; \boldsymbol{\beta})$. Almost invariably, this takes the linear form:

$$g(\boldsymbol{x}; \boldsymbol{\beta}) = \boldsymbol{x}'\boldsymbol{\beta}.$$

Note carefully that for our 2TSF model, we do *not* include a constant term in $g(\boldsymbol{x}; \boldsymbol{\beta})$, while in conventional BR models, it is included and it is good advice to be included. In them, the error term is assumed to be zero mean, and the constant term in $g(\boldsymbol{x}; \boldsymbol{\beta})$ hedges against misspecification if this is not what holds in the data. But in our case, we expect that the error ε will not have a zero mean (since it will be a 2TSF composite error). Then, including a constant term in $g(\boldsymbol{x}; \boldsymbol{\beta})$ would not only be redundant, but it would absorb the mean of our 2TSF error term, and attempting to estimate this constant alongside the distribution parameters would create colinearity and identification problems.

To complete the specification, we need to fix the standard deviation of ε. We will follow what is done in the Probit model, and set $\sigma_\varepsilon = 1$.[3] While this identification

---

[2] This will not be the case in our 2TSF framework.

[3] In the Logit model, σ_ε is a known scale parameter not equal to the standard deviation. In the Probit model, it equals the standard deviation.

issue and the resulting restriction does not cause much heartache in the conventional BR models (because the focus there is in estimating the effects of the regressors $\boldsymbol{x}$ on the probabilities, and this is not affected by this scaling), it hurts more in the 2TSF-BR model. Nevertheless, by setting $\sigma_\varepsilon = 1$, we will be able to obtain meaningful estimation results on the parameters that characterize the latent forces.

13.1.2 The 2TSF Probit Exponential Model

The case for a 2TSF-BR model arises once we believe that there are latent factors that are not included in the regressors that they affect the latent state variable y^ℓ in opposite directions. In such a case, we want to model the error term as a composite 2TSF one, $\varepsilon = v + w - u$. If we make the following statistical assumptions:

$$v \sim \mathrm{N}(0, \sigma_v), \qquad w \sim \mathrm{Exp}(\sigma_w), \qquad u \sim \mathrm{Exp}(\sigma_u),$$
$$\varepsilon \perp\!\!\!\perp \boldsymbol{x}, \qquad (v, w, u) \text{ jointly statistically independent},$$

we have just described the error term from the benchmark 2TSF Normal-Exponential specification (see Chap. 7).[4] Since we assumed v to be Normal as in the Probit model, we will call our model here "2TSF Probit Exponential."[5]

The distribution of ε cannot be treated as symmetric, since in general $E(w) \neq E(u)$, so we cannot compact further the expression for the pmf of y as in conventional models. Moreover, because we have assumed Exponential distributions for w and u, we have imposed

$$\mathrm{Var}(\varepsilon) \equiv \sigma_\varepsilon^2 = \sigma_v^2 + \sigma_w^2 + \sigma_u^2 = 1.$$

This creates a restriction on the moments. In full we have the 2TSF Probit Exponential model:

$$\Pr(y|\boldsymbol{x}) = \left[1 - F_\varepsilon(-\boldsymbol{x}'\boldsymbol{\beta}|\boldsymbol{x})\right]^y \cdot \left[F_\varepsilon(-\boldsymbol{x}'\boldsymbol{\beta}|\boldsymbol{x})\right]^{1-y} \tag{13.2}$$
$$\sigma_w > 0, \quad \sigma_u > 0, \quad \sigma_w^2 + \sigma_u^2 < 1, \quad \sigma_v^2 = 1 - \sigma_w^2 - \sigma_u^2.$$

We have expressed the inequalities as strict rather than as weak, because otherwise the software algorithm would encounter problems if it tried to execute at the exact boundary. If the data indicate that one of the scale parameters is zero, we

[4] And as mentioned there, we avoid here the NHN specification because it does not have a closed-form expression for its distribution function.

[5] The reason we do not venture into Logistic territory is that the Logistic distribution, however mathematically convenient it may be on its own, becomes intractable when we need to combine/convolute it with other distributions.

will understand it by the estimator output because this parameter will be estimated as nearly zero, and the estimator will probably have trouble providing standard errors.

The distribution function F_ε for this 2TSF composite error term is Eq. (7.12):

$$F_\varepsilon(\varepsilon) = \Phi\left(\frac{\varepsilon}{\sigma_v}\right) + \frac{\sigma_u}{\sigma_w + \sigma_u} \exp\{a_1\}\,\Phi(b_1) - \frac{\sigma_w}{\sigma_w + \sigma_u} \exp\{a_2\}\,\Phi(b_2).$$

with

$$a_1 = \frac{\sigma_v^2}{2\sigma_u^2} + \frac{\varepsilon}{\sigma_u}, \quad b_1 = -\left(\frac{\varepsilon}{\sigma_v} + \frac{\sigma_v}{\sigma_u}\right),$$

$$a_2 = \frac{\sigma_v^2}{2\sigma_w^2} - \frac{\varepsilon}{\sigma_w}, \quad b_2 = \frac{\varepsilon}{\sigma_v} - \frac{\sigma_v}{\sigma_w}.$$

We will eventually need also the density of the composite error term, which is Eq. (7.11):

$$f_\varepsilon(\varepsilon) = \frac{\exp\{a_1\}\,\Phi(b_1) + \exp\{a_2\}\,\Phi(b_2)}{\sigma_w + \sigma_u}.$$

In place of ε in all these expressions, we insert $(-\boldsymbol{x}'\boldsymbol{\beta})$. Note that we have removed $\boldsymbol{x}$ from the subscript of the CDF and now we write just F_ε, but we have left it as a conditioning factor in the argument of the function, since irrespective of the independence between the regressors and the error term, the whole expression is still conditional on the regressors since they appear in the argument of this distribution function, which must be non-stochastic for F to be a marginal distribution function. Hence the need to still declare the conditioning on $\boldsymbol{x}$.

As regards stochastic properties, the restriction on the variances does not create any form of statistical dependence between v, w, and u. As regards interpretation and inference in the 2TSF Probit Exponential model, σ_v^2, σ_w^2, and σ_u^2 do not represent the absolute scale of the variables but their relative scales, relative to the total, non-identifiable scale of the composite error term, which has been set equal to unity for the reasons discussed previously.

Essentially, there exist the "true" distributional specifications, say $w^* \sim \text{Exp}(\sigma_w^*)$ which are unreachable, and we are estimating $w = w^*/\sigma_\varepsilon$ that has standard deviation $\sigma_w = \sigma_w^*/\sigma_\varepsilon$.

This is why we said earlier that this identifiability issue that exists in all binary response models with a linear specification is more consequential in the 2TSF context: we can only recover the relative magnitudes of the underlying latent forces. Still this does not render the model useless. Note that in the current framework, the metrics we are going to use will be, as we will see in a while, a subset of the metrics for when the dependent variable is in its original measurement units, since y^ℓ is unknown and it is pointless and useless to argue that it is expressed in logarithms. Then, since these metrics are ratio measures, the implied scaling cancels out, and their values will be the values we would have obtained if we were able to directly estimate with the unscaled variables.

13.1.3 Estimation

The model is estimated by maximum likelihood with the log-likelihood contribution of observation i being

$$\ell_i = y_i \cdot \ln\left[1 - F_\varepsilon(-\boldsymbol{x}_i'\boldsymbol{\beta}|\boldsymbol{x}_i)\right] + (1 - y_i) \cdot \ln F_\varepsilon(-\boldsymbol{x}_i'\boldsymbol{\beta}|\boldsymbol{x}_i).$$

Note that due to the restriction on the variances, we do not estimate σ_v but only σ_w and σ_u, and we obtain σ_v from the constraint. The inequality restrictions can easily be attached to a maximum likelihood estimation script. If during the iterative search of the estimation, the software algorithm treads into forbidden territory, the existence of the inequality restrictions will force it to search elsewhere. It is the same thing as imposing the restriction that variance terms must be non-negative. In addition, because σ_v exists in the likelihood, to avoid writing out every instance of it through the restriction, we can keep the symbol σ_v but include also in the maximum likelihood command/script the relation $\sigma_v^2 = 1 - \sigma_w^2 - \sigma_u^2$ or its square root, and not include σ_v in the vector of parameters designated for estimation.

13.1.4 Inference

Once estimation is completed, we will have available estimates $\widehat{\boldsymbol{\beta}}$ as well as $\hat{\sigma}_w$, $\hat{\sigma}_u$, and so also $\hat{\sigma}_v$. Based on the beta estimates alone, we can perform all the standard analysis that is done in binary response models, which includes computing individual probabilities, examining how good is the model in predicting the binary outcome y (where we endorse the "more probable than not" convention and if $\widehat{\Pr}(y = 1|\boldsymbol{x}_i) > 0.5$ we predict $y = 1$), or forming the empirical distribution of predicted probabilities for a more nuanced picture. Also, we compute the marginal (or "partial") effect of each regressor on the probability per observation. Usually, we compute the marginal effect on $\Pr(y = 1|\boldsymbol{x})$. For regressor x_k, $k = 1 \ldots, K$, this is

$$\frac{\partial \Pr(y = 1|\boldsymbol{x})}{\partial x_k} = \frac{\partial}{\partial x_k}\Big[1 - F_\varepsilon(-\boldsymbol{x}'\boldsymbol{\beta}|\boldsymbol{x})\Big] = f_\varepsilon(-\boldsymbol{x}'\boldsymbol{\beta}|\boldsymbol{x}) \cdot \beta_k. \tag{13.3}$$

But this is standard fare. What is the value-added that the 2TSF formulation brings?

Sample Level The issue here is whether we can use the metrics developed in Chap. 2. To begin with, we look at the metrics when the dependent variable is in levels. The apparent hurdle is that ε and its components w and u relate to the unobservable y^{ℓ}. But if one looks at Table 2.1 where the level metrics are collected, one sees that these metrics use expected values, not realized values. And all expected values involved are computable for the 2TSF Probit Exponential model. Specifically, to map to the symbols of that table,

$$E(y) \to E(y^{\ell}) = E(\boldsymbol{x}')\boldsymbol{\beta} + E(w) - E(u), \quad E(g(\boldsymbol{x})) \to E(\boldsymbol{x}')\boldsymbol{\beta}.$$

But a more important problem is that these metrics were developed for *positive* dependent variables and systematic parts. In the present context, the average systematic part $E(\boldsymbol{x}')\boldsymbol{\beta}$ will often have a negative value, and then, metrics involving $E(y^{\ell})$ will be difficult or even misleading to interpret. It is for this reason that our advice would be to use only metrics M_3 to M_7, and with the absolute value of the systematic part, $|E(\boldsymbol{x}')\boldsymbol{\beta}|$. The expected values $E(w)$ and $E(u)$ are also computable from the estimates obtained.

We present in Table 13.1 this adjusted set of sample-level metrics for the 2TSF Probit Exponential model.

As already mentioned, exactly because these are ratios, the implied division of $\boldsymbol{\beta}$ and of $E(w)$ and $E(u)$ by the unknown and not estimable standard deviation of the 2TSF error term σ_{ε} cancels out from numerator and denominator, and so the above metrics are free from it.

Observation Level At the observation level, we cannot compute the corresponding M-metrics, because we do not have realized values of the dependent variable, nor residuals from the underlying regression. But, in the same spirit that we can derive the marginal effect of each regressor on the probabilities, and compute this per observation, we can also derive and compute the individual marginal effect of the latent forces in an average sense, namely, the effect that changes in the *mean level* of each latent force has on the probabilities, and again compute it per observation.

Table 13.1 Metrics of latent forces for the 2TSF Probit Exponential model

Metric	Formula	Description
M_3	$\dfrac{E(w)}{E(u)}$	Relative strength of w compared to u
M_4	$\dfrac{E(w)+E(u)}{\lvert E(\boldsymbol{x}')\boldsymbol{\beta}\rvert}$	Volatility of systematic part due to w, u
M_5	$\dfrac{E(w)}{\lvert E(\boldsymbol{x}')\boldsymbol{\beta}\rvert}$	Markup effect of w on systematic part
M_6	$\dfrac{E(u)}{\lvert E(\boldsymbol{x}')\boldsymbol{\beta}\rvert}$	Mark-down effect of u on systematic part
M_7	$\dfrac{E(w)-E(u)}{\lvert E(\boldsymbol{x}')\boldsymbol{\beta}\rvert}$	Net effect on systematic part ($\pm$)

In abstract and by the chain rule, using σ_h for $h = w, u$,

$$\frac{\partial \Pr(y|\boldsymbol{x})}{\partial \sigma_h} = \frac{\partial \Pr(y|\boldsymbol{x})}{\partial E(h)} \cdot \frac{\partial E(h)}{\partial \sigma_h} \implies \frac{\partial \Pr(y|\boldsymbol{x})}{\partial E(h)} = \left(\frac{\partial E(h)}{\partial \sigma_h}\right)^{-1} \cdot \frac{\partial \Pr(y|\boldsymbol{x})}{\partial \sigma_h}. \tag{13.4}$$

In the case of the 2TSF Probit Exponential model $E(h) = \sigma_h,\ h = w, u$, so the reciprocal term is equal to unity, and we have only to compute the derivative of the probability with respect to the scale parameter. But in other specifications, this will not be the case.

Further, note that

$$\frac{\partial \Pr(y = 0|\boldsymbol{x})}{\partial \sigma_h} = \frac{\partial F_\varepsilon(-\boldsymbol{x}'\boldsymbol{\beta}|\boldsymbol{x})}{\partial \sigma_h}, \qquad \frac{\partial \Pr(y = 1|\boldsymbol{x})}{\partial \sigma_h} = -\frac{\partial F_\varepsilon(-\boldsymbol{x}'\boldsymbol{\beta}|\boldsymbol{x})}{\partial \sigma_h}.$$

So the one is the negative of the other, and we have only to compute one of them per scale parameter.

But also, we have to respect the constraint $\sigma_v^2 + \sigma_w^2 + \sigma_u^2 = 1$. This implies that during the differentiation, we will not differentiate only with respect to the places where σ_h appears directly, but we will have to express one of the other two scale parameters through the constraint and differentiate that too. In all cases, we choose to substitute for σ_v, so our derivatives can be thought of as changes in the probability when the mean level of one latent force increases, while the other remains constant (and what decreases is the variability of the noise term).

Starting with σ_w, the expression for the derivative of the probability $\Pr(y = 0|\boldsymbol{x})$ is

$$\begin{aligned}\frac{\partial \Pr(y = 0|\boldsymbol{x})}{\partial \sigma_w} &= \frac{\partial F_\varepsilon(-\boldsymbol{x}'\boldsymbol{\beta}|\boldsymbol{x})}{\partial \sigma_w} \\ &= (\sigma_w + \sigma_u)^{-1}\Bigg\{\sigma_u e^{a_1}\left[\Phi(b_1)\frac{\partial a_1}{\partial \sigma_w} + \phi(b_1)\frac{\partial b_1}{\partial \sigma_w}\right] \\ &\quad - \sigma_w e^{a_2}\left[\Phi(b_2)\frac{\partial a_2}{\partial \sigma_w} + \phi(b_2)\frac{\partial b_2}{\partial \sigma_w}\right] - \sigma_u f_\varepsilon(\varepsilon)\Bigg\}\end{aligned} \tag{13.5}$$

The derivatives of the composites are

$$\frac{\partial a_1}{\partial \sigma_w} = \frac{-\sigma_w}{\sigma_u^2}, \qquad \frac{\partial b_1}{\partial \sigma_w} = \frac{\sigma_w}{\sigma_u \sigma_v} - \frac{\sigma_w}{\sigma_v^3}\varepsilon,$$

$$\frac{\partial a_2}{\partial \sigma_w} = \frac{\varepsilon}{\sigma_w^2} - \frac{1 - \sigma_u^2}{\sigma_w^3}, \qquad \frac{\partial b_2}{\partial \sigma_w} = \frac{\sigma_w}{\sigma_v^3}\varepsilon - \frac{1}{\sigma_v} + \frac{\sigma_v}{\sigma_w^2}.$$

For σ_u, we have

$$\frac{\partial \Pr(y=0|\boldsymbol{x})}{\partial \sigma_u} = \frac{\partial F_\varepsilon(-\boldsymbol{x}'\boldsymbol{\beta}|\boldsymbol{x})}{\partial \sigma_u}$$

$$= (\sigma_w + \sigma_u)^{-1}\left\{\sigma_u e^{a_1}\left[\Phi(b_1)\frac{\partial a_1}{\partial \sigma_u} + \phi(b_1)\frac{\partial b_1}{\partial \sigma_u}\right]\right.$$

$$\left. - \sigma_w e^{a_2}\left[\Phi(b_2)\frac{\partial a_2}{\partial \sigma_u} + \phi(b_2)\frac{\partial b_2}{\partial \sigma_u}\right] + \sigma_w f_\varepsilon(\varepsilon)\right\} \tag{13.6}$$

The derivatives of the composites here are

$$\frac{\partial a_1}{\partial \sigma_u} = \frac{\varepsilon}{\sigma_u^2} - \frac{1-\sigma_w^2}{\sigma_u^3}, \qquad \frac{\partial b_1}{\partial \sigma_u} = \frac{1}{\sigma_v} + \frac{\sigma_v}{\sigma_u^2} - \frac{\sigma_u}{\sigma_v^3}\varepsilon,$$

$$\frac{\partial a_2}{\partial \sigma_u} = \frac{-\sigma_u}{\sigma_w^2}, \qquad \frac{\partial b_2}{\partial \sigma_u} = \frac{\sigma_u}{\sigma_w \sigma_v} + \frac{\sigma_u}{\sigma_v^3}\varepsilon.$$

Some notes on the above: first, the above expressions are the partial effects on the probability of the event $\{y = 0|\boldsymbol{x}\}$. Since we usually compute the effect on the probability of the event $\{y = 1|\boldsymbol{x}\}$, remember to put a minus sign in front of the result of each computation. Second, due to nonlinear differentiation, σ_v reappears, and we have kept it for computational compactness. Third, wherever ε appears in the expression for the derivative, we should insert $-\boldsymbol{x}'\boldsymbol{\beta}$. Fourth, the formulas are to be computed per observation, but we have suppressed the observation index i. Fifth, readers should be mindful of the general expression Eq. (13.4), if they decide to apply the Half-Normal variant of the model. Apart from differentiating a different distribution function, they have to adjust the results for the fact that in the Half-Normal case, $\sigma_w \neq E(w)$. Sixth, in order to obtain a single "average" marginal (partial) effect, we can either evaluate the expression at the sample means of the regressors or compute the sample mean of the expression series. Both approaches appear in the literature. Mathematically, what stands between them and makes them not equal is Jensen's inequality. Conceptually, these are not competing estimates of the same thing, but two different measures: the first talks about the marginal effects of the latent forces on the probabilities characterizing *the average individual* (since we take the sample means of the regressors). The second is the actual average marginal effect, or average partial effect. Computing both provides more information about what happens in the sample. In general, with sample means, the effects of Jensen's inequality tend to vanish, so with large samples, the two are expected to give close results.

Interpretation Marginal effects on the probability in the conditional probability models are usually derived by differentiation, even though the regressor may be binary or discrete. In such cases, the obtained magnitude is directly interpretable as the change in probability (measured in probability points) when the regressor changes by one absolute unit (i.e., going from 0 to 1 for the binary variables, or, say, from 3 to 4 for a variable that measures years of education, or it is measured on a Likert scale.)

In interpreting the marginal effects of a change in $E(w)$ or $E(u)$, we must remember that in the 2TSF Probit Exponential model, they are constrained to lie in the (0, 1) interval because they are relative measures as discussed earlier. Then "increasing by one absolute unit" is infeasible. For the marginal effect to be seen as something more tangible than a mathematical slope, a mathematical derivative must be associated with some specific change in magnitude.

In the application of the model in Chap. 6, we chose to consider a change by 0.1. Meaning, that if we have estimated, say, $\widehat{E}(w)$, we determine what the change in probability would be if it went to $\widehat{E}(w)+0.1$. Consequently, the computed marginal effect has to be multiplied by 0.1 to obtain the associated change:

$$\left.\frac{\Delta \Pr(y=1|\boldsymbol{x})}{\Delta E(w)}\right|_{\Delta E(w)=0.1} = \frac{\partial \Pr(y=1|\boldsymbol{x})}{\partial \sigma_w} \times 0.1.$$

13.1.5 Heterogeneity in the 2TSF Probit Exponential Model

If we have available covariates that we believe affect the latent variables, we can enrich the model by making σ_w and σ_u functions of these variables as described in Chap. 8, as in

$$\sigma_{wi} = \sigma_w(z_{wi};\, \boldsymbol{\delta}_w), \qquad \sigma_{ui} = \sigma_u(z_{ui};\, \boldsymbol{\delta}_u).$$

In such a setup, we will not have a single estimate for σ_w or σ_u but a series. What we presented in Chap. 8 for z variables in the context of moment dependence and maximum likelihood estimation holds here too: we can now compute individual metrics, using the conditional expectations given z, which are the unconditional moment expressions of the assumed distributions with the sigma parameters varying. In this way, we sidestep the unavailability of the underlying dependent variable y^ℓ and of the residuals. For example, for the 2TSF Probit Exponential model, we have

$$M_{7i} = \frac{E(w_i|z_{wi}) - E(u_i|z_{wi})}{|E(\boldsymbol{x}_i')\boldsymbol{\beta}|} = \frac{\sigma_w(z_{wi};\, \boldsymbol{\delta}_w) - \sigma_u(z_{ui};\, \boldsymbol{\delta}_u)}{|E(\boldsymbol{x}_i')\boldsymbol{\beta}|}, \qquad i = 1, \ldots, n.$$

We can also compute sample averages:

$$E(\sigma_w) \approx \frac{1}{n}\sum_{i=1}^{n}\sigma_w(z_{wi};\, \boldsymbol{\delta}_w), \qquad E(\sigma_u) \approx \frac{1}{n}\sum_{i=1}^{n}\sigma_u(z_{ui};\, \boldsymbol{\delta}_u),$$

and use them in computing the M-metrics at the sample level.

As regards the marginal effects of the scale parameters on the probability, we can proceed in the same way: we can obtain full series by inserting per observation σ_{wi} and σ_{ui} wherever σ_w and σ_u appear and obtain a fully heterogenized series for the marginal effects. For sample-level measures, we can either compute the sample mean of this series or we can insert in place of the sigmas the above computed sample means (as well as the sample means of the regressors).

With this, we conclude the theoretical development of the 2TSF Probit Exponential model. We remind readers that we have applied the model in a psychology context, in Chap. 6, to measure the strength of latent forces that affect the probability of occurrence of extramarital affairs.

13.2 Decomposing Conflicting Treatment Effects

In this section, we present a 2TSF model to estimate dualistic treatment effects. Every treatment is Rudrian, it has beneficial and undesirable effects, and the conventional way of modeling treatment effects estimates the average net effect. The 2TSF model will separate the two. We presented in Chap. 6 an application of it relating to healthcare; here we provide the formulas and the other necessary tools to actually implement the model.

13.2.1 The 2TSF Random Treatment Effects Model

Denote T_i as the binary variable that indicates whether the treatment has been applied ($T_i = 1$), or not ($T_i = 0$), for subject i. Denote, as in earlier chapters, $z_i = w_i - u_i$, where here w_i and u_i represent the two opposing treatment effects "packed" in the treatment we apply. Notice that in our formulation, the treatment effects are random variables, not constants to be estimated, hence the "random treatment effects" (RTE) name for the model. Note also that which is the "desirable" effect and which the "undesirable" depends on the formulation; for example, in the healthcare empirical application of Chap. 6, the positive effect w_i represented the undesirable effect, due to the way the related dependent variable was defined. Following the setup of the previous section, we postulate the existence of a latent regression, which here takes the form (dropping again the observation index)

$$y^{\ell} = \mathbf{x}'\boldsymbol{\beta} + \varepsilon, \qquad \varepsilon = Tz + v. \tag{13.7}$$

We note that the regression specification *includes* a constant term, contrary to the model in the previous section.

So the latent regression for the treated subjects/observations has a 2TSF error structure, but the untreated observations have a standard error structure:

$$T = 1 \implies \varepsilon = 1 \cdot z + v = w - u + v,$$
$$T = 0 \implies \varepsilon = v.$$

Despite this heterogeneity, we still face an identification issue with the scale of the latent regression, but this time with respect to the standard deviation of v, not of ε. We make the stochastic assumptions:

$$v\Big|\{\boldsymbol{x}, T, z\} \sim \mathrm{N}(0, 1), \qquad w\Big|\{\boldsymbol{x}, T\} \sim \mathrm{Exp}(\sigma_w), \qquad u\Big|\{\boldsymbol{x}, T\} \sim \mathrm{Exp}(\sigma_u)$$
$$z\Big|\{\boldsymbol{x}, T\} = (w - u)\Big|\{\boldsymbol{x}, T\} = w\Big|\{\boldsymbol{x}, T\} - u\Big|\{\boldsymbol{x}, T\}.$$

By assuming (by necessity) that v has a known variance, which we set equal to unity to make it useful, we turn it into a *numéraire*. So here, the estimated σ_w and σ_u will not represent the strength of the latent forces relative to the whole composite error term, but their relative magnitude with respect to the actual σ_v. The ratios σ_w/σ_v and σ_u/σ_v that we will essentially estimate are the meaningful "signal-to-noise" ratios.

The stochastic assumptions say that, conditional on everything else, v follows a Normal distribution, while, conditional on the observables, $z_i = w_i - u_i$ follows the Asymmetric Double Exponential distribution, which we encountered in Chap. 12 as one of the specifications in a 2TSF Tug-of-War model.

In this setup too, we have data on a binary y variable for which

$$\{y = 1\} \Leftrightarrow \{y^\ell \geq 0\}, \qquad \{y = 0\} \Leftrightarrow \{y^\ell < 0\}.$$

We apply the analogous probability argument as in the previous section, but here we will condition (temporarily) on $(\boldsymbol{x}, T, z)$, i.e., on observables and unobservables. The important thing is that the only remaining source of randomness is the Normal component v. We have

$$\begin{aligned}
\Pr(y = 1|\{\boldsymbol{x}, T, z\}) &= \Pr(y^\ell > 0|\{\boldsymbol{x}, T, z\}) \\
&= \Pr(\boldsymbol{x}'\boldsymbol{\beta} + Tz + v > 0|\{\boldsymbol{x}, T, z\}) \\
&= 1 - \Pr(v \leq -\boldsymbol{x}'\boldsymbol{\beta} - Tz|\{\boldsymbol{x}, T, z\}) \\
&= 1 - \Phi\left(-\boldsymbol{x}'\boldsymbol{\beta} - Tz|\{\boldsymbol{x}, T, z\}\right), \\
&= \Phi\left(\boldsymbol{x}'\boldsymbol{\beta} + Tz|\{\boldsymbol{x}, T, z\}\right).
\end{aligned}$$

We have used the symmetry property of the standard Normal distribution function (CDF) Φ. We also have

$$\begin{aligned}\Pr(y = 0|\{\boldsymbol{x}, T, z\}) &= 1 - \Pr(y = 1|\{\boldsymbol{x}, T, z\}) \\ &= 1 - \Phi\left(\boldsymbol{x}'\boldsymbol{\beta} + Tz|\{\boldsymbol{x}, T, z\}\right) \\ &= \Phi\left(-\boldsymbol{x}'\boldsymbol{\beta} - Tz|\{\boldsymbol{x}, T, z\}\right).\end{aligned}$$

Then, we can obtain compactly the full probability distribution of the binary y as

$$\Pr(y|\{\boldsymbol{x}, T, z\}) = \Phi\Big((2y-1)\left(\boldsymbol{x}'\boldsymbol{\beta} + Tz\right)\Big).$$

For the moment, this looks like a typical Probit model. But, crucially, the right-hand side includes the unobservable z, so it cannot be used in estimation. What we want in order to be able to estimate the model is the probability distribution $\Pr(y|\{\boldsymbol{x}, T\})$. By basic laws of probability and conditioning, and our stochastic assumptions, we have the relations

$$\Pr(y|\{\boldsymbol{x}, T\}) = \int_{-\infty}^{\infty} \Pr(y, z|\{\boldsymbol{x}, T\})dz,$$

$$\Pr(y, z|\{\boldsymbol{x}, T\}) = \Pr(y|\{\boldsymbol{x}, T, z\}) \cdot \Pr(z|\{\boldsymbol{x}, T\}).$$

The probability distribution of z and its density $f_z(z)$ are known. Combining all, we have

$$\Pr(y|\{\boldsymbol{x}, T\}) = \int_{-\infty}^{\infty} \Pr(y|\{\boldsymbol{x}, T, z\}) \cdot f_z(z)\, dz.$$

Further, note that because T is also binary, we can write

$$\Pr(y|\{\boldsymbol{x}, T, z\}) = T \cdot \Phi\Big((2y-1)(\boldsymbol{x}'\boldsymbol{\beta} + z)\Big) + (1-T)\cdot \Phi\Big((2y-1)(\boldsymbol{x}'\boldsymbol{\beta})\Big).$$

Inserting this into the previous relation, we have

$$\begin{aligned}\Pr(y|\{\boldsymbol{x}, T\}) &= \int_{-\infty}^{\infty} \Big[T \cdot \Phi\left((2y-1)\left(\boldsymbol{x}'\boldsymbol{\beta} + z\right)\right) \\ &\quad + (1-T)\Phi\left((2y-1)\left(\boldsymbol{x}'\boldsymbol{\beta}\right)\right)\Big] \cdot f_z(z)\, dz \\ &= T \cdot \int_{-\infty}^{\infty} \Phi\left((2y-1)\left(\boldsymbol{x}'\boldsymbol{\beta} + z\right)\right) \cdot f_z(z)\, dz \\ &\quad + (1-T)\Phi\left((2y-1)\left(\boldsymbol{x}'\boldsymbol{\beta}\right)\right).\end{aligned}$$

Given our stochastic assumptions, this integration is feasible. Doing it, we obtain the probability distribution, now written for observation i:

$$
\begin{aligned}
\Pr(y_i | \{\boldsymbol{x}_i, T_i\}) &= \Phi\left((2y_i - 1)\left(\boldsymbol{x}_i'\boldsymbol{\beta}\right)\right) + T_i(\mathcal{D}_{1i} - \mathcal{D}_{2i}), \\
\mathcal{D}_{1i} &= \frac{\sigma_w}{\sigma_w + \sigma_u} \exp\left\{\frac{1}{2\sigma_w^2} + \frac{\boldsymbol{x}_i'\boldsymbol{\beta}}{\sigma_w}\right\} \\
&\quad \times \left\{y_i - \Phi\left[(2y_i - 1)\left(\boldsymbol{x}_i'\boldsymbol{\beta} + \frac{1}{\sigma_w}\right)\right]\right\}, \\
\mathcal{D}_{2i} &= \frac{\sigma_u}{\sigma_w + \sigma_u} \exp\left\{\frac{1}{2\sigma_u^2} - \frac{\boldsymbol{x}_i'\boldsymbol{\beta}}{\sigma_u}\right\} \\
&\quad \times \left\{y_i - \Phi\left[(2y_i - 1)\left(-\boldsymbol{x}_i'\boldsymbol{\beta} + \frac{1}{\sigma_u}\right)\right]\right\}. \qquad (13.8)
\end{aligned}
$$

Before proceeding with estimation and inference, let's pause and contemplate Eq. (13.8). It no longer looks like a typical Probit model. Observe that the component $\Phi((2y_i - 1)\boldsymbol{x}_i'\boldsymbol{\beta})$ is the probability distribution (covering both events for y) of the non-treated subjects. It includes the controls and the characteristics of the random noise element v (its distribution function). The other component $(\mathcal{D}_{1i} - \mathcal{D}_{2i})$ is the *net change* in probability due to the treatment, since it arises only when $T_i = 1$. This additive separation of the treatment probability effect does not exist in conventional conditional probability models, and it makes inference more straightforward and transparent. It is essentially due to the fact that we modeled the treatment effect, z, as a random variable with its own distribution, in combination with the treatment variable being binary. Another important difference is that the treatment effect here is co-determined by the control variables, something absent in the conventional model.

Another equally important note is that, by inspecting the mathematical expressions for $\mathcal{D}_{1i}$ and for $\mathcal{D}_{2i}$ in Eq. (13.8), we should realize that we should *not* interpret $\mathcal{D}_{1i}$ as the positive treatment effect and $\mathcal{D}_{2i}$ as the negative treatment effect (although certainly it would be nice to be able to do so): such a discriminating characterization would bear a causal flavor, and from their expressions, we see that this would be questionable—the σ_u parameter that relates to the negative effect is present in $\mathcal{D}_{1i}$, and the σ_w parameter that relates to the positive effect is present in $\mathcal{D}_{2i}$. These interrelations arose statistically, and they are not supported by any underlying causal mechanism.

Nevertheless, the main advantage and improvement of the 2TSF RTE model over the conventional BR models are that the net treatment effect:

(a) Is varying per observation
(b) Has been obtained as a difference
(c) *Sign* is not constrained

The term $(\mathcal{D}_{1i} - \mathcal{D}_{2i})$ can be positive or negative, possibly changing sign observation per observation. The model does not impose any across-the-board restriction on the direction of influence of the treatment. *But the conventional models do.* Specifically, in these models, the treatment effect is the difference of two probability distributions, which, we remind the reader, are monotonic functions. Suppose we had the conventional Logit model:

$$\Pr(y_i = 1|\boldsymbol{x}_i, T_i) = \Lambda\left(\boldsymbol{x}_i'\boldsymbol{\beta} + \gamma T_i\right).$$

This model would estimate the Treatment Effect on the Treated related to any treated observation m $(T_m = 1)$ as

$$\text{ATT}_m = \Lambda\left(\boldsymbol{x}_m'\boldsymbol{\beta} + \gamma\right) - \Lambda\left(\boldsymbol{x}_m'\boldsymbol{\beta}\right), \quad \forall m.$$

Due to the monotonicity of Λ, we would necessarily have

$$\gamma > 0 \rightarrow \text{ATT}_m > 0, \qquad \gamma < 0 \rightarrow \text{ATT}_m < 0, \quad \forall m.$$

So if the *average* marginal effect factor of the Treatment, γ, is estimated as positive, all individual treatment effects will necessarily be positive. And the analogous result holds for the case where we would estimate $\gamma < 0$. This gives an artificial flavor to such individual treatment effects in the conventional models, since we very well know that the treatment may exert influence in one direction for one subject and in the other direction for another. But the conventional models distort that, assigning to all such individual estimates the same sign as the sign of the average effect. At best, these estimates would be helpful in order to rank the observations, but their level is spurious, and, most importantly, their sign, and hence the direction of influence of the treatment, remains unknown, since what we get necessarily is the sign of the average effect for all treated observations.

In the 2TSF RTE model, this does not happen. So we break important new ground and we can now measure the possibly direction-switching effect of the Treatment in different subjects or in different subgroups of treated subjects.

13.2.2 Estimation and Inference

The logarithm of Eq. (13.8) will be the likelihood contribution ℓ_i for maximum likelihood estimation. There is no need for any further transformations:

$$\ell_i = \ln\left\{\Phi\left((2y_i - 1)\left(\boldsymbol{x}_i'\boldsymbol{\beta}\right)\right) + T_i(\mathcal{D}_{1i} - \mathcal{D}_{2i})\right\}. \tag{13.9}$$

Just remember that σ_v is not a parameter to be estimated, and it has already been set equal to unity in the derived likelihood expression.

13.2.2.1 Probabilities and Treatment Effects

What kind of treatment effects we would want to estimate depends on the particular study we have undertaken. In the current prevailing approach in the treatment effects literature, we examine factuals against counterfactuals (together, "potential outcomes"), either for the whole sample, or for the subsample of treated observations, or that of untreated observations. We mention that the iid sample assumption effectively guarantees the satisfaction of "SUTVA" ("stable unit treatment value assumption"), which is an underlying assumption in benchmark treatment effects estimation.

The other important issue to be examined is the "ignorability" of treatment, namely, the assumption that the potential outcomes are independent of the treatment. Unconditional ignorability exists in randomized control trials, which are rarely available in social sciences, either because they are costly to implement (and when they are implemented, they are very limited in size) or because they face insurmountable ethical issues. Moreover, such kind of ignorability appears to make the whole real-world endeavor pointless: if the outcome is "independent" from the treatment, *why apply the treatment in the first place*?

A modified version of ignorability exists, which also helps clarify its (non)relation with causality: now we assume that the potential outcomes are independent of the treatment *conditional on the control variables*.[6] There is also a weaker version, where we assume only mean independence, not full independence. Still, it would appear that the same question persists: why apply the treatment then?

This brings to the surface the important distinction between statistical dependence and causal dependence. Ignorability conditional on the controls is about the nonexistence of statistical dependence, not about the nonexistence of causal dependence. It is not an assumption about causality, but about information: under ignorability of treatment conditional on the controls, we argue that they contain enough information about *the statistical distribution* of the potential outcomes, so that taking into account the treatment will not change that distribution. We do *not* make an assumption about whether the treatment "affects or not" the outcomes, in the causal sense of the word. And if one finds difficult to imagine how the existence of a causal effect on the outcomes can coexist with an unchanged statistical distribution for them, one can resort to "mean-ignorability": given the controls, what remains unchanged in the presence of the treatment is the mean of this distribution, but its variance or its shape more generally may change.

Whether conditional ignorability can be assumed to exist will crucially depend on each particular study, on the causal nexus of the situation, any temporal ordering, and other such factors. In any case, it is a strong assumption, or even a *very* strong one. But as Papadopoulos and Stark (2021) show, in the conditional probability model, we do not need ignorability of any kind to estimate consistently treatment

---

[6] We note that conditional statistical (in)dependence is not a "weaker" form of unconditional statistical (in)dependence. Neither implies or presupposes the other.

effects, as long as we can establish consistency of the estimator, which can exist regardless of the existence of some version of ignorability. This is where the latent-regression formulation of a conditional probability model becomes indispensable, because it is at this level that consistency of the estimator can be examined with all the rigorous tools of probability theory and statistics.

With this discussion in mind, we proceed to offer the expressions for some basic quantities of interest. We define n_1 to be the size of the treated subsample and n_0 the size of the untreated one, $n_1 + n_0 = n$. We can reorder the sample so that all treated observations are collected together and the same for the untreated ones. Note that for all n_0 observations, $T = 0$, while for all n_1 observations $T = 1$.

For clarity, we provide first the probability expression specifically for the event $\{y = 1\}$:

$$\Pr(y_i = 1|\{\boldsymbol{x}_i, T_i\}) = \Phi\left(\boldsymbol{x}_i'\boldsymbol{\beta}\right) + T_i\left(\mathcal{D}_{1i}|_{y=1} - \mathcal{D}_{2i}|_{y=1}\right),$$

$$\mathcal{D}_{1i}|_{y=1} = \frac{\sigma_w}{\sigma_w + \sigma_u}\exp\left\{\frac{1}{2\sigma_w^2} + \frac{\boldsymbol{x}_i'\boldsymbol{\beta}}{\sigma_w}\right\}\cdot\Phi\left(-\boldsymbol{x}_i'\boldsymbol{\beta} - \frac{1}{\sigma_w}\right),$$

$$\mathcal{D}_{2i}|_{y=1} = \frac{\sigma_u}{\sigma_w + \sigma_u}\exp\left\{\frac{1}{2\sigma_u^2} - \frac{\boldsymbol{x}_i'\boldsymbol{\beta}}{\sigma_u}\right\}\cdot\Phi\left(\boldsymbol{x}_i'\boldsymbol{\beta} - \frac{1}{\sigma_u}\right). \tag{13.10}$$

Probability of the Outcome for the Untreated Sample As already mentioned, the probability of the event to occur ($\{y = 1\}$) for the untreated subpopulation is

$$\Pr(y_j = 1|\{\boldsymbol{x}_j, T_j = 0\}) = \Phi\left(\boldsymbol{x}_j'\boldsymbol{\beta}\right), \qquad j = 1, \ldots, n_0. \tag{13.11}$$

This is exactly what we would get in the standard Probit model, as should be expected, given the additive structure as regards the influence of the treatment. This is a series *over the untreated subsample*, and readers are advised to examine also its statistical properties apart from computing its mean value which is

$$\text{APU} = \frac{1}{n_0}\sum_{j=1}^{n_0}\Phi\left(\boldsymbol{x}_j'\boldsymbol{\beta}\right). \tag{13.12}$$

Probability of the Outcome for the Treated Sample The probability of the event to occur ($\{y = 1\}$) for the treated subpopulation is

$$\Pr(y_m = 1|\{\boldsymbol{x}_m, T_m = 1\}) = \Phi\left(\boldsymbol{x}_m'\boldsymbol{\beta}\right) + \mathcal{D}_{1m|y=1} - \mathcal{D}_{2m|y=1}, \quad m = 1, \ldots, n_1. \tag{13.13}$$

Its mean value is

$$\text{APT} = \frac{1}{n_1}\sum_{m=1}^{n_1}\left[\Phi\left(\boldsymbol{x}_m'\boldsymbol{\beta}\right) + \mathcal{D}_{1m|y=1} - \mathcal{D}_{2m|y=1}\right]. \tag{13.14}$$

Two other interesting measures are the average *excess* probability for the treated over the untreated, before the treatment ("pre") and after the treatment ("post"). The before-treatment average excess probability is

$$\text{AExP}_{pre} = \frac{1}{n_1}\sum_{m=1}^{n_1}\Phi\left(\boldsymbol{x}_m'\boldsymbol{\beta}\right) - \text{APU}. \tag{13.15}$$

The first component is a counterfactual: what the probability for the treated would be if they have not been treated. It essentially characterizes the situation between the treated and the untreated subpopulations in probabilistic terms before treatment was applied.

The after-treatment average excess probability is simply

$$\text{AExP}_{post} = \text{APT} - \text{APU}. \tag{13.16}$$

Nothing prevents these two measures from being negative.

Treatment Effect on the Treated The familiar "Treatment Effects on the Treated," which we will denote "TTE," is defined as the difference between the related factual and its counterfactual, i.e., the actual probability of the treated minus the counterfactual probability that would have been if they had not been treated:

$$\begin{aligned}\text{TTE}_m &\equiv \Pr(y_m = 1|\{\boldsymbol{x}_i, T_i = 1\} - \Pr(y_m = 1|\{\boldsymbol{x}_i, T_i = 0\} \\ &= \mathcal{D}_{1m|y=1} - \mathcal{D}_{2m|y=1}, \; m = 1, \ldots, n_1.\end{aligned}$$

This is the series that may take on positive or negative values, depending on the complex interaction of the dependent variable and the controls for each observation and the distribution parameters of w and u. It is a proper predictor of *individual* treatment effects and makes the assessment of the treatment much more effective, beyond the estimation of the well-known Average Treatment Effect on the Treated which is

$$\text{ATT} = \frac{1}{n_1}\sum_{m=1}^{n_1}(\mathcal{D}_{1m|y=1} - \mathcal{D}_{2m|y=1}).$$

13.2.2.2 Marginal Effects

The marginal effects of the regressors take a clear intuitive form:

$$\frac{\partial \Pr(y_i = 1|\{\boldsymbol{x}_i, T_i\})}{\partial x_{ki}} = \left[(1 - T_i)\phi\left(\boldsymbol{x}_i'\boldsymbol{\beta}\right) + T_i\left(\frac{1}{\sigma_w}\mathcal{D}_{1i} + \frac{1}{\sigma_u}\mathcal{D}_{2i}\right)\right] \cdot \beta_k. \tag{13.17}$$

In the 2TSF RTE model, the marginal effects are fully disjoint between the treated and the non-treated observations. It then makes more sense to compute the series and average over it, rather than compute the expression at the sample averages of the regressors. Additionally, one could look at the two subsamples separately and compute the average partial effects of the regressors for each sample, computed at the subsample means of the regressors.

But we are very much interested in the marginal effects of our latent forces also and individually so. First, we stress that w and u operate at the level of the latent variable y^ℓ, so they do not relate directly to the effect on the probability. Second, we should remember that in reality we are estimating not the "true" σ_w, say, but the associated signal-to-noise ratio σ_w/σ_v. Combined, these imply that we can meaningfully compute the ratio

$$\frac{\widehat{\sigma}_w}{\widehat{\sigma}_u} \to_p \frac{\sigma_w/\sigma_v}{\sigma_u/\sigma_v} = \frac{\sigma_w}{\sigma_u},$$

since the denominators will cancel out, and this will give us the relative magnitude of the average force of w over u as regards their impact on the underlying latent y^ℓ (we could of course consider also the reciprocal expression, depending on the interpretational focus of the model). This is another important new offering from the 2TSF RTE model compared to the conventional models, but it does not yet speak about their effects on the probability, effects that will be intermediated by the control variables.

To compute them, we need to compute the derivative of the probability with respect to the sigmas, noting that here the variance restriction does not make matters more complicated because it does not constrain either σ_w or σ_u (in contrast to what happened in the 2TSF Probit Exponential model). But we need to remember that the estimated sigmas are still relative magnitudes, and so we should adjust the computed number to represent an effect for a plausible change, say for a 0.1 change in σ_w, which here would mean that σ_w increases by 10% relative to σ_v. The mathematical expressions are

$$\frac{\partial \Pr(y_m = 1|\{\boldsymbol{x}_m, T_m = 1\})}{\partial \sigma_w} = \frac{\partial(\mathcal{D}_{1m|y=1} - \mathcal{D}_{2m|y=1})}{\partial \sigma_w} =$$

$$\frac{1}{\sigma_w(\sigma_w + \sigma_u)}\left[\left(\sigma_u - \frac{1}{\sigma_w} - \boldsymbol{x}_m'\boldsymbol{\beta}\right)\mathcal{D}_{1m|y=1} + \sigma_w\mathcal{D}_{2m|y=1} + \phi(\boldsymbol{x}_m'\boldsymbol{\beta})\right], \tag{13.18}$$

$$\frac{\partial \Pr(y_m = 1|\{\boldsymbol{x}_m, T_m = 1\})}{\partial \sigma_u} = \frac{\partial(\mathcal{D}_{1m|y=1} - \mathcal{D}_{2m|y=1})}{\partial \sigma_u} =$$

$$\frac{-1}{\sigma_u(\sigma_w + \sigma_u)} \left[\sigma_u \mathcal{D}_{1m|y=1} + \left(\sigma_w - \frac{1}{\sigma_u} + \boldsymbol{x}'_m \boldsymbol{\beta} \right) \mathcal{D}_{2m|y=1} + \phi(\boldsymbol{x}'_m \boldsymbol{\beta}) \right]. \tag{13.19}$$

These metrics are relevant for the subsample of treated observations, since in the untreated sample, the marginal effect of σ_w and σ_u is zero.

To properly compute the *net* marginal effect of the two, we start with the total differential of the treatment effect $(\mathcal{D}_{1m|y=1} - \mathcal{D}_{2m|y=1})$ with respect to σ_w and σ_u, which is

$$d(\mathcal{D}_{1m|y=1} - \mathcal{D}_{2m|y=1}) = \frac{\partial(\mathcal{D}_{1m|y=1} - \mathcal{D}_{2m|y=1})}{\partial \sigma_w} d\sigma_w + \frac{\partial(\mathcal{D}_{1m|y=1} - \mathcal{D}_{2m|y=1})}{\partial \sigma_u} d\sigma_u.$$

To be able to arrive at a single expression, we set the sigma differentials equal, say $d\sigma_w = d\sigma_u = \bar{d}_\sigma$. Then we can obtain

$$d(\mathcal{D}_{1m|y=1} - \mathcal{D}_{2m|y=1}) = \left[\frac{\partial(\mathcal{D}_{1m|y=1} - \mathcal{D}_{2m|y=1})}{\partial \sigma_w} + \frac{\partial(\mathcal{D}_{1m|y=1} - \mathcal{D}_{2m|y=1})}{\partial \sigma_u} \right] \bar{d}_\sigma.$$

Note that we are *adding* the two marginal effects while remembering that the one related to σ_u has a minus sign in front of it. This expression allows us to take the net marginal effect and scale it by a common $\bar{d}_\sigma$ to get an approximation on the net change of the treatment effect if the average strength of both factors w and u changes by the same amount, say by 0.1.

With this additional new insight, we conclude the presentation of the 2TSF Random Treatment Effects BR model.

Chapter 14
Extensions III: Honey Pots and Epilogue

Early on, we decided to not make this book an extended theoretical exposition or a dry look-up reference, but a guide for two-tier stochastic frontier applications across the social sciences. This necessitated choice: to focus on and present, in detail pertinent to empirical analysis, a rather tight selection of the tools in the 2TSF armamentarium, while also leaving aside many interesting doors to open and promising paths to explore. And this is what we did (hopefully). But eventually, things accumulated, so much so that we could not but acknowledge their existence. This last chapter of the book is here to provide leads to existing complementary approaches, but also to motivate future research.

14.1 More Estimation Methods

The vast majority of published research on estimation of the 2TSF model has been undertaken in the parametric maximum likelihood framework, initially using the specification of Polachek and Yoon (1987) and, in the last years, its more tidy and transparent formulation by Kumbhakar and Parmeter (2009). This is for good reason. The likelihood has a simple closed-form solution and is, relatively speaking, easy to implement in a variety of statistical software. However, this does not mean that other approaches should not be considered.

As we discussed in Chap. 9, 2TSF panel data models and applications are virtually nonexistent. The same is true when it comes to alternative estimation approaches for the model. While a variety of proposals exist in the single-tier literature, few, if any, of these suggestions have permeated the 2TSF literature. This offers many potential avenues for applied researchers to enhance their work. Here we discuss some rudiments of these approaches and how they have (or have not) so far been adapted in the 2TSF world.

A. Papadopoulos, C. F. Parmeter, *Two-Tier Stochastic Frontier Analysis for the Social Sciences*, https://doi.org/10.1007/978-3-031-81513-3_14

14.1.1 Nonparametric Estimation

A key concern in applied work is the impact that various parametric assumptions have on the corresponding findings. How much does my linear specification impact the hedonic regression? How much does the Normal assumption for v or the Exponential for either w or u play a role in what we learn about these unobservables? It is natural to ask these questions and it is also natural to seek to insulate oneself from having to make these types of assumptions. Enter nonparametric smoothing methods. While they are becoming quite popular in virtually all applied economic subfields, their use remains far behind their more common parametric counterparts. This could be expected. Parameters are the colors and pigments that give life to models and make it easier for the researcher to "see" what is happening. Nonparametric methods, however, while having as much if not more vibrancy and color, make it more challenging (though not impossible) to oversee what is happening in a model. As we will discuss briefly here, nonparametric estimation has quite close connections to standard parametric estimation.

To understand key distinctions between parametric and nonparametric estimation, it helps to distinguish between *local* and *global* estimation. Parametric modeling falls under the "global" umbrella. A researcher specifies a model, say linear, and then uses *all* of the data to estimate the parameters of that particular functional form. With a "local" model, it is nearly the same except that we typically use *some* of the data for each point we are interested in estimating. The "some" here is important. Which "some" do we use and how do we know if we have used enough "some"? These are important issues over which both theorists and the consumers of nonparametric methods have spent a lot of intellectual capital.

Nearly all nonparametric methods rely on the concept of "localness," and this is measured through a user-defined parameter (or parameters). For example, with kernel smoothing, localness (the "some" above) is dictated by a bandwidth, the size of which controls how much data to use to evaluate the unknown function at a specific point. With splines and series, it is almost the same but knots are used to construct a flexible function—the more knots, the more localness is imparted. For k-nearest neighbor methods it is nearly identical, the more neighbors, the less localness and vice versa. It is not uncommon to hear criticisms levied against the use of localness parameters; however, there is a vast literature discussing the theoretical underpinnings of their choice (Fan & Gijbels, 1995; Henderson & Parmeter, 2015; Li and Racine, 2007; Pagan & Ullah, 1999).

In kernel smoothing, estimators are constructed via a kernel function (which is just a density function—the choice of which has been shown to be irrelevant in empirical work) and a bandwidth (the localness parameter). A typical setup is

$$k_h(x_i, x) = k\left(\frac{x_i - x}{h}\right), \tag{14.1}$$

where h is the smoothing parameter and $k(\cdot)$ is the kernel function. It is common to further abbreviate the kernel notation to k_{ix} (where the dependence upon h has been suppressed). When we have multiple variables over which to smooth, it is common to invoke the product kernel. Here, with s variables, we would have

$$K_{ix} = \prod_{j=1}^{s} k\left(\frac{x_{ij} - x_j}{h}\right).$$

The ordinary least squares regression estimator of the function of interest, $g(\boldsymbol{x}; \beta)$, can be written as follows (Henderson and Parmeter, 2016):

$$\widehat{m}(\boldsymbol{x}, \beta) = \sum_{i=1}^{n} w_i(\boldsymbol{x}) y_i, \tag{14.2}$$

where the weights will depend on the specific functional form (say linear in $\boldsymbol{x}$) chosen by the user. A local-constant least squares regression estimator (which is nonparametric) simply replaces the weights in Eq. (14.2) with kernel weights:

$$\widehat{m}(\boldsymbol{x}) = \sum_{i=1}^{n} w_i(\boldsymbol{x}) y_i, \quad \text{with } w_i(\boldsymbol{x}) = \frac{k_h(\boldsymbol{x}_i, \boldsymbol{x})}{\sum_{i=1}^{n} k_h(\boldsymbol{x}_i, \boldsymbol{x})}. \tag{14.3}$$

The intuition behind the construction of this estimator is as follows: if we were estimating the logarithm of wages for an individual with 16 years of education, we could simply look at the average earnings for all individuals with 16 years of education in the data. However, this fails to recognize that individuals with 15 or 17 years of education may be similar to an individual with 16 years of education. Rather, by (weighted) averaging over all individuals who have education close to 16 years of education, the averaging mitigates some of the noise imparted through the use of only the "16-years-of-education" observations. By judicious choice of the smoothing parameter, we can hope to eliminate concerns that closeness was in some sense arbitrary.

Now to place this general discussion on nonparametric estimation in our context, consider the benchmark 2TSF model for a sample of units $i = 1, \ldots, n$:

$$y_i = g(\boldsymbol{x}_i) + v_i + w_i - u_i. \tag{14.4}$$

The key condition required for consistent estimation of the production frontier in a regression setting is $E(\varepsilon|\boldsymbol{x}) = 0$. However, given the one-sided nature of u and w, it is unlikely that this condition will be satisfied, rather $E(\varepsilon|\boldsymbol{x}) = \gamma$. While this is enough in a parametric setting to estimate consistently the regression slope coefficients, the constant term, namely, the level of the regression model, cannot be identified, as it is the case that

$$y_i = g(\boldsymbol{x}_i) + \varepsilon_i = g(\boldsymbol{x}_i) + \gamma + (\varepsilon_i - \gamma) \equiv g^*(\boldsymbol{x}_i) + \varepsilon_i^*. \tag{14.5}$$

Fan et al. (1996, FLW hereafter), in the single-tier setting, proposed a solution to correct the bias in the estimation of $g(\boldsymbol{x})$ by retaining standard distributional assumptions from the SFA literature (e.g., Normal noise, Half-Normal inefficiency) and estimating the corresponding distributional parameters via maximum likelihood on the nonparametrically estimated residuals from a local-constant regression estimator, as just described above. Once these distributional parameters have been estimated, the estimated conditional mean can be shifted (bias-corrected) by the estimated mean of inefficiency (the mean correction factor).

FLW's procedure can be easily adapted to the 2TSF setting with only slight modification. To our knowledge, nonparametric estimation in the 2TSF setting has yet to appear in the applied literature. This is troubling since it is likely that many empirical models suffer from misspecification and deploying methods of this sort could alleviate some of these issues. Under fairly weak conditions, FLW show that the parameters of the composed error distribution can be consistently estimated at the parametric $\sqrt{n}$ rate.[1]

The approach just described is what is known as "plug-in (PI) likelihood." The PI approach of FLW begins by making assumptions on the error components in the 2TSF model. We assume that noise follows a Normal distribution (as per usual, with zero mean and constant, but unknown, homoskedastic variance, σ_v) and that the latent forces follow Exponential distributions, with unknown finite and homoskedastic variances to be estimated, σ_u and σ_w, respectively.

Given these distributional assumptions and the (biased) nonparametric estimator of the frontier, one would estimate the semiparametric two-tier stochastic frontier model and the unknown distributional parameters as follows:

Step 1: Estimate the conditional expectation of (14.5), $E(y_i|\boldsymbol{x}_i)$, using nonparametric methods; FLW deploy kernel smoothing but a regression spline or sieve estimator would work equally well. Call this $\widehat{g}^*(\boldsymbol{x}_i)$ and let the residuals be denoted $\widehat{\varepsilon}_i^* = y_i - \widehat{g}^*(\boldsymbol{x}_i)$.

Step 2: Estimate the three distributional parameters, $\boldsymbol{\theta} = (\sigma_v, \sigma_u, \sigma_w)$ by maximizing the log-likelihood function consistent with the Normal-Exponential distributional assumptions (see Chap. 7), which is

$$\widehat{\boldsymbol{\theta}} = \underset{\boldsymbol{\theta}}{\text{argmax}} \left\{ -n \cdot \ln(\sigma_w + \sigma_u) + \sum_{i=1}^{n} \ln \left(e^{a_{1i}} \Phi(b_{1i}) + e^{a_{2i}} \Phi(b_{2i}) \right) \right\}, \tag{14.6}$$

where we set $\varepsilon_i = \widehat{\varepsilon}_i^* - \gamma$.[2] The mean correction factor is $\gamma = E(\varepsilon_i) = \sigma_w - \sigma_u$ and so we have $\widehat{\gamma} = \widehat{\sigma}_w - \widehat{\sigma}_u$.

[1] Martins-Filho and Yao (2015, also MY hereafter) later showed that these parameter estimators are biased and inefficient unless undersmoothing is undertaken.

[2] The expressions for the composite terms a_{1i}, b_{1i}, a_{2i}, and b_{2i} can be found in Chap. 7.

Step 3 The two-tier stochastic frontier $g(\boldsymbol{x}_i)$ is consistently estimated by

$$\widehat{g}(\boldsymbol{x}_i) = \widehat{g}^*(\boldsymbol{x}_i) - \widehat{\gamma}. \tag{14.7}$$

If alternative distributional assumptions were made on u, for example, Half-Normal, a different likelihood function and mean correction should be deployed.

Martins-Filho and Yao (2015) demonstrated that while the kernel estimator of the frontier detailed in FLW is consistent, the estimator for the parameters of the density of the composite error produces an asymptotic bias when normalized by $\sqrt{n}$. As an alternative, MY proposed an estimator that jointly estimates the distributional parameters and the unknown frontier.

More specifically, adapting their approach to the 2TSF model, estimation of the parameters of the distributions and the unknown conditional mean is accomplished through the likelihood function:

$$\mathcal{L}_n(\boldsymbol{\theta}, g) = n^{-1} \sum_{i=1}^{n} \ln f_\varepsilon \left(y_i - g^*(\boldsymbol{x}_i; \boldsymbol{\theta}) - \gamma(\boldsymbol{\theta}); \boldsymbol{\theta}\right), \tag{14.8}$$

where $E(y|\boldsymbol{x}) = g^*(\boldsymbol{x}_i; \boldsymbol{\theta}, g) = g(\boldsymbol{x}) + \gamma(\boldsymbol{\theta})$. FLW's estimator of $\boldsymbol{\theta}$ is based on the idea that if $g^*(\boldsymbol{x}_i; \boldsymbol{\theta}, g)$ were known, a standard parametric ML estimator for $\boldsymbol{\theta}$ could be obtained in routine fashion by maximizing $\mathcal{L}(\boldsymbol{\theta}, g)$. However, since $g(\cdot)$ is unknown, the exact likelihood function is replaced with the approximation:

$$\tilde{\mathcal{L}}_n(\boldsymbol{\theta}, \widehat{g}) = n^{-1} \sum_{i=1}^{n} \ln f_\varepsilon \left(y_i - \widehat{g}^*(\boldsymbol{x}_i; \boldsymbol{\theta}, g) - \gamma(\boldsymbol{\theta}); \boldsymbol{\theta}\right) \tag{14.9}$$

and the parametric estimator of $\boldsymbol{\theta}$ is $\widehat{\boldsymbol{\theta}} = \underset{\boldsymbol{\theta}}{\operatorname{argmax}}\, \tilde{\mathcal{L}}_n\left(\boldsymbol{\theta}, \widehat{g} + \gamma(\boldsymbol{\theta})\right)$. The bias associated with estimating the two-tier frontier in the first stage is **independent** of the structure of f_ε, and MY demonstrated that this produces an asymptotic bias in the ML estimates of $\boldsymbol{\theta}$. There are two remedies to handle the bias in $\widehat{\boldsymbol{\theta}}$. First, one can use a suboptimal bandwidth when estimating $g^*(\boldsymbol{x}_i; \boldsymbol{\theta}, g)$; this suboptimal bandwidth will allow the bias in $\widehat{g}$ to decay at an appropriate rate such that $\widehat{\boldsymbol{\theta}}$ no longer carries an asymptotic bias when normalized by $\sqrt{n}$. Second, a joint estimation procedure that explicitly connects estimation of $\boldsymbol{\theta}$ and $g(\boldsymbol{x}_i)$ can be deployed; this procedure is termed "profile likelihood estimation" in MY.

MY's joint estimation approach relies on local-likelihood estimation except that the parameters of the distribution of the composed error term are held fixed. In their setup, this is represented as

$$\check{\mathcal{L}}_n(\boldsymbol{\theta}, \widehat{g}) = n^{-1} \sum_{i=1}^{n} \ln f_\varepsilon(y_i - \tilde{g}(\boldsymbol{x}_i); \boldsymbol{\theta}) K_{i\boldsymbol{x}}, \tag{14.10}$$

where $\tilde{g}(\boldsymbol{x}_i) = \alpha(\boldsymbol{x}) + (\boldsymbol{x}_i - \boldsymbol{x})'\beta(\boldsymbol{x})$ is the first-order Taylor approximation of $g(\cdot)$ such that $\alpha(\boldsymbol{x})$ captures the function $g(\cdot)$ and $\beta(\boldsymbol{x})$ captures the derivative of function $m(\cdot)$, $\partial g(x)/\partial x_j$ for $j = 1, \ldots, k$.

Estimation involves two repeated steps.

Step 1: For a fixed $\boldsymbol{x}$ and $\boldsymbol{\theta}$, $\widehat{\alpha}_{\boldsymbol{\theta}}(\boldsymbol{x})$ and $\widehat{\beta}_{\boldsymbol{\theta}}(\boldsymbol{x})$ are estimated as

$$\left(\widehat{\alpha}_{\boldsymbol{\theta}}(\boldsymbol{x}), \widehat{\beta}_{\boldsymbol{\theta}}(\boldsymbol{x})\right) = \underset{\alpha(\boldsymbol{x}),\beta(\boldsymbol{x})}{\text{argmax}}\ \check{\mathcal{L}}_n(\boldsymbol{\theta}, g).$$

Step 2: Using $\widehat{\alpha}_{\boldsymbol{\theta}}(\boldsymbol{x})$, an estimator of θ is determined as

$$\widehat{\boldsymbol{\theta}} = \underset{\boldsymbol{\theta}}{\text{argmax}}\ \check{\mathcal{L}}_n(\boldsymbol{\theta}, \widehat{\alpha}_{\boldsymbol{\theta}}(\boldsymbol{x})).$$

The estimator for $g(\boldsymbol{x})$ is $\widehat{g}(\boldsymbol{x}) = \widehat{\alpha}_{\widehat{\boldsymbol{\theta}}}(\boldsymbol{x})$.

This two-step procedure requires iteration to obtain the final estimates. As starting values, the estimated parameters from simple application of FLW's estimator can be used. From here, the two-step approach is implemented for each point of interest. Keep in mind that if you evaluate the frontier at 100 data points, then 101 optimizations need to be solved at each iteration, one for every evaluation point and a final optimization to determine $\boldsymbol{\theta}$. The iterations can be terminated once there is sufficiently small movement in $\widehat{\boldsymbol{\theta}}$ from iteration to iteration.[3] As a last step, the final estimate of θ is then used to re-estimate $g(\boldsymbol{x})$ for each of the evaluation points.

An important implementation hurdle is the selection of the bandwidth vector to smooth $\boldsymbol{x}$. Martins-Filho and Yao (2015) note that $g(\boldsymbol{x}) = E(y|\boldsymbol{x}) + \gamma(\boldsymbol{\theta})$, where $\gamma(\boldsymbol{\theta})$ is non-stochastic. This implies that an appropriate level of smoothing for $g(\boldsymbol{x})$ is the same as the appropriate level of smoothing for $E(y|\boldsymbol{x})$. Given this, they recommend the use of a plug-in type bandwidth for local-linear regression (Ruppert et al., 1995). Alternatively, a data-driven method, such as least squares cross-validation (LSCV), could also be applied.

14.1.2 Fast Fourier Transform

To motivate the use of the Fast Fourier Transform (FFT), we start with the single-tier Normal-Exponential setup of Aigner et al. (1977) and Meeusen and van den Broeck (1977). Certainly, in this case, we have a closed-form solution for $f(\varepsilon)$, but this will allow us to understand how the FFT works in more complicated settings.

[3] MY use a tolerance of 0.001 in their Monte Carlo simulations.

The composite error density here is

$$f(\varepsilon) = \sigma_u^{-1} \exp\left\{\frac{\varepsilon}{\sigma_u} + \frac{\sigma_v^2}{2\sigma_u^2}\right\} \Phi\left(-\frac{\varepsilon}{\sigma_v} - \frac{\sigma_v}{\sigma_u}\right). \tag{14.11}$$

Now, imagine that we did not know this. If we assume that $v \perp\!\!\!\perp u$, a common assumption in stochastic frontier models, then we know that the characteristic function of $f(\varepsilon)$ is simply the product of the individual characteristic functions. Denote a generic characteristic function of a random variable X as $\varphi_X(t) = E\left(e^{\imath tx}\right) = \int e^{\imath tx} f_X(x)dx, \forall t \in \mathbb{R}$, where $\imath = \sqrt{-1}$ is complex unity.

You may be asking why we care about characteristic functions. From Levy's inversion theorem,

$$f_\varepsilon(\varepsilon) = \int_{-\infty}^{\infty} e^{-\imath\varepsilon t}\varphi_\varepsilon(t)dt. \tag{14.12}$$

Thus, the **Fourier transform** of the characteristic function can be used to construct the unknown density. Typically, this integral will not be available in closed form, but it can be approximated efficiently and accurately using the FFT. Moreover, characteristic functions for a range of densities are known. For our purposes, here $\varphi_v(t) = \exp\{-0.5\sigma_v^2 t^2\}$ (which does not include the complex unity because the mean is zero) and $\varphi_u(t) = (1 - \sigma_u \imath t)^{-1}$.

To explain how we will leverage characteristic functions to estimate the unknown density, we first rewrite the integral in Eq. (14.12) using the change of variables $t = 2\pi\omega$, so that $dt = 2\pi d\omega$. Then we have

$$f_\varepsilon(\varepsilon) = \int_{-\infty}^{\infty} 2\pi \exp\{-\imath\varepsilon 2\pi\omega\}\varphi_\varepsilon(2\pi\omega)d\omega. \tag{14.13}$$

Further, define a sequence $x_k = (k - 1 - N/2)h$ where $h > 0$ is a user-defined step length,[4] and $N = 2^J$ for J some positive integer. With this sequence, the integral can be approximated using the trapezoidal rule as

$$f_\varepsilon(x_k) \approx s \sum_{j=1}^{N} \exp\{-\imath 2\pi\omega_j x_k\}\varphi_\varepsilon(2\pi\omega_j), \tag{14.14}$$

where $\omega_j = s(j - 1 - N/2)$ and $s = (hN)^{-1}$. Keeping in mind Euler's identity $e^{\imath\pi} + 1 = 0$, this produces

[4] Tsionas (2012) recommended use of $h = 2\bar{e}/N$ where $\bar{e} = \max |e_i|$.

$$
\begin{aligned}
f_\varepsilon(x_k) &\approx s\sum_{j=1}^{N} e^{-\imath 2\pi s(j-1-N/2)(k-1-N/2)h}\varphi_\varepsilon(2\pi\omega_j) \\
&= s\sum_{j=1}^{N} e^{-\imath 2\pi(j-1-N/2)(k-1-N/2)/N}\varphi_\varepsilon(2\pi\omega_j) \\
&= s\sum_{j=1}^{N} e^{-\imath 2\pi(j-1)(k-1)/N}e^{-\imath 2\pi(j-1)(-1/2)} \\
&\quad\times e^{-\imath 2\pi(-1/2)(k-1)}e^{-\imath 2\pi(N/2)(N/2)/N}\varphi_\varepsilon(2\pi\omega_j) \\
&= s\sum_{j=1}^{N} e^{-\imath 2\pi(j-1)(k-1)/N}e^{\imath\pi(j-1)}e^{\imath\pi(k-1)}e^{\imath\pi(-N/2)}\varphi_\varepsilon(2\pi\omega_j) \\
&= \frac{(-1)^{k-1-(N/2)}}{hN}\sum_{j=1}^{N}(-1)^{j-1}e^{-\imath 2\pi}\varphi_\varepsilon(2\pi(j-1-N/2)/hN).
\end{aligned}
\tag{14.15}
$$

This sequence can be computed efficiently applying the FFT to the sequence $(-1)^{j-1}\varphi_\varepsilon(2\pi(j-1-N/2)/hN)$ for $j = 1, \ldots, N$. Since our observations of ε_i will not correspond to this sequence, we will need to use interpolation. That is, when we perform maximum likelihood estimation, we will evaluate $f_\varepsilon(\cdot)$ at $\varepsilon_i = y_i - \boldsymbol{x}_i\boldsymbol{\beta}$, but the FFT evaluates $f_\varepsilon(\cdot)$ at the points $x_k = (k-1-N/2)h$. Both the FFT and linear/cubic interpolation are trivially handled in `R` using the `fft()` and `approx()` commands, respectively.

While this may seem computationally frightening, the benefit is that by moving to imaginary space, convolutions are much easier to work with (under independence). This was the key insight of Tsionas (2012) when he proposed a Normal-Gamma two-tier stochastic frontier. In real space, this density is intractable and either simulation or numerical integration would be needed. However, in imaginary space, this density is easy to construct. In fact, this estimation approach readily lends itself to trying all manner of densities (but has yet to make a serious appearance in applied work).

For example, using the fact that the characteristic function of a Gamma distributed random variable with shape k and scale θ is $\varphi_X(t) = (1-\theta\imath t)^{-k}$, we can easily construct $\varphi_\varepsilon(t) = e^{-0.5\sigma_v^2 t^2}(1-\theta_u\imath t)^{-k_u}(1+\theta_w\imath t)^{-k_w}$. The NE model stems directly from this by setting $k_u = k_w = 1$.

14.1.3 Bayesian Estimation of the 2TSF Model

Another area that has substantially lagged behind the corresponding inroads in the single-tier frontier setting is Bayesian estimation and inference of the 2TSF model. The use of Bayesian methods has a long history in single-tier SFA. Beginning with van den Broeck et al. (1994) and through today, there are a great many researchers and proposals for estimating a single-tier frontier through the argot of Bayesian statistics. However, when casting an eye for similar research for the 2TSF model, there is only one, Zhao and Losak (2024), who developed a Markov Chain Monte Carlo sampling algorithm to estimate the posterior distributions of all the parameters in the model.

They further enhance their model by allowing dependence, one of the first applications to do so beyond those examples discussed here in the book. To elicit draws from the posterior, Zhao and Losak (2024) rely on Gibbs sampling, which allows sequential draws of the Markov chain through larger and larger conditional distributions. Technically, the 2TSF is not a simple model, having at its core a convolution density of three random variables, sometimes including also a copula density, as we saw in Chap. 10. Being able to invoke the unstoppable march of the MCMC army is comforting, as long as, of course, we feel confident that there is indeed a destination.

14.1.4 Regressor Endogeneity

Regressor endogeneity is not an estimation method, but a potential situation that threatens the validity of all estimation methods in a regression setup. In short, if regressors have statistical dependence with the regression error term, the validity and reliability of estimation are questioned. It is true that regressors can be statistically dependent with the error term in a way that does not affect the properties of the estimators: if the error term is "mean-independent" of the regressors, i.e., $E(\varepsilon|\boldsymbol{x})$ is a constant, or even if it is just orthogonal, $E(\varepsilon \cdot \boldsymbol{x}) = 0$, or at least has zero covariance $\text{Cov}(\varepsilon, \boldsymbol{x}) = 0$, most estimation methods maintain their desirable asymptotic properties, while higher-order statistical dependence is allowed to exist. For example, the standard "benign" assumption of dependence between error term and regressors is conditional heteroskedasticity, in parallel with mean independence. But this may not be always, or even frequently, the case, and so the issue of how a regression model handles possible endogeneity comes up routinely in assessing the validity of a study.

The most widespread and studied method is instrumental variables, where we replace suspected endogenous regressors with other available covariates that have zero correlation with the error term but enough correlation with the dependent variable to provide asymptotically consistent estimates (sacrificing efficiency), or, in a more sophisticated approach, where we use these untainted covariates to create an orthogonal projection of the endogenous regressors, and then run the regression using these orthogonal projections instead of the original endogenous variables (this is the "two-stage-least-squares" methodology). The problems with instrumental variables are, first, that it is not easy to acquire good quality instruments, and second, actually verifying that these instruments are indeed uncorrelated with the error term is rarely possible, and we mostly rely on theoretical arguments to support it.

In the single-tier stochastic frontier domain, quite a bit of attention has been given to handling endogeneity. These methods range from extensions of COLS (Amsler et al., 2016), using instrumental variables of course, to full maximum likelihood estimation (Amsler et al., 2017; Karakaplan & Kutlu, 2017; Kutlu, 2010) where we model the dependence between the error term and the endogenous covariates as well as using the moment conditions from the endogeneity in a GMM setup (Amsler et al., 2016; Tran & Tsionas, 2013). Other approaches have also been developed that focus on various economic restrictions, such as appending the first-order conditions from cost minimization to estimation of the frontier itself (for a full discussion see Kumbhakar et al., 2022).

There are no theoretical tools developed to handle regressor endogeneity specifically for the 2TSF setup. Given that COLS and/or GMM estimation can be applied to the 2TSF model (and we have presented such estimators in Chap. 7), it is likely that the methods referenced above could be adjusted for 2TSF estimation with endogenous regressors. But there is another general approach, that of using Copulas to model statistical dependence, which can be applied to any setting, and we have indeed used it in certain applications in this book.[5] In practice, because in the Copula formulation all dependence is encapsulated in the copula density which multiplies the joint density under independence (i.e., the product of the marginal densities), when we consider the joint log-likelihood of the sample, the copula-enhanced log-likelihood is just the log-likelihood under no endogeneity plus the log-copula density.

The one ad hoc aspect of using Copulas to handle endogeneity is the choice of the Copula function and the resulting copula density. In Chap. 10, we presented three Copulas that can be used to model dependence between the two latent forces. But they can also be used to capture the dependence between the error term and the regressors, although, if the regressors suspected for endogeneity are more than one, then only elliptical Copulas like the Gaussian are really practical to use, since Copulas of dimension higher than two are cumbersome to work with.

Papadopoulos (2022) presents in an accessible manner the fundamental statistical and likelihood theory underlying the approach along with a step-by-step guide on how to use Copulas to handle endogeneity in a general regression setup.

[5] This has been proposed for the single-tier SF model by Tran and Tsionas (2015).

14.2 More Distributional Specifications

While many things developed for single-tier stochastic analysis have not been applied for the 2TSF framework, some have. Still, we decided not to stuff the core chapters of the book with too many alternatives, in order to avoid inducing "choice bewilderment" to our readers. In this subsection, we present summarily three alternative 2TSF models, providing the basic reference for each for detailed guidance.

14.2.1 *The Generalized Exponential Specification*

Both the 2TSF Normal-Exponential and Normal-Half-Normal models have latent forces whose distribution has its mode at zero and is monotonically decreasing, incorporating, in a sense, a "probabilistic bias" in favor of the realization of values of low magnitude. In a setting of economic efficiency and inefficiency, this has theoretical support: competition pushes the firms to be as efficient as they can, and, although this may not be enough to fully eliminate inefficiency, it results in keeping it at low levels. But two-tier stochastic frontier analysis is a general statistical tool that can be applied to all social sciences, and it is not necessarily the case that such zero-mode distributions will be the best representatives of all phenomena. Papadopoulos (2021c) presents single-tier and 2TSF stochastic frontier models based on a version of the Generalized Exponential (GE) distribution that now characterizes each of two latent forces and is accompanied as usual by Normal noise. This distribution has its mode *always away* from zero. Its marginal density function is, say for the latent force u,

$$f_u(u) = \frac{2}{\theta_u} \exp\{-u/\theta_u\} \left(1 - \exp\{-u/\theta_u\}\right), \quad \theta_u > 0. \tag{14.16}$$

If one looks closely, this is also the density of the maximum of two iid Exponential random variables. So conceptually, the Generalized Exponential represents a factual and a counterfactual. For *each* of our latent forces, there are two underlying forces that compete to take this one place—and the bigger of the two gets to be realized: a "high-intensity" situation on both sides, that could, for example, find interesting applications in psychology, or in situations of social conflict.

The basic moments/cumulants of this distribution are

$$E(u) = \kappa_1(u) = \frac{3}{2}\theta_u, \ \mathrm{Var}(u) = \kappa_2(u) = \frac{5}{4}\theta_u^2,$$

$$\kappa_3(u) = \frac{9}{4}\theta_u^3, \ \kappa_4(u) = \frac{51}{8}\theta_u^4,$$

$$\text{Skewness}(u) = \gamma_1(u) = \frac{\kappa_3(u)}{\kappa_2^{3/2}(u)} = \frac{18}{\sqrt{125}} \approx 1.61,$$

$$\text{Ex. Kurtosis}(u) = \gamma_2(u) = \frac{\kappa_4(u)}{\kappa_2^2(u)} = \frac{102}{25} = 4.08.$$

The GE distribution has lower skewness and excess kurtosis compared to the Exponential distribution (for which they are 2 and 6, respectively), but exceeds both for the Half-Normal (0.995 and 0.869, respectively).

Papadopoulos (2021c) contains the full toolkit for the 2TSF Generalized Exponential model, with two qualifications: he provides the formulas only for the model variant where the dependent variable is in logarithms; and he uses the conditional *mode*, rather than the conditional expectation, to form individual metrics. This requires maximizing the conditional density (which is provided) per observation by a numerical solver, something that with the current sophistication of computational tools is a rather simple task.

The conditional mode as an alternative predictor at the individual level was proposed very early in the stochastic frontier literature, but it did not really catch on. Still, it is the optimality concept that aligns with likelihood inference, and it could be better suited to represent one-off decisions or outcomes, compared to the conditional expectation, which is the best predictor in the Mean-Square-Error sense, a criterion that is appropriate mainly for optimizing prediction as a repetitive activity, or when a single outcome represents implicitly the aggregation of many distinct ones. For example, in the World Happiness application of Chap. 5, the dependent variable is the national average of the subjective assessments made by numerous individuals, which, in turn, evaluate ("predict") many different aspects of their life and aggregate them into a single measure.

14.2.2 Heterogeneous Distributions for the Latent Forces

Yet one other common aspect of all 2TSF models presented up to now is that the opposing latent forces follow the same distribution (just with different values for the distribution parameters). But this is not necessarily the case in the real world: the opposing forces may be two very different entities, not necessarily identical in nature. For example, in the first 2TSF application of this book (Chap. 2), we estimated a hedonic 2TSF regression on the products of a *single* firm in the early days of the personal computer market. In setting its prices, we had the firm wanting to "exploit the hype" on the one hand, while at the same time, it was "fearing the competition." While both forces live inside the same decision-maker unit, they have no reason to follow the same distribution, when viewed as random variables.

A first step toward heterogeneous distributions for the latent forces is the models presented in Papadopoulos (2024b). There the author develops a Gamma-

Exponential and an Exponential-Gamma model, allowing one of the two components to have a non-zero mode. The model has a non-closed-form density, and the author proposes simulated maximum likelihood with the use of Halton numbers, which reduce considerably computation time. Alternatively, one could apply COLS/GMM estimation (see Chap. 7) or the just described FFT method. The article contains the full package of formulas to implement the two models either with a dependent-variable-in-levels-formulation or with a logarithmic formulation, and it also applies the Vuong specification test that we presented in Chap. 7, which illustrates how two models whose log-likelihood values are really close may fare very differently under that test.

14.2.3 Latent Class Modeling

In Chap. 8, we presented methods to incorporate heterogeneity in a cross-sectional 2TSF model, by the use of additional data, variables that affect the latent forces. But what can we do when heterogeneity is believed to exist, but there are no such data available? One solution would be the latent class model (LCM).

The LCM is characterized by its use and estimation of a *mixture* of densities as opposed to a single density that is assumed to have produced all of the data. In the standard finite mixture model, there is a pre-specified number of groups (though we do not know to which group each observation belongs), the proportion of which is assumed to be fixed (a parameter to be estimated), see, e.g., Caudill (2003). In essence, the LCM operates like the subsample approach of Polachek and Yoon (1987) but without a specific variable dictating how the subsamples are created, and so without knowing a priori which observations belong to which class. Rather, the data itself inform the model on which group an observation is likely to come from.

The use of LCM has become popular in the single-tier stochastic frontier setting: Caudill (2003) introduces an expectation-maximization (EM) algorithm to estimate a mixture of two stochastic cost frontiers in the presence of no sample separation information; Orea and Kumbhakar (2004) use brute force maximization of the likelihood function to allow differences across banks, while Greene (2005b) deploys maximum simulated likelihood to obtain the maximum likelihood estimator in which the (prior) probability of adopting a particular technology is assumed to depend on firm characteristics. However, to our knowledge, use of the LCM in a two-tier stochastic frontier setup has yet to appear. While the LCM implemented by these authors are all developed around a panel, our discussion here will focus on the cross-sectional case.

To properly construct the latent class 2TSF model, we assume that there is a latent, unobservable sorting of the data into C classes. For an observation from class c, the conditional density for observation i can be characterized as

$$f(y_i|\boldsymbol{x}_i, c) = f(\Theta_c, y_i, \boldsymbol{x}_i), \qquad c = 1, \ldots, C \tag{14.17}$$

where Θ_c is the collection of parameters, specific to class c for the two-tier stochastic frontier model. However, all the data are used to estimate all of the parameters as well as the probability of belonging to a particular class.

To provide a clear exposition, we will assume that $f(\cdot)$ is the same across the classes; specifically, we will use the NE model. Of course one could specify a different distribution per class. Given this the class-specific vector of parameters to be estimated is $\Theta_c = \left(\boldsymbol{\beta}_c, \sigma_{uc}, \sigma_{vc}, \sigma_{wc}\right)$ and our class-specific density is

$$f(\varepsilon_{ci}|c) = \frac{e^{a_{1ci}}}{\sigma_{uc} + \sigma_{wc}} \Phi\left(b_{1ci}\right) + \frac{e^{a_{2ci}}}{\sigma_{uc} + \sigma_{wc}} \Phi\left(b_{2ci}\right), \tag{14.18}$$

where the composite terms are as shown in Sect. 7.2.1 (with a "c" appended to the various subscripts) and $\varepsilon_{ci} = y_i - \boldsymbol{x}_i'\boldsymbol{\beta}_c$. $f(\varepsilon_{ci}|c)$ represents observation i's contribution to the conditional on c likelihood. The unconditional log-likelihood for observation i is the weighted average over all C classes:

$$\ell_i = \ln\left[\sum_{c=1}^{C} \Pi(\varepsilon_{ci}, c) f(\varepsilon_{ci}|c)\right]. \tag{14.19}$$

Here, $\Pi(\varepsilon_{ci}, c)$ is the prior probability of observation i's membership in class c. In this framework, $\Pi(\varepsilon_{ci}, c)$ reflects the uncertainty residing around class membership for a given observation.[6] $\Pi(\varepsilon_{ci}, c)$ can be parameterized based on a set of covariates (similar to the inclusion of determinants of information deficiency), but it must be kept in mind that $0 \leq \Pi(\varepsilon_{ci}, c) \leq 1$, so a logit or probit parameterization is appropriate. For $C > 2$, the multinomial logit form would parameterize $\Pi(\varepsilon_{ci}, c)$ as

$$\Pi(\varepsilon_{ci}, c) = \frac{e^{\boldsymbol{z}_{l,i}'\boldsymbol{\pi}_c}}{1 + \sum_{m=1}^{C-1} e^{\boldsymbol{z}_{l,i}'\boldsymbol{\pi}_m}}, \tag{14.20}$$

where $\boldsymbol{z}_{l,i}$ and the subscript l captures the fact that here $\boldsymbol{z}$ is a variable reflecting inclusion into a specific class as opposed to influencing information deficiency directly (see Chap. 8). Note also that $\Pi(\varepsilon_{ci}, c)$ does not have to be dependent upon a set of covariates; this is just an option. In this case, $\Pi(\varepsilon_{ci}, c) = \Pi(c)$ is a constant across observations, namely, all observations have the same probability of belonging to class c.

As noted by Greene (2005b), the log-likelihood of the sample can be optimized either through the EM algorithm or by maximum simulated likelihood.

---

[6] In the subsample approach of Polachek and Yoon (1987), $\Pi(\varepsilon_{ci}, c)$ is essentially set equal to 0 or 1.

The EM algorithm has intuitive appeal. Define the posterior probability of class c given observation i as

$$w(c|i) = \frac{f(\varepsilon_{ci}|c)\Pi(\varepsilon_{ci}, c)}{\sum_{m=1}^{C} f(\varepsilon_{ci}|m)\Pi(\varepsilon_{mi}, m)}. \tag{14.21}$$

Then, maximum likelihood estimates can be found by iterating between

$$\widehat{\boldsymbol{\Theta}}_c = \max \sum_{i=1}^{n} w(c|i) \ln f(\varepsilon_{ci}|c), \quad c = 1, \ldots, C, \tag{14.22}$$

and

$$(\widehat{\boldsymbol{\pi}}_1, \widehat{\boldsymbol{\pi}}_2, \ldots, \widehat{\boldsymbol{\pi}}_C) = \max \sum_{i=1}^{n} \sum_{c=1}^{C} w(c|i) \ln \Pi(\varepsilon_{ci}, c), \tag{14.23}$$

making sure to update $w(c|i)$ after each iteration. The EM algorithm essentially acts as though there is some initial guess at the class probabilities, and then the class parameters are estimated through maximization of the log-likelihood function of the c-th class. Finally, with these class parameters fixed, new multinomial coefficients are estimated and the whole process is repeated. If the class membership probabilities are fixed across observations, then the estimates of π_c represent posterior probabilities. Once a solution is reached, the $w(c|i)$ provide estimates of the class probabilities for a given observation. Class assignment is typically based on the largest posterior probability since it is likely that no estimated π_c will be equal to 1.

Individual measures for the latent forces can be calculated either using the highest class membership probability or by calculating the conditional mean. To calculate the conditional mean value of each measure we aggregate over the posterior probabilities as

$$\widehat{M}_s = E(\widehat{M}_s|c) = \sum_{i=1}^{n} \sum_{c=1}^{C} \widehat{w}(c|i)\widehat{M}_{sc} \tag{14.24}$$

which is the sample average metric, or

$$\widehat{M}_{si} = E(\widehat{M}_{si}|c) = \sum_{c=1}^{C} \widehat{w}(c|i)\widehat{M}_{sci} \tag{14.25}$$

which is the observation specific metric of interest. In both cases we would have $s = 1, \ldots, 10$ based on the specific metrics we wished to use to explain deficiency and $\widehat{M}_{sc}$ is the group c specific metric and $\widehat{M}_{sci}$ is the group c specific observation metric.

Lastly, as Greene (2005b) mentions, one cannot treat C as a parameter to be estimated. C is assumed known at the start of the analysis. However, the setup with $C-1$ classes is nested in the C class structure and so one can select a model with C classes based on traditional model selection criteria such as the Akaike or Schwarz information measures (this is advocated in Orea & Kumbhakar, 2004).

14.2.4 Sample Selection Bias in the 2TSF Model

A common data set studying wages might have many individuals who have zero wages. This could be because those individuals are unemployed and currently seeking employment, or they could have simply chosen to not work. It might be tempting to remove these zero observations from the data prior to estimation. However, this induces the well-known "sample selection bias." In essence, the decision not to work, or to be unemployed, is not random, but the dropping from the data set acts as though this were the case. Thus, the data that is actually used is not a random sample from the population of interest, and ignoring sample selection can lead to biased and inconsistent parameter estimates, as the analysis only reflects the selected subset of the population. This problem and its many solutions have a long history in applied econometrics. However, the issue of sample selection in frontier models (either single or two-tier) is much less studied.

More specifically, there is commonly a choice or decision variable, call it d_i that must be made in order for a given output to occur. So in the labor example, $d_i = 1$ if an individual decides to work. This decision is random and is commonly modeled as

$$d_i = I\{z_i\gamma + \zeta_i > 0\}. \tag{14.26}$$

With sample selection, the concern is that the randomness in decision-making, ζ_i, is correlated with observables or the error in the main model specification. The correlation with observables is trivial to handle since they can be made a part of z_i in which case this correlation is purged from ζ_i.

Kumbhakar et al. (2009) were the first to address sample selection in the stochastic frontier model. Their application focused on the decision of dairy farmers in Finland to switch from traditional farming to organic farming, via the presence of subsidies from the Finnish government. The concern is that those farmers who switched may have done so with knowledge of their efficiency levels, in which case low (or high) efficiency farmers were the ones switching (i.e., the switching from conventional to organic dairy farming was not random). They developed a sample selection single-tier stochastic frontier model where selection was based on the level of inefficiency. Greene (2010) also developed a sample selection single-tier stochastic frontier approach. His model differed from that of Kumbhakar et al. (2009) in that he allowed selection based on noise, rather than inefficiency.

Extensions of these ideas to the 2TSF currently are discussed only in Blanco (2017), who looks at a 2TSF earnings frontier with sample selection. The likelihood function of Blanco (2017) involves two major changes to the standard Normal-Exponential specification discussed in Chap. 7. First, the ζ_i in the decision is correlated with the noise v_i in the wage frontier. These are modeled jointly as Bivariate Normal. Second, an additional piece is added to account for those individuals who did not participate in a training program prior to working. The likelihood in this case is intractable and Blanco (2017) recommends simulated maximum likelihood to solve for the parameters. In fact, the simulated maximum likelihood approach of Blanco (2017) is nearly identical to that of Greene (2010) except that Greene (2010) approach is for the Normal-Half Normal single-tier framework and Blanco (2017) is for the Normal-Half-Normal two-tier framework. But moving to the 2TSF setup requires nothing more than adding an additional Half-Normal random variate from which to sample for the simulated part of the estimation.

14.3 Terra Incognita

While we have highlighted some recent (or perhaps not so recent) attempts at extending the 2TSF model, there are other areas where no exploration currently exists, but it might prove fruitful. As with all exploration, we cannot know what we will find unless we seek that which we do not know.

14.3.1 Non-Normal and Asymmetric Noise Component

A small but robust literature has recently cropped up in the single-tier SFA world exploring the impacts and consequences of having a non-Normal noise component v in the error term. More specifically, Horrace and Parmeter (2018) propose using a Laplace distribution, which has interesting insights regarding the nature of conditional inefficiency. Bonanno et al. (2016) proposed a Logistic Distribution (even earlier was a proposal by Smith (2008) to use this for the noise distribution). Badunenko and Henderson (2024) suggested using Skew Normal, while Horrace et al. (2024) investigated connections with the Asymmetric Laplace. Wheat et al. (2019) proposed the Student's-t distribution to control for outliers and studied robustness properties associated with the maximum likelihood estimator of this specification.

Any one of these distributions could be deployed in the 2TSF setup, but naturally, the likelihood function is likely to not possess a closed-form solution (nearly all of the aforementioned papers were able to obtain closed-form solutions, but going from two to three error components is a big step). However, with either simulation or the Fast Fourier Transform, researchers have at their disposal tools that can estimate the likelihood function efficiently, as well as the corresponding metrics.

14.3.2 Quantile Methods for 2TSF Models

Quantile methods have started to make an appearance in the single-tier stochastic frontier literature. However, their use has been somewhat scattershot and is summarized by Papadopoulos and Parmeter (2022) as not really capturing the notion of a frontier unless implemented in a correct manner. Intuitively, quantile estimation of the single-tier frontier hinges on what is the cumulative probability at the point where the composed error is exactly 0 (prior to a mean shift). It is not obvious at writing if this relationship implies the same outcome with two distinct one-sided components working in opposite directions, and if so, how. There is most certainly scope for investigating the benefits of quantile estimation in the 2TSF setting.

14.3.3 Dynamic Models

Our last insight for extensions to the 2TSF model is dynamics. For the panel chapter, the time varying information deficiency terms, while allowed to vary over time, did so in a random manner. It might prove scientifically interesting to have these terms depend in some formal manner on their past values. Further, there also exist dynamics in many outcomes of interest, such as environmental pollution. As has perhaps been overstated in this closing chapter, the single-tier stochastic frontier literature has seen a wide variety of modeling choices as it pertains to various forms of dynamics (Lai & Kumbhakar, 2020; Minviel & Sipiläinen, 2021; Tsionas, 2006), but nothing of the sort has arisen for the 2TSF model. It is envisioned that the methods applied in the single-tier framework can be adapted here. However, we caution the reader again that, as in Chap. 9, the lack of closedness of the Normal-Exponential density in the 2TSF setting means that alternative approaches are likely necessary to make headway.

14.4 Afterword

Full disclosure: the PhD theses of both authors were about the two-tier stochastic frontier model. And in the noosphere of science, that was enough to bridge some 9,000 kilometers/6,000 miles of distance between them. Both authors have spent countless hours thinking about and deriving and extending the 2TSF model, as well as implementing it in as many and as diverse empirical applications as possible.

In this book, we try to distill our experience and knowledge with two-tier stochastic frontier analysis, holding back as much as possible on technical expositions but making sure that scientific robustness is never compromised. The main purpose of the book is to allow social scientists to apply empirical two-tier stochastic frontier analysis in their fields in a self-contained way. This required providing all the technical tools needed, together with practical guidance for their implementation, but also a frequent referral back to the real world and a discussion on how the models and the tools can be mapped to all these diverse phenomena under study, and so be used in a meaningful and illuminating way.

But two-tier stochastic frontier analysis is not an off-the-shelf finished product line. As this last chapter clearly shows, there are many interesting topics on the estimation and modeling front that have been touched only lightly, or not at all. Our hope is that this book will also contribute to scholars picking up the mantle and developing these new areas for the 2TSF model to maturity.

References

Abowd, J., Kramarz, F., & Margolis, D. N. (1999). High wage workers and high wage firms. *Econometrica, 67*(2), 251–333.

Aigner, D. J., Lovell, C. A. K., & Schmidt, P. (1977). Formulation and estimation of stochastic Frontier production functions. *Journal of Econometrics, 6*(1), 21–37.

Alvarez, A., Amsler, C., Orea, L., & Schmidt, P. (2006). Interpreting and testing the scaling property in models where inefficiency depends on firm characteristics. *Journal of Productivity Analysis, 25*(2), 201–212.

Amsler, C., Prokhorov, A., & Schmidt, P. (2014). Using copulas to model time dependence in stochastic frontier models. *Econometric Reviews, 33*(5–6), 497–522.

Amsler, C., Prokhorov, A., & Schmidt, P. (2016). Endogeneity in stochastic frontier models. *Journal of Econometrics, 190*, 280–288.

Amsler, C., Prokhorov, A., & Schmidt, P. (2017). Endogenous environmental variables in stochastic frontier models. *Journal of Econometrics, 199*, 131–140.

Anderson, T. W. (1989). Linear latent variable models and covariance structures. *Journal of Econometrics, 41*(1), 91–119.

Anglin, P. M., & Gençay, R. (1996). Semiparametric estimation of a hedonic price function. *Journal of Applied Econometrics, 11*, 633–648.

Arnold, B. C., Beaver, R. J., Groeneveld, R. A., & Meeker, W. Q. (1993). The nontruncated marginal of a truncated bivariate normal distribution. *Psychometrika, 58*(3), 471–488.

Badunenko, O., & Henderson, D. J. (2024). Production analysis with asymmetric noise. *Journal of Productivity Analysis, 61*, 1–18.

Balakrishnan, N., & Lai, C.-D. (2009). *Continuous bivariate distributions*. Springer.

Belotti, F., & Ilardi, G. (2018). Consistent inference in fixed-effects stochastic frontier models. *Journal of Econometrics, 202*(2), 161–177.

Bigsten, A., Isaksson, A., Söderbom, M., Collier, P., Zeufack, A., Dercon, S., & Gunning, J.W., Teal, M. F., Appleton, S., Gauthier, B., Oduro, A., Oostendorp, R., & Pattillo, C. (2000). Rates of return on physical and human capital in Africa's manufacturing sector. *Economic Development and Cultural Change, 48*(4), 801–827.

Blanco, G. (2017). Who benefits from job placement services? A two-sided analysis. *Journal of Productivity Analysis, 47*(1), 33–47.

Bloom, N., Genakos, C., Sadun, R., & Van Reenen, J. (2012). Management Practices across Firms and Countries, Working Paper w17850, National Bureau of Economic Research.

Bollen, K. A. (1989). *Structural equations with latent variables*. John Wiley & Sons.

A. Papadopoulos, C. F. Parmeter, *Two-Tier Stochastic Frontier Analysis for the Social Sciences*, https://doi.org/10.1007/978-3-031-81513-3

Bonanno, A., Bimbo, F., Costanigro, M., Oude Lansink, A., & Viscecchia, R. (2019). Credence attributes and the quest for a higher price–a hedonic stochastic frontier approach. *European Review of Agricultural Economics, 46*(2), 163–192.

Bonanno, G., De Giovanni, D., & Domma, F. (2017). The 'wrong skewness' problem: A re-specification of stochastic frontiers. *Journal of Productivity Analysis, 47*(1), 49–64.

Bradley, E. H., Curry, L., Horwitz, L. I., Sipsma, H., Wang, Y., Walsh, M. N., Goldmann, D., White, N., Pina, I. L., & Krumholz, H. M. (2013). Hospital strategies associated with 30-day readmission rates for patients with heart failure. *Circulation: Cardiovascular Quality and Outcomes, 6*, 444–450.

Bronfenbrenner, M. (1956). Potential monopsony in labor markets. *ILR Review, 9*(4), 577–588.

Brown, J., & Goolsbee, A. (2002). Does the internet make markets more competitive? Evidence from the life insurance industry. *Journal of Political Economy, 110*(3), 481–507.

Caudill, S. B. (2003). Estimating a mixture of stochastic frontier regression models via the EM algorithm: A multiproduct cost function application. *Empirical Economics, 28*(1), 581–598.

Clayton, D. G. (1978). A model for association in bivariate life tables and its application in epidemiological studies of familial tendency in chronic disease incidence. *Biometrika, 65*(1), 141–151.

Colombi, R., Kumbhakar, S., Martini, G., & Vittadini, G. (2014). Closed-skew normality in stochastic frontiers with individual effects and long/short-run efficiency. *Journal of Productivity Analysis, 42*(2), 123–136.

Cooper, J. C. (2006). Prices and Price Dispersion in Online and Offline Markets for Contact Lenses, Technical Report 283, Federal Trade Commission, Bureau of Economics.

Cornwell, C., & Rupert, P. (1988). Efficient estimation with panel data: An empirical comparison of instrumental variables estimators. *Journal of Applied Econometrics, 3*(2), 149–155.

Curtiss, J., Jelínek, L., Medonos, T., Hruška, M., & Hüttel, S. (2021). Investors' impact on Czech farmland prices: A microstructural analysis. *European Review of Agricultural Economics, 48*(1), 97–157.

Das, T., & Polachek, S. W. (2017). Estimating labor force joiners and leavers using a heterogeneity augmented two-tier stochastic frontier. *Journal of Econometrics, 199*(2), 156–172.

Deltas, G., Stengos, T., & Zacharias, E. (2011). Product line pricing in a vertically differentiated oligopoly. *Canadian Journal of Economics/Revue canadienne d'économique, 44*(3), 907–929.

Deltas, G., & Zacharias, E. (2004). Sampling frequency and the comparison between matched-model and hedonic regression price indexes. *Journal of Business & Economic Statistics, 22*(1), 94–106.

Doornik, J. A., & Hansen, H. (2008). An omnibus test for univariate and multivariate Normality. *Oxford Bulletin of Economics and Statistics, 70*, 927–939.

Dutang, C., & Savicky, P. (2023). *Randtoolbox: Generating and testing random numbers.* R package version 2.0.4.

El Mehdi, R., & Hafner, C. M. (2014). Inference in stochastic frontier analysis with dependent error terms. *Mathematics and computers in simulation, 102*, 104–116.

Fair, R. C. (1978). A theory of extramarital affairs. *Journal of Political Economy, 86*(1), 45–61.

Fan, J., & Gijbels, I. (1995). Data-driven bandwidth selection in local polynomial fitting: Variable bandwidth and spatial adaptation. *Journal of the Royal Statistical Society. Series B, 57*(2), 371–394.

Fan, Y., Li, Q., & Weersink, A. (1996). Semiparametric estimation of stochastic production frontier models. *Journal of Business & Economic Statistics, 14*(4), 460–468.

Ferona, A., & Tsionas, E. G. (2012). Measurement of excess bidding in auctions. *Economics Letters, 116*(3), 377–380.

Frank, M. J. (1979). On the simultaneous associativity of $f(x, y)$ and $x+y-f(x, y)$. *Aequationes mathematicae, 19*(1), 194–226.

Freund, J. E. (1961). A bivariate extension of the exponential distribution. *Journal of the American Statistical Association, 56*(296), 971–977.

Fried, H., & Tauer, L. (2019). Efficient wine pricing using stochastic frontier models. *Journal of Wine Economics, 14*(2), 164–181.

Gaynor, M., & Polachek, S. W. (1994). Measuring information in the market: An application to physician services. *Southern Economic Journal, 60*(4), 815–831.

Genest, C. (1987). Frank's family of bivariate distributions. *Biometrika, 74*(3), 549–555.

Godfrey, L. G., & Orme, C. D. (1991). Testing for skewness of regression disturbances. *Economics Letters, 37*(1), 31–34.

Greene, W. (2005a). Fixed and random effects in stochastic frontier models. *Journal of Productivity Analysis, 23*, 7–32.

Greene, W. H. (2003). Simulated likelihood estimation of the normal-gamma stochastic frontier function. *Journal of Productivity Analysis, 19*(2), 179–190.

Greene, W. H. (2005b). Reconsidering heterogeneity in panel data estimators of the stochastic frontier model. *Journal of Econometrics, 126*(2), 269–303.

Greene, W. H. (2010). A stochastic frontier model with correction for sample selection. *Journal of Productivity Analysis, 34*(1), 15–24.

Griliches, Z. (1971).Price indexes and quality change. In Z. Griliches (ed.) *Studies in new methods of measurement*. Harvard University Press.

Groot, W., & Oosterbeek, H. (1994). Stochastic reservation and offer wages. *Labour Economics, 1*(3), 383–390.

Groot, W., & van den Brink, H. (2007). Optimism, pessimism and the compensating income variation of cardiovascular disease: A two-tiered quality of life stochastic frontier model. *Social Science and Medicine, 65*(7), 1479–1489.

Gumbel, E. J. (1960). Bivariate exponential distributions. *Journal of the American Statistical Association, 55*(292), 698–707.

Harding, J. P., Rosenthal, S. S., & Sirmans, C. F. (2003). Estimating bargaining power in the market for existing homes. *Review of Economics and Statistics, 85*(1), 178–188.

Häring, J. (2003). Different Prices for Identical Products? Market Efficiency and the Virtual Location in B2C E-Commerce, ZEW Discussion Papers 03–68, ZEW - Leibniz Centre for European Economic Research.

Harrell, F. E. J. (2015). *Regression modeling strategies: With applications to linear models, logistic and ordinal regression, and survival analysis*. Springer.

Helliwell, J. F., Layard, R., Sachs, J., & De Neve, J.-E. (Eds.) (2020). *World happiness report 2020*. Sustainable Development Solutions Network, New York.

Henderson, D. J., & Parmeter, C. F. (2015). *Applied nonparametric econometrics*. Cambridge University Press.

Henderson, D. J., & Parmeter, C. F. (2016). Teaching nonparametric econometrics to undergraduates. *Journal of Econometric Methods, 5*(1), 179–193.

Hofler, R., & Polachek, S. W. (1982). Ignorance in the labor market: A new approach for measuring information content. In *Proceedings of the American Statistical Association* (pp. 422–425).

Hofler, R. A., & Polachek, S. W. (1985). A new approach for measuring wage ignorance in the labor market. *Journal of Economics and Business, 37*, 267–276.

Horrace, W. C. (2005). Some results on the multivariate truncated normal distribution. *Journal of Multivariate Analysis, 94*(2), 209–221.

Horrace, W. C., & Parmeter, C. F. (2018). A Laplace stochastic frontier model. *Econometric Reviews, 37*(3), 260–280.

Horrace, W. C., Parmeter, C. F., & Wright, I. A. (2024). On asymmetry and quantile estimation of the stochastic frontier model. *Journal of Productivity Analysis, 61*, 19–36.

Hosmer, D. W., & Lemeshow, S. (2000). *Applied logistic regression* (2nd ed.). Wiley.

Jamalizadeh, A., Pourmousa, R., & Balakrishnan, N. (2009). Truncated and limited Skew-Normal and Skew-t distributions: Properties and an illustration. *Communications in Statistics - Theory and Methods, 38*(16–17), 2653–2668.

Kaplan, G., Menzio, G., Rudanko, L., & Trachter, N. (2019). Relative price dispersion: Evidence and theory. *American Economic Journal: Microeconomics, 11*(3), 68–124.

Karakaplan, M. U., & Kutlu, L. (2017). Handling endogeneity in stochastic frontier analysis. *Economics Bulletin, 37*, 889–901.

Kim, J.-O., & Mueller, C. W. (1978). *Factor analysis: Statistical methods and practical issues.* Sage.

Kinukawa, S., & Motohashi, K. (2010). Bargaining in technology markets: An empirical study of biotechnology alliances, Discussion Paper Series 10-E-200, RIETI.

Kinukawa, S., & Motohashi, K. (2016). What determines the outcome of licensing deals in market for technology? Empirical analysis of sellers and buyers in biotechnology alliances. *International Journal of Technology Management, 70*(4), 257–280.

Kivedal, K. (2024). *Applied statistics and econometrics. basic topics and tools with Gretl and R.* Springer.

Klaassen, C., & Wellner, J. (1997). Efficient estimation in the bivariate Normal Copula model: Normal margins are least favourable. *Bernoulli, 3*(1), 55–77.

Kleiber, C., & Zeileis, A. (2008). *Applied econometrics with R.* Springer. ISBN: 978-0-387-77316-2. https://CRAN.R-project.org/package=AER

Kotz, S., Balakrishnan, N., & Johnson, N. L. (2000). *Continuous multivariate distributions vol 1: Models and Applications* (2nd ed.). John Wiley and Sons.

Kumbhakar, S. C., Lien, G., & Hardaker, J. B. (2014). Technical efficiency in competing panel data models: A study of Norwegian grain farming. *Journal of Productivity Analysis, 41*, 321–337.

Kumbhakar, S. C., & Parmeter, C. F. (2009). The effects of match uncertainty and bargaining on labor market outcomes: Evidence from firm and worker specific estimates. *Journal of Productivity Analysis, 31*(1), 1–14.

Kumbhakar, S. C., & Parmeter, C. F. (2010). Estimation of hedonic price functions with incomplete information. *Empirical Economics, 39*(1), 1–25.

Kumbhakar, S. C., Parmeter, C. F., & Zelenyuk, V. (2022). Stochastic frontier analysis: Foundations and advances I. In S. Ray, R. Chambers & S. C. Kumbhakar (Eds.), *Handbook of production economics* (Vol. 1). Springer.

Kumbhakar, S. C., Tsionas, E. G., & Sipiläinen, T. (2009). Joint estimation of technology choice and technical efficiency: An application to organic and conventional dairy farming. *Journal of Productivity Analysis, 31*(2), 151–161.

Kutlu, L. (2010). Battese-Coelli estimator with endogenous regressors. *Economics Letters, 109*, 79–81.

Lach, S. (2002). Existence and persistence of price dispersion: An empirical analysis. *The Review of Economics and Statistics, 84*(3), 433–444.

Lai, H.-P., & Kumbhakar, S. C. (2020). Estimation of a dynamic stochastic frontier model using likelihood-based approaches. *Journal of Applied Econometrics, 35*(2), 217–247.

Lei, P., Cai, Q., & Jiang, F. (2024). Assessing the impact of environmental regulation on enterprise high-quality development in China: A two-tier stochastic frontier model. *Energy Economics, 133*, 107502.

Leppin, A. L., Gionfriddo, M. R., Gionfriddo, M. R., Brito, J. P., Mair, F. S., Gallacher, K., Wang, Z., Erwin, P. J., Sylvester, T., Boehmer, K., Ting, H. H., Murad, M. H., Shippee, N. D., & Montori, V. M. (2014). Preventing 30-day hospital readmissions: A systematic review and meta-analysis of randomized trials. *JAMA Internal Medicine, 174*(7), 1095–1107.

Li, Q., & Racine, J. (2007). *Nonparametric econometrics: Theory and practice.* Princeton University Press

Lian, Y., Liu, C., & Parmeter, C. F. (2023). Two-tier stochastic frontier analysis using Stata. *The Stata Journal, 23*(1), 197–229.

Lu, H., Lian, Y., & Lu, S. (2011). Measuring the degree of information asymmetry in China's medical service market. *Economic Research Journal, 4*.

Maćkowiak, B., Matějka, F., & Wiederholt, M. (2023). Rational inattention: A review. *Journal of Economic Literature, 61*(1), 226–73.

Madigan, E. A. (2008). People with heart failure and home health care resource use and outcomes. *Journal of Clinical Nursing, 17*(7b), 253–259.

Madigan, E. A., Gordon, N. H., Fortinsky, R. H., Koroukian, S. M., Pina, I., & Riggs, J. S. (2012). Rehospitalization in a national population of home health care patients with heart failure, 47(6). *Health Services Research, 47*(6), 2316–2338.

Martins-Filho, C., & Yao, F. (2015). Semiparametric stochastic frontier estimation via profile likelihood. *Econometric Reviews, 34*(4), 413–451.

Meeusen, W., & van den Broeck, J. (1977). Efficiency estimation from Cobb-Douglas production functions with composed error. *International Economic Review, 18*(2), 435–444.

Minviel, J. J., & Sipiläinen, T. (2021). A dynamic stochastic frontier approach with persistent and transient inefficiency and unobserved heterogeneity. *Agricultural Economics, 52*(4), 575–589.

Moran, P. A. P. (1967). Testing for correlation between non-negative variates. *Biometrika, 54*(3–4), 385–394.

Mortensen, D. T. (2003). *Wage dispersion: Why are similar workers paid differently?* MIT.

Mundlak, Y. (1961). Empirical production function free of management bias. *Journal of Farm Economics, 43*(1), 44–56.

Mundlak, Y. (1978). On the pooling of time series and cross section data. *Econometrica, 46*(1), 69–85.

Nash, J. F. J. (1950). The bargaining problem. *Econometrica, 18*(2), 155–162.

Nelsen, R. (1986). Properties of a one-parameter family of bivariate distributions with specified marginals. *Communications in Statistics: Theory and Methods, 15*(11), 3277–3285.

Nelsen, R. (2006). *An introduction to copulas* (2nd ed.). Springer.

Olson, J. A., Schmidt, P., & Waldman, D. A. (1980). A Monte Carlo study of estimators of stochastic frontier production functions. *Journal of Econometrics, 13*, 67–82.

Orea, L., & Kumbhakar, S. C. (2004). Efficiency measurement using a latent class stochastic frontier model. *Empirical Economics, 29*(1), 169–183.

Pagan, A., & Ullah, A. (1999). *Nonparametric econometrics*. Cambridge University Press.

Papadopoulos, A. (2015a). The double half normal distribution and its extensions, Technical report, Athens University of Economics and Business, Athens, Greece. http://www.pyxida.aueb.gr

Papadopoulos, A. (2015b). The half normal specification for the two-tier stochastic frontier model. *Journal of Productivity Analysis, 43*(2), 225–230.

Papadopoulos, A. (2018). The two-tier stochastic frontier framework: Theory and applications, models and tools, PhD thesis, Athens University of Economics and Business. https://www.didaktorika.gr/eadd/handle/10442/44888?locale=en

Papadopoulos, A. (2021a). The effects of management on production: A survey of empirical studies. In S. Ray, C. R.G., & S. Kumbhakar (Eds.), *Handbook of production economics* (Vol. 2, pp. 1651–1697). Springer.

Papadopoulos, A. (2021b). Measuring the effect of management on production: A two-tier stochastic Frontier approach. *Empirical Economics, 60*, 3011–3041.

Papadopoulos, A. (2021c). Stochastic frontier models using the Generalized Exponential distribution. *Journal of Productivity Analysis, 55*(1), 15–29.

Papadopoulos, A. (2021d). The two-tier stochastic frontier framework (2TSF): Measuring frontiers wherever they may exist. In C. Parmeter & R. Sickles (Eds.), *Advances in efficiency and productivity analysis* (pp. 163–194). Springer.

Papadopoulos, A. (2022). Accounting for endogeneity in regression models using Copulas: A step-by-step guide for empirical studies. *Journal of Econometric Methods, 11*(1), 127–154.

Papadopoulos, A. (2024a). The Nash bargaining two-tier stochastic frontier model. In *Essays in honor of Subal Kumbhakar* (Vol. 46, pp. 439–476). Emerald Publishing Limited.

Papadopoulos, A. (2024b). Two-tier stochastic frontier analysis: Heterogeneous error distributions and model selection. *Journal of Productivity Analysis*. https://doi.org/10.1007/s11123-024-00740-4

Papadopoulos, A., & Parmeter, C. F. (2022). Quantile methods for stochastic frontier analysis. *Foundations and Trends® in Econometrics, 12*(1), 1–120.

Papadopoulos, A., Parmeter, C. F., & Kumbhakar, S. C. (2021). Modeling dependence in two-tier stochastic frontier models. *Journal of Productivity Analysis, 56*(2–3), 85–101.

Papadopoulos, A., & Stark, R. B. (2021). Does home health care increase the probability of 30-day hospital readmissions? Interpreting coefficient sign reversals, or their absence, in binary logistic regression analysis. *The American Statistician, 75*(2), 173–184.

Park, S., & Gupta, S. (2012). Handling endogenous regressors by joint estimation using copulas. *Marketing Science, 31*(4), 567–586.
Parmeter, C. F. (2006). *Two-tier frontier and generalized kernel estimation of hedonic price indices*. State University of New York at Binghamton.
Parmeter, C. F. (2018). Estimation of the two-tiered stochastic frontier model with the scaling property. *Journal of Productivity Analysis, 49*(1), 37–47.
Parmeter, C. F., & Kumbhakar, S. C. (2014). Efficiency analysis: A primer on recent advances. *Foundations and Trends® in Econometrics, 7*(3–4), 191–385.
Pennerstorfer, D., Schmidt-Dengler, P., Schutz, N., Weiss, C., & Yontcheva, B. (2020). Information and price dispersion: Theory and evidence. *International Economic Review, 61*(2), 871–899.
Petersen, T., Tatic, M., Hartmann, M., & Hirsch, S. (2023). Meat and meat substitutes—a hedonic-pricing model for the German market. *Journal of the Agricultural and Applied Economics Association, 2*(4), 668–685.
Phelps, E. S. (1972). The statistical theory of racism and sexism. *The American Economic Review, 62*(4), 659–661.
Poggi, A. (2010). Job satisfaction, working conditions and aspirations. *Journal of Economic Psychology, 31*(6), 936–949.
Polachek, S. W., & Yoon, B. J. (1987). A two-tiered earnings frontier estimation of employer and employee information in the labor market. *The Review of Economics and Statistics, 69*(2), 296–302.
Polachek, S. W., & Yoon, B. J. (1996). Panel estimates of a two-tiered earnings frontier. *Journal of Applied Econometrics, 11*(2), 169–178.
Pope, J. C. (2008). Do seller disclosures affect property values? Buyer information and the hedonic model. *Land Economics, 84*(4), 551–572.
Pratt, J. W., Wise, D. A., & Zeckhauser, R. (1979). Price differences in almost competitive markets. *The Quarterly Journal of Economics, 93*(2), 189–211.
Prokhorov, A. (2024). *Efficiency and productivity analysis: Using copulas in stochastic frontier models*. Routledge.
Prokhorov, A., & Schmidt, P. (2009). Likelihood-based estimation in a panel setting: Robustness, redundancy and validity of copulas. *Journal of Econometrics, 153*(1), 93–104.
Rajapaksa, D. P. (2015). Floods and property values: A hedonic property and efficiency analysis, PhD thesis, Queensland University of Technology.
Rosen, S. (1974). Hedonic prices and implicit markets: Product differentiation in pure competition. *Journal of Political Economy, 82*(1), 34–55.
Roth, A. E. (1979). *Axiomatic models of bargaining*. Springer.
Rothschild, M. (1973). Models of market organization with imperfect information: A survey. *Journal of Political Economy, 81*(6), 1283–1308.
Ruppert, D., Sheather, S., & Wand, M. P. (1995). An effective bandwidth selector for local least squares regression. *Journal of the American Statistical Association, 90*, 1257–1270.
Samuelson, P. (1947). *Foundations of economic analysis*. Harvard University Press. Enlarged edition 1983.
Seifert, S., Kahle, C., & Hüttel, S. (2021). Price dispersion in farmland markets: What is the role of asymmetric information? *American Journal of Agricultural Economics, 103*(4), 1545–1568.
Smith, M. D. (2008). Stochastic frontier models with dependent error components. *The Econometrics Journal, 11*(1), 172–192.
Song, P. (2000). Multivariate dispersion models generated from Gaussian copula. *Scandinavian Journal of Statistics, 27*(2), 305–320.
Sorensen, A. T. (2000). Equilibrium price dispersion in retail markets for prescription drugs. *Journal of Political Economy, 108*(4), 833–850.
Stengos, T., & Zacharias, E. (2006). Intertemporal pricing and price discrimination: A semiparametric hedonic analysis of the personal computer market. *Journal of Applied Econometrics, 21*(3), 371–386.
Stigler, G. J. (1961). The economics of information. *Journal of Political Economy, 69*(3), 213–225.

Stigler, G. J. (1962). Information in the labor market. *Journal of Political Economy, 70*(5, Part 2), 94–105.

Tauer, L. W. (2021). Determining the accuracy of holstein bull semen prices with a two-tier efficiency model. *Journal of Applied Farm Economics, 4*(1), 5.

Tomini, S., Groot, W., & Pavlova, M. (2012). Paying informally in the Albanian health care sector: A two-tiered stochastic frontier model. *The European Journal of Health Economics, 13*, 777–788.

Tran, K. C., & Tsionas, E. G. (2013). GMM estimation of stochastic frontier models with endogenous regressors. *Economics Letters, 118*, 233–236.

Tran, K. C., & Tsionas, E. G. (2015). Endogeneity in stochastic frontier models: Copula approach without external instruments. *Economics Letters, 133*, 85–88.

Tsionas, E. G. (2006). Inference in dynamic stochastic frontier models. *Journal of Applied Econometrics, 21*(5), 669–676.

Tsionas, E. G. (2012). Maximum likelihood estimation of stochastic frontier models by the Fourier transform. *Journal of Econometrics, 170*, 234–248.

van den Broeck, J., Koop, G., Osiewalski, J., & Steel, M. F. (1994). Stochastic frontier models: A Bayesian perspective. *Journal of Econometrics, 61*(2), 273–303.

Vuong, Q. H. (1989). Likelihood ratio tests for model selection and non-nested hypotheses. *Econometrica, 57*(2), 307–333.

Wang, H.-J., & Schmidt, P. (2002). One-step and two-step estimation of the effects of exogenous variables on technical efficiency levels. *Journal of Productivity Analysis, 18*, 129–144.

Wang, S., & Lei, H. (2017). Measurement of medical bargaining power of the elderly population in China. *Scientific Research on Aging, 5*, 35–45.

Wang, Y. (2016). The effect of bargaining on US economic aid. *International Interactions, 42*(3), 479–502.

Wang, Y. (2018). Bargaining matters: An analysis of bilateral aid to developing countries. *Journal of International Relations and Development, 21*, 1–21.

Wang, Y. (2021). Bargaining power and US military aid in the post-cold war era. *Security and Defence Quarterly, 35*(3), 7–23.

Weinberger, M., Oddone, E. Z., & Henderson, W. G. (1996). Does increased access to primary care reduce hospital readmissions? *New England Journal of Medicine, 334*(22), 1441–1447.

Wheat, P., Stead, A. D., & Greene, W. H. (2019). Robust stochastic frontier analysis: A Student's *t*-Half Normal model with application to highway maintenance costs in England. *Journal of Productivity Analysis, 51*(1), 21–38.

White, H. (1994). *Estimation, inference, and specification analysis*. Cambridge University Press.

Xiaojun, L., Yujing, Z., & Xiaotong, G. (2018). The analysis of difference on lessor-tenant bargaining power in housing rental market of China. *Price: Theory & Practice* p. 03.

Xu, C., Yang, L., Zhang, B., & Song, M. (2021). Bargaining power and information asymmetry in China's water market: An empirical two-tier stochastic frontier analysis. *Empirical Economics, 61*, 2395–2418.

Xu, W., Ding, Y., & Liu, Y. (2013). Measurement about information asymmetry degree of Chinese medical service market. *Computer Modelling and New Technologies, 17*, 277–288.

Yang, X. (2018). Path selection of China's medical reform based on empirical measurement of asymmetric information. *Health, 10*, 1406–1430.

Yee, T. W. (2015). *Vector generalized linear and additive models: With an implementation in R*. Springer.

Zeileis, A. (2004). Econometric computing with HC and HAC covariance matrix estimators. *Journal of Statistical Software, 11*(10), 1–17.

Zeileis, A. (2006). Object-oriented computation of sandwich estimators. *Journal of Statistical Software, 16*(9), 1–16.

Zhang, H., Zhang, J., Yang, Y., & Zhou, Q. (2018). Bargaining power in tourist shopping. *Journal of Travel Research, 57*(7), 947–961.

Zhao, S., & Losak, J. (2024). Two-tiered stochastic frontier models: A Bayesian perspective. *Journal of Productivity Analysis, 61*(2), 85–106.

MIX
Papier aus verantwortungsvollen Quellen
Paper from responsible sources
FSC® C105338

If you have any concerns about our products,
you can contact us on
ProductSafety@springernature.com

In case Publisher is established outside the EU,
the EU authorized representative is:
Springer Nature Customer Service Center GmbH
Europaplatz 3, 69115 Heidelberg, Germany

Printed by Libri Plureos GmbH
in Hamburg, Germany